Cytopreparation

AF760232

ESSENTIALS IN CYTOPATHOLOGY SERIES

Dorothy L. Rosenthal, MD, FIAC, Series Editor

Editorial Board

Syed Z. Ali, MD
Douglas P. Clark, MD
Yener S. Erozan, MD

For further volumes:
http://www.springer.com/series/6996

Gary W. Gill, BA, CT(ASCP)
Independent Cytological Consultant, Indianapolis, IN, USA

Cytopreparation

Principles & Practice

Springer

Gary W. Gill
Independent Cytological Consultant
Indianapolis, IN, USA

ISSN 1574-9053 ISSN 1574-9061 (electronic)
ISBN 978-1-4614-4932-4 ISBN 978-1-4614-4933-1 (eBook)
DOI 10.1007/978-1-4614-4933-1
Springer New York Heidelberg Dordrecht London

Library of Congress Control Number: 2012947399

© Springer Science+Business Media New York 2013
This work is subject to copyright. All rights are reserved by the Publisher, whether the whole or part of the material is concerned, specifically the rights of translation, reprinting, reuse of illustrations, recitation, broadcasting, reproduction on microfilms or in any other physical way, and transmission or information storage and retrieval, electronic adaptation, computer software, or by similar or dissimilar methodology now known or hereafter developed. Exempted from this legal reservation are brief excerpts in connection with reviews or scholarly analysis or material supplied specifically for the purpose of being entered and executed on a computer system, for exclusive use by the purchaser of the work. Duplication of this publication or parts thereof is permitted only under the provisions of the Copyright Law of the Publisher's location, in its current version, and permission for use must always be obtained from Springer. Permissions for use may be obtained through RightsLink at the Copyright Clearance Center. Violations are liable to prosecution under the respective Copyright Law.
The use of general descriptive names, registered names, trademarks, service marks, etc. in this publication does not imply, even in the absence of a specific statement, that such names are exempt from the relevant protective laws and regulations and therefore free for general use.
While the advice and information in this book are believed to be true and accurate at the date of publication, neither the authors nor the editors nor the publisher can accept any legal responsibility for any errors or omissions that may be made. The publisher makes no warranty, express or implied, with respect to the material contained herein.

Printed on acid-free paper

Springer is part of Springer Science+Business Media (www.springer.com)

Dedicated to Doris and Buddy, my late parents; they gave me life.

Dedicated to my family: Marianne, stepson BJ, David, Megan, Michael, daughter-in-law Kristen, and grandchildren Jackson Thomas and Mariah MacKenzie; they are my life.

Foreword

At last! A textbook with the accumulated knowledge of Gary Gill on that most tangled web of complexity, cytopreparation. Don't be deceived by the title. This is not a textbook solely for cytotechnologists, cytopreparatory technicians, and educators; it should become every pathologist's companion. Anyone who uses a microscope to study tissue is cognizant of the importance of perfect tissue preparation and stain and its impact on reaching an accurate diagnosis. Staining principles apply to both cytology and histology. However, this textbook explores much more than cytopreparation. It provides useful advice for improving processes that extend beyond cytopathology, highlighted with unexpected tints of humor that will make you laugh out loud. Gary Gill has long been an advocate of optimizing cytopathology for the benefit of the patient, whether in specimen submission, preparation, staining, screening, or reporting. This textbook fills a long-vacant knowledge gap and provides a succinct explanation of the chemical processes involved in staining and processing in a format that is familiar to all laboratories: the principles and practices of standard operating procedures.

Nearly everyone in pathology is familiar with the ubiquitous "Gill hematoxylin," the formulation for which he is famous. From his observations and experience as a newly minted cytotechnologist graduating from The Johns Hopkins Hospital's School of Cytotechnology, Gary Gill concluded that the Papanicolaou stain suffered from a serious lack of standardization and began his investigations on the optimal stain by researching hematoxylin preparations. He discovered a method of progressive staining that did not require differentiation (the extraction of stain using weak acid rinses), thereby preventing overstaining of cellular chromatin. His progressive hematoxylin has been universally embraced by both histologists and cytologists and has replaced almost all other hematoxylin formulations in most laboratories. He continued his journey by examining all of the chemical reagents and reactions in staining and eventually streamlined Papanicolaou staining to an environmental-friendly processing line that he coined "Enviro-Pap." His career culminated as the corporate compliance officer for a large cytology laboratory in Indiana, where he ensured excellence and adherence to regulatory standards. His career has touched nearly every aspect of cytopathology. You can read more about his illustrious career in Appendix J.

The cytopathology community knows Gary Gill as its foremost authority on cytology specimen processing, preparation, staining, and screening. In fact, several years ago, he presented "Managing Cytology Information Overload: a Glimpse into Gary Gill's Brain," an invited lecture at a Program Faculty Seminar during the American Society of Cytopathology's Annual Scientific Meeting in 2005. The lecture was intended to answer the question: how do you keep track of your extensive cytological files? He has been nicknamed the "cytogoogle" of the cytopathology community. Just ask Gary, and he will know the answer. When I first met Gary at a national meeting, I was appropriately in awe of this icon, but was immediately placed at ease by his down-to-earth, self-effacing, and approachable personality. Since then, I have not ceased needling him to compile his knowledge into one definitive text, and he hasn't let me down.

His 1 TB of electronic information and warehouse of paper information have only now made their way into print. It is an opportunity too good to pass up, an endeavor long urged by all his friends and colleagues, and possibly the only time the medical community has been able to rein him in from his personal intellectual pursuits long enough to leave a professional legacy.

If you, too, want to get inside Gary Gill's brain, then do not skip the appendices—for here lies a true treasure trove of ideas and explanations. My favorite is Appendix I, "Screening and

CPR." All of those who have ever screened cytology slides and have had a dog will appreciate the analogy and the problems encountered as elucidated in this section.

Bethesda, MD, USA Barbara A. Crothers

Foreword

In my view, the most important aspect of any book on cytopreparation is that it has practical application to the cytopathology laboratory. In that sense, this book delivers! It has just the right mix of practical application with sound scientific principles. The reader is not only given instructions on what to do, but also explanation as to why they are doing it. This book has everything for the individual thirsting for the best information concerning cytopreparation and does not distract with unneeded embellishments. The author's desire to inspire us to a greater awareness of the importance of excellent cytopreparatory technique and its direct relationship to good patient care is evident throughout the book.

Cytopreparation Principles and Practices is the result of the author's exploration of all the ins-and-outs that comprise good cytopreparatory technique. He meticulously walks us through the steps to good specimen preparation and then challenges us to consider what it really means to screen a cytologic sample. Having known Gary for many years, it is not surprising that his book reflects this practical methodology, which is how he approaches life in general. He's a very rational man who makes decisions based upon proven constructs with sound information.

I suspect that over the years, there are many of us, and I would venture to say most of us in cytopathology who have been touched in some way by Gary's expertise in cytopreparation. I initially met Gary in my first year out of Cytotechnology School while attending his Cytopreparation Workshop at an American Society of Cytopathology annual meeting. At the time, I was in charge of the urine processing for our lab, and we were using membrane filters for all of our urine preparations. I was having some difficulty getting reproducibly good preparations, and the pathologist I worked with suggested I attend Gary's workshop.

My first impression of Gary was that he possessed a serious demeanor which was, initially, a little intimidating. Dressed in his characteristic double-breasted blazer, he appeared a little stiff, but clearly had an intense interest and grasp of all things related to cytopreparation. By the end of the workshop, I got the information I needed for the preparation of exquisite urine samples, and my pathologist and I were very pleased with our investment in Gary's expertise.

Since that first encounter, it has been my distinct pleasure to know Gary, not only as an icon of cytopreparation, but also as a good friend. His quick wit and dry sense of humor are generally not only entertaining but also informative. These characteristics shine through periodically in the writing of this book as well.

Upon sharing with several individuals that Gary was writing a book on cytopreparation, I was amazed often by their reaction, which was usually, "I thought he had already done that."

In fact, it happened so often that it really got my attention. The man whom everyone assumed had written the book on cytopreparation had actually never written it…quite a testimony to his reputation in the field.

It is with much appreciation that we thank Gary for writing this extraordinary work that will allow all who read it to have a part in carrying forward the Gill legacy: to apply due vigilance when considering the best approach in preparing the patient's cytologic sample set before us.

Indianapolis, IN, USA William N. Crabtree

Series Preface

When Springer asked me to be Editor-in-Chief of Essentials in Cytopathology, the year was 2004. Even though there was already a plethora of adequate texts on cytomorphology, most were hard-bound and weighty. My editorial board agreed with Springer's concept of small format paper backs that were inexpensive to produce and therefore accessible for the prospective buyer. Each volume has concentrated on the cellular patterns from a particular organ site or related complex, emphasizing diagnostic criteria and pitfalls in a simple format style with abundant high-quality color plates and graphic illustrations. The series has been an unqualified success by any measure, already publishing 12 volumes in 7 years, one volume being a second revised edition with 6 more volumes under contract.

This latest volume (number 13) in the Essentials in Cytopathology series is perhaps the most important. It clearly is unique, as it barely has any photographs of cells and does not define cellular criteria. It does, however, instruct the observant laboratorian in how to achieve optimal cellular samples for microscopic interpretation. Not only is it being published in the year of the 60th anniversary meeting of the American Society of Cytopathology and the author's 70th birthday, but also the 25th anniversary year of the Wall Street Journal articles that earned the 1988 Pulitzer Prize in investigative journalism for the author, Walt Bogdanich. The resultant Clinical Laboratory Amendments of 1988 are addressed in this volume as each of the principles is delineated that responds to a regulation.

Regulatory compliance is not the intent of this work, although it is a side benefit. I coerced Gary into writing it for posterity. My bribe was naming The Johns Hopkins Cytopreparatory Laboratory in honor of him. No one knows cytopathology preparatory techniques and their scientific bases better than Gary. In fact, many of the chapters apply to histopathology samples as well. Therefore, anyone in Anatomic Pathology, from Laboratory Directors and Managers, to the preparatory technicians, can benefit from this unique manual.

I urge everyone who reads this seminal work to consider where we've come as a medical specialty in the quarter century following these journalistic exposes of laboratory incompetence. Much time is spent by laboratory managers attending to the "paper work" necessary to prove compliance to laboratory inspectors. But have we actually been able to prevent deaths from cervical cancer by these measures? Quoting Gary, "quality begins here." The final interpretation of a sample is dictated by the quality of the preparation. Until we get that right, we will not reach our goal for cervical cancer nor provide optimal care for all our patients with the other diseases that come to us at the light microscope.

Dorothy L. Rosenthal, M.D., FIAC
Baltimore, MD, USA

Preface

"I have no special talents. I am only passionately curious."
Albert Einstein

Having recently graduated from Western Maryland College with a baccalaureate degree in premed, I was looking for employment. An ad in the Baltimore Sun newspaper caught my eye. Someone was looking for a person with experience in a variety of biological subjects. I wrote to the box number provided, telling the unidentified source that I had coursework in the advertised areas but no job experience. To shorten a long story, the prospective employer turned out to be Dr. John K. Frost in the Division of Cytopathology of The Johns Hopkins Hospital. The advertised position had been filled internally, but he had a School of Cytotechnology, and there were student stipends available that would cover the cost. Would I be interested in enrolling?

That was 1963, and the rest—as is often said—is history. Confucius was right: find a job you love and you'll never work a day in your life. By pure dumb luck, I had stumbled onto a profession and career path that fuelled the passion that resulted in my writing this book. After graduating on October 9, 1964, with a certificate in "Medical Cytotechnology" from The Johns Hopkins Hospital, I remained employed there until January 16, 1987.

Parenthetically, the formal name of the institution is The Johns Hopkins Hospital. The word "The" is capitalized, and Johns Hopkins is the name of the Quaker philanthropist who donated $7 million to construct the hospital in 1875 and several other famous

Baltimore-based institutions that bear his name. Johns is a family name; it is not John and is not followed by apostrophes (i.e, not John's or Johns' and not Hopkin's or Hopkins'). He died Christmas Eve 1873 at age 78.

The first research project in which I participated was circulating cancer cells in the blood. Nine years earlier, Engell had published a review about the subject that sparked enormous interest.[1] Note that the year was 1955. We didn't know then what we didn't know.

Our small research team's initial charge was to gather from the published medical literature papers about processing peripheral blood, evaluate each method, identify the most promising one, improve it as needed, and apply it to real-life specimens. One thing above all became abundantly clear: we didn't know what we were doing. Among other things, for example, we couldn't get erythrocytes or leukocytes to stick to glass slides when wet-fixed (i.e, plunged into alcohol). We learned that normal saline—contrary to expectation—destroys cells in vitro. We also learned that the Pap stain was not standardized. In short, almost everything we had been taught about cytopreparation was insufficiently reliable to be useful. No one was to blame. After all, it was the 1960s.

Since we were unable to get blood cells to stick to glass slides that were wet-fixed for cytology, instead of having been air-dried for hematology, we began collecting them on Millipore filters. I observed that cells near the boundary of the cell collection area of one preparation in particular were well preserved, while neighboring cells were not. One of our early "successes" is pictured in this photomicrograph of what we believed to be megakaryocyte. Megakaryocytes ordinarily don't circulate in peripheral blood.

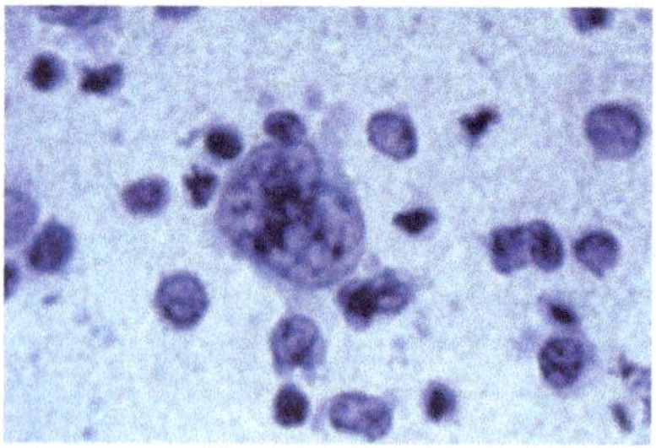

That one observation made me think that if we could identify the contributing factors responsible for this isolated success, we could take the guesswork out of making filter preparations of well-preserved cells. Thus began the unending questions and answers that are embodied in this book.

Readers will note that most of the cited references are in journals unrelated to cytology as we know it, and they're *old*. Many were published in the first half of the twentieth century and, occasionally, the seventeenth century. These reflect the fact that I had questions, and they had answers. I had no recourse but to visit the musty dusty stacks of the Welch Medical Library of The Johns Hopkins Medical Institutions. In those days, now nearly 50 years ago, there was no easy way to research topics of interest. PubMed and Google were far into the future. Volumes and volumes of Index Medicus on tables and tables in the library's reading room do not inspire serious scholarship. To facilitate focused searches, I learned that reading the lists of references in published papers that were useful to me often revealed titles of articles of likely interest and the names of journals outside those commonly associated with diagnostic cytopathology. I would often go to the non-air-conditioned stacks, select the last issue of these unfamiliar journals for each year, and read the titles of articles published for the entire year. The library's policy allowed me to check out the journals and copy the articles, which I still have.

This volume can be used to teach cytopreparation and help students:

- *Understand the principles that underlie the various procedures and practices*
- *Appreciate that everything done to a specimen makes a microscopically appreciable difference*
- *Encourage observations that may elicit suggestions for improvements*
- *Discourage potential shortcuts that cost more than they gain*
- *Promote curiosity* (e.g, How do you know that? Are you sure? Show me the citation.)

These lectures are needed because cytopreparation for technicians is not taught anywhere as a formal program. While part of every cytotechnologist's education, it is a relatively small part and often not taught well. Nationwide, the need for high-quality cytopreparation is great.

This book covers the entire range of processes that contribute to a useful cytologic preparation, from specimen collection thru microscopy. Since "the Pap test is cytopathology," I have also included an approach to screening Pap tests and data analysis. I have tried to provide sufficient details throughout the book so that others outside this country may benefit.

I want to acknowledge with gratitude my first teachers in cytopathology: Dr. John K. Frost, Arline K. Howdon, and Sue T. Shutt. Their unbridled enthusiasm was infectious; their encouragement, unflagging. Pre-everything regulatory, nothing slowed my researches or dampened my curiosity. Others at Hopkins I want to acknowledge include the following: Dr. Yener S. Erozan, Dr. Prabodh K. Gupta, Dr. William M. Howdon, and Dr. Norman J. Pressman; cytotechnologists Fran Burroughs, Sue Ermatinger, Gene Ford, Deirdre Kelly, Jack Kirby, Ellen Patz, and Karen Plowden; cytopreparatory technicians Dianna Farrar, Villa Gardner, Darlene Ratajczak, and Linda Reynolds; and Secretary Shirley Long. The named individuals were the core staff during my 23-year tenure. I remember them all fondly.

Lastly, I want to thank Dr. Dorothy Rosenthal, Series Editor of *Essentials in Cytopathology*, for inviting me to write this book. She also spearheaded the November 7, 2011 dedication in my name of the Cytopreparatory Laboratory in the Pathology Building of The Johns Hopkins Hospital. The dedication recognizes the fundamental soundness of my contributions to cytopreparation, which have stood the test of time since my 1987 departure.

Will there always be a place for cytopreparation in a world of molecular medicine? I think so as long as humans are curious to see things otherwise invisible. On the other hand, however, "prediction is very difficult, especially about the future"—Neils Bohr.

In the 1981 movie On Golden Pond, Henry Fonda portrays Norman Thayer, an 80-year old curmudgeon who is celebrating his 80th birthday. When presented with a cake ablaze with 80 candles, he says: "I've been trying all day to draw some…

profound conclusions about living fourscore years. Haven't thought of anything. Surprised it got here so fast." It's the latter statement I remember and a sentiment I now understand.

Reference

1. Engell HC. Cancer cells in the circulating blood; a clinical study on the occurrence of cancer cells in the peripheral blood and in venous blood draining the tumour area at operation. Acta Chir Scand Suppl. 1955;201:1–70.

Contents

Part II The Image

Part III Everything Else

Part I
The Object

Chapter 1
Introduction

I am therefore I think.

Cytopreparation is the science of optimizing and standardizing the collection, preparation, and analysis of cytologic samples in ways that promote the detection of cells-of-interest and accurate interpretation of nuclear morphology. Cytopreparation as a science is based on a single overarching principle: when we make a microscopical preparation for cytopathology, we should try to understand what we are doing and why. Otherwise, we are examining cells that have been treated in unknown ways that may diminish their usefulness. The Elements of Style, the well-known little book of English style by Strunk and White, stated famously, "Make every word tell."[1] In the context of cytopreparation, the goal is to "make every cell tell."

This book is divided into three major parts:

1. The Object
2. The Image
3. Everything Else

The Object includes all materials and methods that interact with the specimen itself—from specimen collection through staining. The Image includes those materials and methods that impact the light that forms the image itself—the clearant, the mounting medium, cover glass, and microscope illumination. The distinction is entirely practical, as we are looking at images of cells, and not the cells themselves. If we know what to expect in terms of

G.W. Gill, *Cytopreparation: Principles & Practice*,
Essentials in Cytopathology 12, DOI 10.1007/978-1-4614-4933-1_1,
© Springer Science+Business Media New York 2013

quality, we will recognize its absence, and know how to identify the cause, and fix it once and for all. Everything Else includes how we find abnormal cells when screening Pap tests and what we do with those findings in terms of evaluating cytotechnologist screening performance and of managing the laboratory's potential risk due to false negatives.

The principles underlying the overarching one are based on relevant laws of biology, chemistry, physics, and optics. From specimen collection through microscopic examination, these principles are the following:

1. Fresh specimens facilitate specimen processing and cell flattening.
2. Make preparations that represent the sample.
3. Flatten cells to enhance chromatin display.
4. Fix preparations immediately to maintain morphology.
5. Stain preparations to facilitate cell visibility, detection, and interpretation.
6. Mount preparations to optimize microscope objective's performance.
7. Examine with a clean microscope and Köhler illumination to promote highest resolution.
8. Screen preparations in ways that facilitate abnormal cell detection.

The materials and methods that reduce the principles of cytopreparation to practice are those that interact with cells from the time the specimen is collected to the time of microscopic examination. All these materials and methods impact visibly and measurably on the resulting preparation. Collectively, they determine the quality and quantity of cells available for examination and affect preparation features and properties such as cell number, mix, flattening, chromatin distribution and particle size, biochemical makeup, penetrability by biological dyes, optical density, color, texture, refractive index, refractility, and microscopic resolution.

The laboratory determines the quality of the preparations to be examined cytomorphologically, and it assesses the quality of the outcomes. Poor quality preparations can challenge the interpretive skills of the best morphologists (Fig. 1.1).

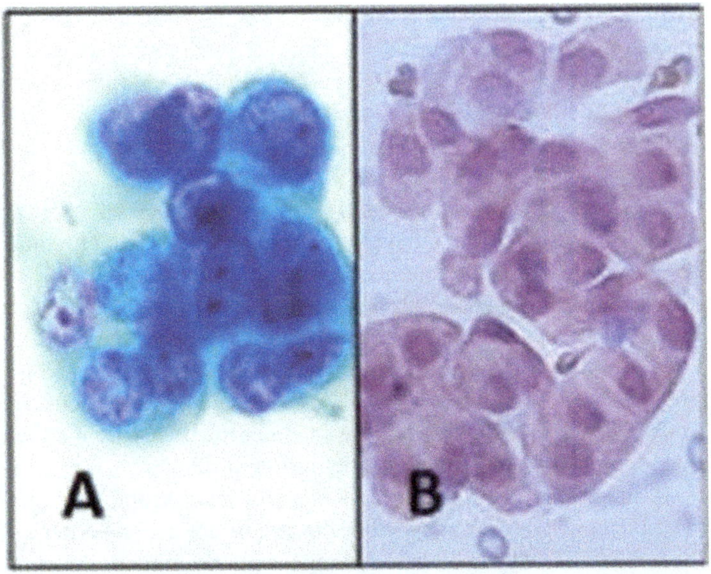

Fig. 1.1. (**A**) Well-prepared samples are useful for their intended purpose, in contrast with those that are not (**B**).

And poor data analysis of Pap test screening outcomes can artificially inflate a laboratory's impression of its screening sensitivity and lessen its perception of risk related to false-negative outcomes. This book attempts to impose objective standards on inherently subjective processes and by so doing strives to improve the overall performance of cytology laboratories.

This book addresses fundamentals of cytopreparation, including microscopy, screening, and data analysis. It is intended to be entirely pragmatic. It provides not only phenomenological descriptions of the most common materials and methods as they apply to gyn, non-gyn, and FNA preparations but also the underlying mechanistic bases.

It is not intended to be encyclopedic; readers will not find answers to their every question. Not everything published is worth reading or merits mention and citation. Everything that is discussed makes a visible difference. Nothing is abstract. Quality cytopreparation is all about controlling cellular artifacts. It is my intention to provide information that will enable and empower readers to review and improve their laboratories' cytopreparatory techniques as they apply to the vast majority of specimens.

This book includes much, but not all, of what I've published previously but is not readily available. See Appendix J. Astute readers will recognize substantial portions that have appeared in larger formats.[2] I have attempted to put in one relatively small book things I've learned that contribute to quality results.

While intended primarily for everyone involved directly or indirectly in making cytologic preparations for cytopathology, this book will benefit anyone who prepares cytologic and histologic preparations for any purpose. All microscopic preparations are more alike than different: they're intended to be useful for their intended purpose—the working definition of quality.

If one has never prepared specimens for microscopic examination, how does one know whether the results are useful? After all, the number of ways to prepare specimens poorly exceeds the number to prepare them well. By sheer numbers of possible substandard alternative materials and methods, one is more likely to get it wrong than right. This book will guide new practitioners "down the path of righteousness."

I've always found it curious that cytopreparation, which can make or break a specimen as it begins its journey through the laboratory, is entrusted to "cytoprep techs," those with the least formal training. Cytoprep techs are supervised by cytotechnologists who receive some, but not enough, technical training. Well-supervised cytoprep techs can do excellent work. However, cytotechnologists and pathologists are neither taught nor tested on technique well. Even today, more than 60 years after the first cytotechnology school was established, there is not even a published SOP for screening a Pap test. If that doesn't qualify as curious, I don't know what does.

According to CLIA '88 (Clinical Laboratory Improvement Amendments of 1988), "Facilities only collecting or preparing specimens (or both) or only serving as a mailing service and not performing testing are not considered laboratories." In the view of Centers for Medicare and Medicaid Services, a "*Laboratory* means a facility for the biological, microbiological, serological, chemical, immunohematological, hematological, biophysical, cytological, pathological, or other examination of materials derived from the human body for the purpose of providing information for the diagnosis, prevention, or treatment of any disease or impairment of, or the assessment of the health of, human beings. These examinations also include procedures to determine, measure, or otherwise describe the presence or absence of various substances or organisms in the body."[3]

CLIA '88 leaves the quality of the preparations on which diagnoses are based on the professionals who do the diagnosing and ignores the reality that cytologic preparations in particular are not routinely as useful as needed. Interpreting poorly stained Pap tests sometimes contributes to false-negative results. For example, "While slide staining (preparation) is seldom the direct target of litigation, it is in fact one of the things repeatedly mentioned when slides from cytology claims are reviewed by experts in the course of evaluating and defending these claims."[4]

This book is intended for education coordinators in cytotechnology and histotechnology programs, cytotechnology and histotechnology students, cytotechnologists, histotechnicians, histotechnologists, pathologists, and for lack of a better all-embracing term, biologists who examine poorly prepared specimens and don't know it. The book includes information that is more in-depth than some might prefer. Such inclusion is deliberate. If I don't memorialize it in print, it may be lost forever to future generations.

Historically, the methods for preparing cells for microscopic examination and diagnostic interpretation have not always been grounded on sound principles. Indeed, often the logic behind a particular methodology is not obvious. The ultimate consequence can be preparations that are functionally inadequate for their intended purpose. The problem is further compounded by the variety of body sites from which biologic specimens come, the different cellular compositions and suspension mediums, and the impact of these differences in cytopreparation.

"In the middle forties, when the cytologic method of diagnosis was not widely accepted as valid, Dr. Papanicolaou presented a paper on its applicability in uterine cancer. One speaker who discussed this paper presented figures to show that it was impossible to distinguish individual cancer cells. Dr. Papanicolaou was distressed by his statements and quite bewildered by such divergent conclusions. To find out why their conclusions varied so widely, Dr. Papanicolaou canceled his train reservation, stayed overnight, and the next morning went to the man's laboratory. They sat together at a microscope with a box of slides. After a few minutes, Dr. Papanicolaou said, 'Sir, I am happy to state we are in full agreement. I couldn't make a cytologic diagnosis from such poor technical preparations, either'."[5] Truth is timeless.

Cytopreparation is a one-time investment that pays multiple dividends with each successive microscopic examination. Cytopreparation is relatively inexpensive; microscopical examination time is expensive. It takes no more time or money to prepare a specimen correctly than it does to prepare it incorrectly. Marginally satisfactory cytologic preparations, however, will cost the laboratory in insidious and sometimes dramatically obvious ways. CLIA '88 and proficiency testing notwithstanding, techniques to improve the quality of specimen preparation and microscope usage will contribute greatly to improving the screening and cytomorphological interpretive skills of the observer.

Readers should be aware of two classic books by John R. Baker (1900–1984) that are still relevant today: Cytological Technique—The Principles Underlying Routine Methods[6] and Principles of Biological Microtechnique—A Study of Fixation and Dyeing.[7] Dr. Baker's scholarship is plainly evident; his writing influenced mine. I never met Dr. Baker, but I corresponded once with him about hematoxylin. His framed handwritten note, dated 29 November 1972, hangs on the wall in front of where I'm sitting as I prepare this manuscript.

Readers will note that throughout this book there are no ® or ™ marks with product names: "*Use of Trademarked Names in Publication.* Under the US Federal Trademark Dilution Act, restricted use of trademark names applies mainly to commercial use of trademarks, not to editorial use in publication. For example, a photography magazine may not place the word "Kodak®" as part of its cover design and a computer manufacturer may not place the word "Kodak"—without the trademark symbol—in an article about cameras and film development without risking trademark infringement.

The symbol ®, or letters ™ or ℠, should not be used in scientific articles or references, but the initial letter of a trademarked word should be capitalized.

On occasion, a trademark owner will request that its trademark or trade name appear in all capital letters or a combination of capital and lowercase letters often with the trademark symbol. Authors and editors are not required by law to follow such request."[8]

References

1. Strunk Jr W, White EB. The elements of style. 2nd ed. New York: Macmillan; 1972.
2. Gill GW. The laboratory. In: DeMay RM, author. The art & science of cytopathology, vol. 3. 2nd ed. Chicago: ASCP Press; 2011. p. 1539–92.
3. Part 493—Laboratory Requirements. 42 CFR Ch. IV (10–1–04 Edition): 967–1087. http://wwwn.cdc.gov/clia/regs/toc.aspx. Accessed 15 Jan 2012.
4. Scott MD. Liability issues with the Papanicolaou smear—an insurance industry perspective. Arch Pathol Lab Med. 1997;121(3):239–40.
5. Graham RM. Technique, chapter XXXI. In: The cytologic diagnosis of cancer, 2nd ed. Philadelphia: WB Saunders; 1963. p. 335
6. Baker JR. Cytological technique–the principles underlying routine methods. 5th ed. London: Methuen; 1966. Edition 4 available http://www.archive.org/details/cytologicaltechn00bake. Accessed 12 Jan 2012.
7. Baker JR. Principles of biological microtechnique—a study of fixation and dyeing. London: Methuen; 1958. http://www.archive.org/details/principlesofbiol01bake. Accessed 14 Jan 2012.
8. Iverson C (Chair). AMA Manual of Style—A guide for authors and editors, 10th ed. Oxford: University Press; 2007. p. 213.

Chapter 2
Quality Control and Quality Assessment

A difference to be a difference must make a difference.

Gertrude Stein

Quality

The word "quality" is thrown about so frequently that is has lost meaning in general and in cytology in particular.[1] It's not what you say, it's what people hear.[2] Examples abound: "death tax" instead of "estate tax," "affordable health care act," "drilling for oil" instead of "exploring for energy," "quality outcomes." I have zero tolerance for loose language (e.g., referring to random rescreening as QC).

"Quality" is derived from the Latin *qualitas*, meaning "of what sort." The set of attributes that allows a product to be used for its intended purpose *defines* its quality. In short, therefore, quality means useful for its intended purpose. That having been said, one must define the purpose, and go on from there, one logical step after another.

Cytopreparation allows cytotechnologists and cytopathologists to get the right answer by not being the limiting factor. "Right answer" means finding abnormal cells when present and interpreting them in ways that guide the clinician in patient management. Note, I did *not* say reproducibly and reliably interpreting the cytologic changes so they correlate with the underlying histology.

G.W. Gill, *Cytopreparation: Principles & Practice*,
Essentials in Cytopathology 12, DOI 10.1007/978-1-4614-4933-1_2,
© Springer Science+Business Media New York 2013

Variation is normal.[3–6] After all, clinicians want to know whether cytology has identified a lesion that requires follow-up.

To make cytology specimens useful for their intended purpose, cytopreparation must include and display the cells-of-interest, so they make sense visually. Period. Simply, cytology preparations must include cells that are a representative sample of the raw specimen, well-preserved, flattened, fixed, stained to promote the visibility of nuclear chromatin and differentiation of cell types, and coverslipped to promote optimal imaging by microscope objectives.

Quality Control

Quality control activities look *forward*. They define the product's quality, imparting to it the credibility needed for its intended purpose. QC activities are the result of planning and are applied prospectively to everything that contributes to the final product, thereby impacting the outcome. QC activities are deterministic (i.e., lead to expected results when followed). Quality control is mentioned 26 instances in CLIA '88, but it is not defined and is not mentioned once in the context of cytology.

Quality Assessment

On January 24, 2003, CLIA '88 was finalized, which was 5,198 days after President Reagan signed it into law on October 31, 1988. That's more than 14 years! Among the changes, quality assurance became known as quality assessment. Properly implemented, quality assessment leads to quality assurance.

Quality assurance is mentioned 6 times in CLIA '88; quality assessment, 23 times. Unlike quality control, quality assessment is defined: "The laboratory's quality systems must include a quality assessment component that ensures continuous improvement of the laboratory's performance and services through ongoing monitoring that identifies, evaluates, and resolves problems." Quality

TABLE 2.1. Quality control is any material or method used routinely to promote useful outcomes.

Differential feature	Quality control	Quality assessment
Purpose	Defines quality	Measures success
Timing	Prospective	Retrospective
Application	All processes	Sample outcomes
Impact	Outcomes	Processes
Nature	Deterministic	Probabilistic

Quality assessment samples outcomes to see whether they "measure up," and if not, why not?

assessment is defined identically in preanalytic, analytic, and postanalytic systems. Note that CLIA '88 tells laboratories what they must do, but not how to do it. Therefore, implementation is unavoidably uneven, and at times, questionably effective. The word quality, not paired with control or assessment, is not mentioned once among the 1,327 words that constitute **§ 493.1274 Standard: Cytology** in CLIA '88.

In contrast to QC, QA looks *backward*. QA measures the degree to which desired outcomes are successful (i.e., their impact). QA activities, therefore, retrospectively sample outcomes. Discrepant findings should be investigated to learn the cause(s), if possible. The findings should be incorporated into the processes that contribute to the final product in an effort to prevent recurrences of the same types of discrepant results (e.g., did the patient have cancer as reported,and if not, why?). As a practical matter, quality assessment activities are probabilistic (i.e., have attendant uncertainty relative to reliability), as it not possible to review all product outcomes.

To decide whether an activity qualifies as QC or QA, see Table 2.1:

Differential Features of Quality Control and Quality Assessment

To implement an *effective* QC/QA program, laboratory personnel must first understand the *differences* between the two sets of activities. Otherwise, documentation of such activities to meet regulatory requirements becomes primarily an exercise in paperwork compliance, rather than one that makes a real difference in how work is done. Judging by how often QC and QA are used interchangeably in conversations, quality control and quality assessment appear to be considered synonymous. Usually, it's "I'm going to QC this or QC that," and never "I'm going to QA this." When the terms are used as though interchangeable, the user obviously perceives no difference. When a distinction between the two terms is perceived, it is often applied incorrectly. In either case, the recipient of such information is misinformed. As a result, the planning of QC/QA activities is often confused; the implementation, suboptimal.

Is 10% Random Review of Negative Pap Tests QC?

No, it's QA. The random rescreening of at least 10% of negative gyn cases as required by CLIA '88 is universally referred to as "QC." While performed prospectively relative to the final reporting, rescreening is performed retrospectively relative to the activity it is intended primarily to measure, that is, the performance of the cytotechnologist. "(c) *Control procedures.* The laboratory must establish and follow written policies and procedures for a program *designed to detect errors in the performance of cytologic examinations* [italicized for emphasis] and the reporting of results."

The rescreening samples outcomes; the findings impact the process of screening. The 10% of negative gyn cases that are rescreened is a random sample, which means it is probabilistic. Such a set of differential features is characteristic of quality assessment. On the other hand, routinely rescreening all high-risk gyn cases as a matter of laboratory policy is quality control, as it is applied prospectively to all such cases, and is intended to prevent false negatives.

Total Quality Management

QC activity without associated QA activity is half-action. Documentation per se simply constitutes paper compliance with regulations that fails to satisfy the intent. QC and QA activities must be practiced continuously to monitor and maintain the performance of the two sets of contributory processes, recognize problems as they arise, identify corrective actions to be taken, and improve quality. Taken together, these two sets of activities constitute a program of total quality management.

Analyzing Quality Control and Quality Assessment Activities

In the broadest possible sense, QC activities cease and QA activities begin when the laboratory product, the cytological interpretation or consultation, is complete. In other words, everything that precedes sign-out is quality control and everything that follows is quality assessment. Specifically, that point is the moment in time when the cytological interpretation is committed to the laboratory report. That definition is too broad, however, to be instructive at the levels where QC/QA activities are most useful.

Cytopreparation constitutes the *processes* that determine the outcome. Successfully detecting abnormal cells is the outcome of a series of interdependent samplings of successively diminishing size. The specimen collection technique samples the biologic process, the cytopreparatory technique samples the specimen, the screening process samples the preparation, and the diagnostic interpretation samples the cellular features. A quality laboratory increases the sensitivity of its cytological method by optimizing and standardizing its materials and methods of specimen collection and preparation.

The relation of cytopreparation to the whole process of detecting abnormal cells is depicted as the left side of the CytoTect Triangle (Fig. 2.1).

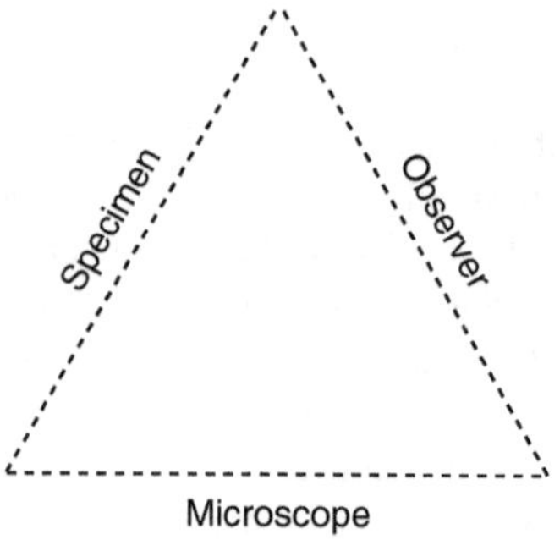

FIG. 2.1. "CytoTect Triangle" is a portmanteau for "cytodetection triangle." The CytoTect Triangle relates the interdependent roles of the specimen, observer, and microscope usage in the detection of abnormal cells.

Numerous physical, psychological, and physical factors must converge in time and space to promote the likelihood of perceiving the presence of abnormal cells. Relating this model to familiar language used in electronics, the specimen is the signal; the observer, the receiver; and the microscope, the transmitter. The many variables that impact the process can introduce noise. By optimizing and standardizing the three processes, the signal is strengthened and the noise is reduced.

Optimized processes increase the probability of abnormal cell detection (i.e., high sensitivity) and reduce the incidence of missed abnormal cells (i.e., false negatives). In Fig. 2.1, the probabilistic nature of the entire process is represented by dashed lines, rather than solid lines, as would be the case for a deterministic process such as the fire triangle.

The fire triangle, also known as the combustion triangle, illustrates simply the relationship among three elements essential to starting and sustaining fire: heat, fuel, and atmospheric oxygen. When present in suitable proportions, these elements will *always* result in combustion. To extinguish a fire, take away any 1 of the elements. The probabilistic CytoTect Triangle connotes the concept that abnormal cells will usually, but not always, be detected during the complex process of screening.

References

1. Krieger PA, McCoogan E, Vooijs GP, et al. Quality assurance/control issues. IAC Task Force Summary. Acta Cytol. 1998;42(1):133–40.
2. Luntz FL. Words that work. New York: Hyperion; 2007.
3. Cooper K. Errors and error rates in surgical pathology. Arch Pathol Lab Med. 2006;130:607–9.
4. Ismail SM, Colclough AB, Dinnen JS, et al. Observer variation in histopathological diagnosis and grading of cervical intraepithelial neoplasia. BMJ. 1989;298(6675):707–10.
5. Llewellyn H. Observer variation, dysplasia grading, and HPV typing: a review. Am J Clin Pathol. 2000;114(Suppl):S21–35.
6. Stelow EB, Skeate R, Wahi MM, et al. Pap test discrepancies and follow-up histology. Who's right and does it help to know? Diagn Cytopathol. 2003;29(2):111–5.

Chapter 3
Specimen Collection

There's no there, there.

Gertrude Stein

PRINCIPLE NO. 1

Fresh specimens facilitate processing and cell flattening.

PRACTICE

Communicate your recommended collection practices to clinicians via intranet, print manuals, and verbally. Help them, help you, help their patients.

Nongynecologic Specimens

Specimen collection is sampling cells from a body site for submission to the cytopreparatory laboratory for processing. Specimen is rooted in the Latin *specimen* meaning "indication, mark, example, sign, and evidence," from *specere* "to look at," meaning "single thing regarded as typical of its kind" first recorded in the mid-seventeenth century.[1] If any cytologic specimen is to be useful, therefore, it must be a representative (i.e., typical) of the body site sampled, whether normal or abnormal. Unlike gynecologic specimens, *all* nongynecologic specimen preparations are examined microscopically by a pathologist before being signed out. If a specimen

G.W. Gill, *Cytopreparation: Principles & Practice*,
Essentials in Cytopathology 12, DOI 10.1007/978-1-4614-4933-1_3,

© Springer Science+Business Media New York 2013

does not contain cells representative of an actual underlying lesion, and is a clinically-based false negative, the patient will not be harmed. The patient's physician will continue working up the patient until a diagnosis is established and treatment is initiated.

Regardless of body site and sampling technique, the specimen will be a cell suspension. Indeed, in the context of cytopreparation, cytologic specimens—regardless of body site—are more alike than different. They may be cellular or not, large or small in volume, and possess native fluid adhesive properties or not. In any case, the challenge is to maintain the viability of fresh cells until arrival into the laboratory or prevent cytologic degeneration by collecting the specimen in preservative. Either way, the cells are transferred in the laboratory from suspension onto a transparent surface—whether slide or filter—for subsequent processing.

Specimen collection is in the hands of the clinician, and therefore, outside the laboratory's direct control. It is absolutely essential that the laboratory educate its clinicians in specimen collection for cytology. It is unwise to assume that all clinicians know what works best. I have encountered, for example, the occasional specimen submitted in formalin or even plain water. At one time, the Cytopathology Laboratory at The Johns Hopkins Hospital provided its clinicians copies of Clinical Cytopathology Techniques for Specimen Preparation. It was a companion volume to Laboratory Cytopathology Techniques for Specimen Preparation that I authored.

The Hopkins paper guide to specimen collection has been replaced by an intranet website that is easily accessed internally but contains less detailed information. Thousands of cytology specimen collection guidelines posted by others on the Internet are available for reference. For example, Googling "specimen collection" and cytology yields 200,000+ results. Regardless, all cytology laboratories should advise their clinicians in proper specimen collection techniques to assure specimens that will serve their intended purpose. Laboratories should provide constructive feedback to clinicians who consistently submit inadequate specimens.

Collection circumstances allowing, fresh non-gyn specimens are best for cytomorphology. "Fresh" means the specimen is in its natural fluid, without added preservative. When applied to a glass surface using proper technique, fresh cells flatten in the same way as a fresh hen egg flattens on a skillet. I use word "technique" to mean doing something in a particular way that contributes to the intended outcome time and time again. Technique will appear in subsequent chapters.

"Fresh" is used in the context of nongynecologic—not FNA or gyn—specimens and includes specimens such as body cavity fluids, brushings, sputum, and urine. They are processed within a few hours following collection to promote cytomorphology that meets user expectations. If the specimen must be collected in, or suspended in, a salt solution before it is ultimately fixed in alcohol, the composition of the salt solution can make or break a specimen. Not all salt solutions are created equal. Salt solutions are covered in Chap. 4.

Collecting specimens in alcoholic preservatives has become the norm as a matter of convenience. The practice eliminates the challenge of maintaining cell viability by refrigeration or unexpected lengthy exposures to room temperatures. Nowadays, FDA-approved liquid-based cytology processing systems require the use of patented (i.e., CytoLyt) or proprietary (i.e., CytoRich) preservatives for gyn and non-gyn specimens. The action of preservatives and consequences of using them will be discussed later in Chap. 8.

Fresh Body Cavity Fluids: To Clot or Not?

Properly prepared body cavity fluids can be a thing of joy and beauty forever. Naturally present proteins protect cells for many hours.[2] However, unless the specimen is collected to avoid clotting, cytopreparation can be problematical. Body cavity fluids frequently contain greater concentrations of thrombin and fibrinogen than blood. If not counteracted by heparin, the thrombin will act on fibrinogen to form fibrin (i.e., a clot). Fibrin, in turn, may entrap potentially diagnostic cells. In plain English, clots are a nuisance to process, if for no other reason to avoid them.

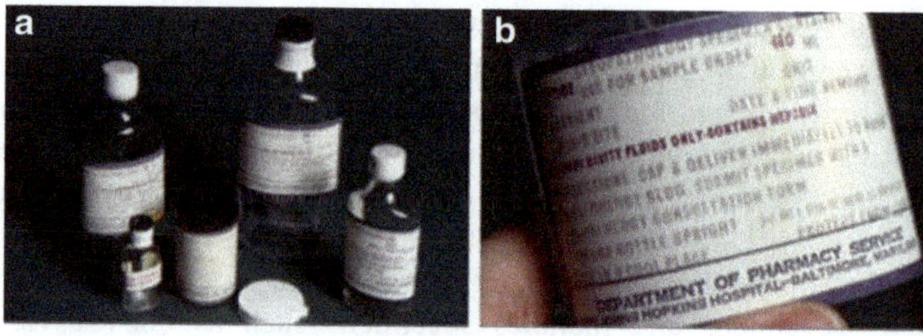

FIG. 3.1. Heparinized bottles in 4 sizes: 15, 120, 240, and 480 mL. The 15-mL bottle is used for cerebrospinal fluids. The bottle is too small to accommodate a label, so the bottle is provided in a larger medication vial that can be labeled.

At the Cytopathology Laboratory of The Johns hopkins Medictal Institutions in Baltimore, prehaparinized sterile bottles were prepared in four different sizes by Pharmacy Services.[3] See Fig. 3.1. The concentration of heparin is 3 units per mL capacity. A serendipitous benefit was that these bottles stopped clinicians from sending gallons of body cavity fluids to the laboratory for disposal. Preheparinized bottles are not used at Hopkins any more. Today, body cavity fluid specimens are submitted in blood collection bags. Heparinized hypodermic needles prevent clotting during the paracentesis procedure.

If blood collection bags are not available, clotting can be prevented by adding 3 units of heparin per cc capacity of the collection container *before* the specimen is added. If added after a specimen has been collected in a glass container, heparin won't prevent clotting. For example, "**Effusions-Ascites, Pleural OR Pericardial:** If a specimen can be transported promptly to the lab, we prefer the fresh fluid. If it cannot be brought immediately, add 3 units of heparin per mL of fluid as precaution against clotting, and place in refrigerator until it can be delivered."[4]

Heparin is rooted in a Greek word for liver. One unit of heparin is the quantity required to keep 1 mL of cat's blood fluid for 24 h at 0° C (i.e., unitized dosing). It is available commercially as 1,000 units per mL. A unit is equivalent to about 0.002 mg pure heparin. Therefore, a very small volume is needed to prevent clotting of a body cavity fluid.

Not all body cavity fluids contain the requisite chemicals that will cause clotting. Those that do, however, often contain higher concentrations and require 3 units of heparin per mL instead of the usual 1 unit. Since one cannot know in advance which will clot, it is wise to assume that all will and to add heparin routinely. To prevent clotting of a 100-mL body cavity fluid, add 0.3 mL heparin (1,000 units per mL) to the collection container before adding the specimen.

Gynecologic Specimens

Pap tests are applied to populations of symptomless women to detect cervical cancer and its precursors. Under such circumstances, the Pap test is a screening technique. A small proportion of these women have underlying lesions that, if left undetected, can develop into cervical cancer and be fatal. For this reason, it is essential that clinicians exercise effective specimen collection techniques. To appreciate today's Pap test, it helps to understand yesterday's Pap test.

Background [5–7]

The cervicovaginal smear as a means to detect cervical cancer and its precursors was introduced independently about a year apart by Romanian physician Aurel A. Babeş and Greek-American physician George N. Papanicolaou in the late 1920s. They were age cohorts. Papanicolaou was born in 1883, 3 years before Babeş, and died at age 79 in 1962, 1 year later than Babeş.

In 1928, Babeş published his only major contribution to cytopathology.[8] It accurately described cells of squamous cell carcinoma in cervical smears. The cells had been taken using a platinum loop to transfer cells from the cervix to the slides, which were then air-dried and Giemsa-stained.

In January 1928, Papanicolaou presented his findings of cancer cells in vaginal aspirates at the Third Race Betterment Conference, which was sponsored by the Kellogg Foundation, in Battle Creek, Michigan.[9] The initial response was underwhelming, to say the least. Most believed at that time that cancer could be diagnosed only by examining tissue. It is fair to say that no one thought early cancer could be detected in symptomless women.

Discouraged by the lukewarm response to his 1928 presentation, Papanicolaou abandoned his uterine cancer researches for nearly 10 years. He resumed his researches at the suggestion of the chairman of Cornell University's Department of Anatomy in New York City. In 1941, Papanicolaou and Herbert Traut, professor of gynecology, published an article[10] and, in 1943, a monograph[11] that heralded the dawn of the modern era of cytopathology.

The Pap test was introduced into clinical practice in the mid-1940s and is now heralded as the most successful cancer screening test in medical history.[12] There are two types of Pap tests: conventional, yesterday's Pap test, and liquid-based preparations, today's Pap test. The conventional Pap test was the standard of practice for decades; there was no competitive alternative.

Conventional Pap Test

The conventional Pap test does not require the approval of the Food and Drug Administration (FDA). The FDA is the oldest comprehensive consumer protection agency in the US federal government. Its origins can be traced back to the Patent Office around 1848. Although it was not known by its present name until 1930, FDA's modern regulatory functions began with the passage of the 1906 Pure Food and Drugs Act, a law a quarter-century in the making that prohibited interstate commerce in adulterated and misbranded food and drugs. FDA is responsible for protecting the public health by assuring the safety and efficacy of products sold interstate commerce. The conventional Pap test is not a commercial product and so is exempt from FDA approval.

In the beginning, only plain glass slides without a frosted label end were available. Papanicolaou wrote, "*Materials needed.* 1. Clean slides, labeled beforehand with a piece of index card 1 inch square fastened to one end of the slide by a paper clip. The patient's name and the date and type of smear should be written on the label with pencil. It is advisable to etch the smear type on the slide with a diamond pencil."[13] The Hopkins cytoprep techs were using diamond pencils when I began as a student in 1963. The practice was discontinued in 2011 because of the availability of better labeling systems.

The original Pap tests were either vaginal aspirates or "swab smears" that were taken with either a nonabsorbent swab or a wooden spatula such as Ayre's. Parenthetically, James Ernest Ayre, a Canadian physician, was granted a patent for his eponymous spatula in 1949.[14]

About 60 million Pap tests are performed annually in America. No one knows for sure how many, but it's a big number. I've seen estimates as low as 35 million and as high as 80 million. In this section of specimen collection, I want to emphasize the role of proper technique in transferring cells from the endocervical brush into liquid preservative and its demonstrable contribution to improved LBP (liquid-based preparation) performance.

Liquid-Based Preparation

The term "liquid-based preparation" was introduced in 1998. "A variety of terms have been proposed to describe the methodology for rinsing sampling implements in liquid preservative, which is then transported to the laboratory, where the specimen is partially homogenized (usually) and a subsample, thin-layer preparation is made. Monolayer conveys a visual image of the process, but most preparations are not true monolayers. Thin layer is more accurate than monolayer, but some conventional smears can be thin layer, too. The workshop participants suggest the term liquid-based preparation, which avoids confusion with proprietary names."[15]

Two FDA-approved liquid-based preparations are marketed today: (1) ThinPrep Pap test (TPPT)[16] and (2) SurePath Pap test (SPPT).[17] They have virtually replaced the conventional Pap test as the standard of practice in a relatively short time. Conventional Paps were introduced in the mid-1940s; the first FDA-approved LBP, 1996.

The inherent limitations of the conventional Pap test and unethical practices by so-called Pap mills were brought to light by the now infamous 1987 Wall Street Journal exposés by Walt Bogdanich.[19–21] These articles were undoubtedly responsible—directly or indirectly—for CLIA '88,[22] the development, FDA-approval[16, 17] and commercialization of LBP, the Bethesda nomenclature system,[23] improved endocervical sampling devices,[25, 29] and who knows what else.

- 1985—Majority of false-negative conventional Pap tests were due to clinical sampling errors (62%); the balance, screening errors (16%), or interpretation errors (22%).[24]
- 1988—Endocervical brush patented.[25]
- 1994—"Counts of epithelial cells on conventional smears showed that only a fraction of the available epithelial cells on the sampling devices (medians, 6.5% to 62.5%) was actually deposited on the slides. In all 27 cases studied with the ThinPrep method, equivalent diagnostic material was obtained on each of the replicate slides prepared per specimen. This identifies a new source of error, preparation error, in conventional smears."[26]
- 1996—"The one case where an abnormality was detected only on the third slide suggests the possibility of nonrandom subsampling and of discarding the majority of the cellular material from the patient."[27]
- 1996—ThinPrep Pap test FDA-approved for use with broom.[16]
- 1997—ThinPrep Pap test FDA-approved for use of the combination of endocervical brush and plastic spatula as an alternative to the broom-like sampling device.[17]
- 1999—SurePath Pap test (*née* AutoCyte Pap test) FDA-approved.[18]
- 2000—*"I think that's the most common complaint of physicians when they switch to liquid-based sampling. They start getting a substantial increase in the number of smears labeled 'inadequate because of no endocervical component'... The brush can hang on to those cells tenaciously. To avoid this problem, the operator should not only swish the brush in the fluid but also*

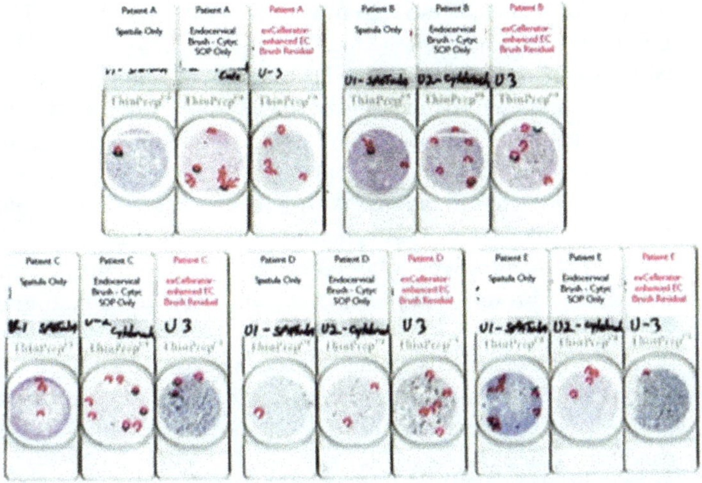

FIG. 3.2. These five sets of 3-part ThinPrep Pap tests were part of a pilot study that demonstrated that residual cellular material remains entrapped in the bristles of the endocervical brush when handled as recommended: "Rinse the brush as quickly as possible in the PreservCyt Solution by rotating the device in the solution ten times while pushing against the PreservCyt vial wall. Swirl the brush vigorously to further release material. Discard the brush."[34] An online video is available.[35]

use the spatula to strip off the mucus from the brush into the fluid. Collection of endocervical cells in that way, and the rate of 'satisfactory but limited by lack of endocervical cells' will drop.'[28] See Fig. 3.2.

- 2002—Apparatus [exCellerator] and method for removing cells from an endocervical brush in a liquid-based Pap test system patented.[29] The device was developed to ensure that virtually 100% of the sample was indeed transferred into the liquid preservative. Pilot studies had demonstrated that about half of the sample remained in the endocervical brush when handled as recommended by the manufacturer.
- 2003—"Globally, 37% of cellular material is lost when the collecting device is discarded. Loss of material is different among gynecologists. The more intense the rinsing process, the less the

loss, but the latter is never zero and is poorly predictable. The discarded subsample often contains a greater amount of endocervical clusters. *Conclusions*. Discarding collecting in liquid-based cytology reproduces one of the flaws of the conventional smear technique. Losing cellular material may have an impact on cervical cancer detection, but this still has to be evaluated with further investigations."[30] Authors did not investigate further.

- 2004—FDA approves alternative collection device (i.e., detachable broom head) for use with TriPath Imaging's PrepStain[31]
- 2006—"Abnormal cells would have been discarded with the TP broom and the amount varied between the colposcopists. This finding could have an affect [*sic*] on adequacy of the sample used for routine diagnosis."[32]
- 2008—TPPT specimens collected with exCellerator that are examined with ThinPrep Imaging System shown to positively impact clinical outcomes.[33] Unless and until FDA-approved, the exCellerator can only be sold and distributed by lab to client clinicians under the category of off-label use. If FDA approval is obtained it could then be fully commercialized, making it available for sale to other laboratories.

The positive impact of enhanced transfer of cellular material from the endocervical brush into the liquid preservative is shown in Fig. 3.3.

Detecting abnormal cells is the outcome of a series of interdependent samplings of successively diminishing size: (1) the specimen collection technique samples the biologic process, (2) the cytopreparatory technique samples the specimen, (3) the screening process samples the preparation, and (4) the morphological interpretation samples the cellular features. It has been demonstrated "that 'rare events' may be overlooked when limited sample aliquots are analyzed by ABLC [automated liquid based cytology] instruments."[36] It is obviously essential, therefore, that as many representative, well-preserved cells as possible be harvested.

Since gyn specimens are not collected by laboratory personnel, the laboratory should educate its clinicians in all aspects of proper technique. This includes reminding them not only about instructions to the patient but also tips on proper sampling technique.[37–39]

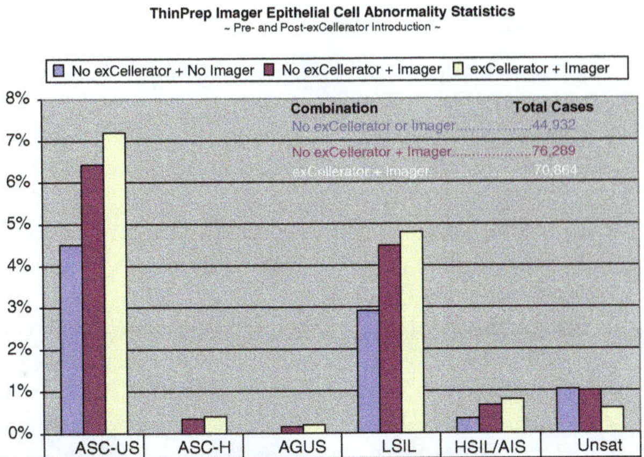

	ASC-US	ASC-H	AGUS	LSIL	HSIL/AIS	Unsat
No exCellerator + No Imager	4.51%	0	0	2.92%	0.34%	1.04%
No exCellerator + Imager	6.42%	0.35%	0.15%	4.48%	0.66%	1.01%
exCellerator + Imager	7.20%	0.40%	0.20%	4.80%	0.80%	0.58%

FIG. 3.3. The increased detection of epithelial cell abnormalities is statistically significant for all categories except AGUS.

The majority of women don't know that the Pap test is intended to detect cancer and its precursors.[40] There is lack of consensus about the best sampling devices, how best to use them, whether the presence or absence of endocervical cells impacts cervical disease detection,[41] the impact of blood and lubricant,[42–45] and on and on. I believe the laboratory is best qualified to educate its client physicians and to advocate for the patient. The Pap test is sometimes characterized as a screening test, implying that one should not rely too heavily on a single Pap test results. However, it is often a woman's best and only chance to avoid developing invasive cervical cancer.

The laboratory assesses microscopically whether specimens are satisfactory or unsatisfactory, and in my view, is ethically obligated to educate its clients when they are submitting specimens that are not useful for their intended purpose. Putting patients first is always best.

References

1. Online Etymology Dictionary. http://etymonline.com/?term=specimen. Accessed 21 Feb 2012.
2. Manosca F, Schinstine M, Fetsch PA, et al. Diagnostic effects of prolonged storage on fresh effusion samples. Diagn Cytopathol. 2007;35(1):6–11.
3. Roffe BD, Wagner FH, Derewicz HJ, Gill GW. Heparinized bottles for the collection of body cavity fluids in cytopathology. Am J Hosp Pharm. 1979;36(2):211–4.
4. Shands at UF Medical Laboratory. Specimen Collection and Test Catalog. PS108542. 8/1/11. http://www.shands.org/professionals/labs/specimenreport.pdf. Accessed 21 Jan 2012.
5. Naylor B, Tasca L, Bartziota E, Schneider V. In Romania it's the Méthode Babeş-Papanicolaou. Acta Cytol. 2002;46(1):1–12.
6. Naylor B. The century for cytopathology. Acta Cytol. 2000;44(5):709–25.
7. Naylor B. Perspectives in cytology. From Battle Creek to New Orleans. Acta Cytol. 1988;32(5):613–21.
8. Babeş A. Diagnostic du cancer du col utérin par les frottis. Presse Med. 1928;36:451–4.
9. Papanicolaou GN. New cancer diagnosis. Proceedings of 3rd Race Betterment Conference, Battle Creek, MI; 1928. p. 528–34.
10. Papanicolaou GN, Traut HF. The diagnostic value of vaginal smears in carcinoma of the uterus. Am J Obstet Gynecol. 1941;42:193–205.
11. Papanicolaou GN, Traut HF. Diagnosis of uterine cancer by the vaginal smear. New York, NY: Commonwealth Fund; 1943.
12. Carmichael DE. The Pap smear: Life of George N. Papanicolaou. Springfield, IL: Charles C. Thomas; 1973.
13. Papanicolaou GN. Atlas of exfoliative cytology. Cambridge, MA: Published for The Commonwealth Fund by Harvard University Press; 1954.
14. Ayre JE, inventor. Cervical scraper. US patent 2471088. May 24; 1949.
15. McGoogan E, Colgan TJ, Ramzy I, et al. Cell preparation methods and criteria for sample adequacy. IAC Task Force Summary. Acta Cytol. 1998;42(1):25–32.
16. Hologic, Inc. ThinPrep Processor, Model TP 2000. PMA Number P950039. Decision date: 20 May 1996.
17. Hologic, Inc. ThinPrep Processor, Model TP 2000. PMA Number P950039, Supplement 002. Decision date: 25 Feb 1997.

18. TriPath Imaging. AutoCyte Prep System. PMA Number P970018. Decision date: 17 Jun 1999.
19. Bogdanich W. False negative. Medical labs, trusted as largely error-free, are far from infallible. Haste, misuse of equipment, specimen mix-ups afflict even best labs at times. Regulation; weak and spotty. Wall St J. 2 Feb 1987.
20. Bogdanich W. Lax laboratories. The Pap test misses much cervical cancer through labs' errors. Cut-rate 'Pap mills' process slides with incentives to rush. Misplaced sense of security? Wall St J. 2 Nov 1987.
21. Bogdanich W. Physicians' careless with Pap tests is cited in procedure's high failure rate. Wall St J. 29 Dec 1987.
22. Part 493—Laboratory Requirements. 42 CFR Ch. IV (10–1–04 Edition):967–1087. http://wwwn.cdc.gov/clia/regs/toc.aspx. Accessed 15 Jan 2012.
23. Solomon D, Davey D, Kurman R, et al. The 2001 Bethesda system: terminology for reporting results of cervical cytology. JAMA. 2002;287(16):2114–9.
24. Gay JD, Donaldson LD, Goellner JR. False-negative results in cervical cytologic studies. Acta Cytol. 1985;29(6):1043–6.
25. Stormby N, inventor. Endocervical sampling brush and smear method. US patent 4759376. 26 Jul 1988.
26. Hutchinson ML, Isenstein LM, Goodman A, Hurley AA, Douglass KL, Mui KK, Patten FW, Zahniser DJ. Homogeneous sampling accounts for the increased diagnostic accuracy using the ThinPrep Processor. Am J Clin Pathol. 1994;101(2):215–9.
27. Goodman A, Hutchinson ML. Cell surplus on sampling devices after routine cervical cytologic smears. A study of residual cell populations. J Reprod Med. 1996;41(4):239–41.
28. Richart RM, Ashfaq R, Cox JT, Ferenczy A. Current understanding and management of glandular lesions of the cervix. Contemporary OB/GYN. April 15, 2000 Supplement.
29. Gill GW, Snyder JA, inventors; Diagnostic Cytology Laboratories Inc, assignee. Apparatus and method for removing cells from an endocervical brush in a liquid-based Pap test system. US patent 6394966. 28 May 2002.
30. Bigras G, Rieder MA, Lambercy JM, et al. Keeping collecting device in liquid medium is mandatory to ensure optimized liquid-based cervical cytologic sampling. J Low Genit Tract Dis. 2003;7(3):168–74.
31. TriPath Imaging. PrepStain system: Approval for modified labeling of the prepstain system for use with an alternate specimen collection device. PMA Number P970018, Supplement Number S007. Decision date: 28 May 2004.

32. Umana A, Kubba AA, Dunsmore H, Kokhan AR, Herbert A. A study to determine whether there is significant cell loss on the discarded ThinPrep broom. Cytopathology. 2006;17 Suppl 1:15–6.
33. Gill GW, Eisenhut CC, Snyder JA. Excellerator™: a new tool to improve specimen adequacy in ThinPrep Pap tests. Cancer Cytopathol (Supplement). 2008;114(5):386. Poster 57.
34. Cytyc Corporation. ThinPrep Pap test quick reference guide: Endocervical Brush/Spatula Protocol. Part No. 85217-001 Rev. G.; 2003.
35. Hologic. ThinPrep Pap test specimen collection video (7′45″). http://www.thinprep.com/hcp/specimen_collection/view_video.html. Accessed 25 Feb 2012.
36. Maksem JA, Dhanwada V, Trueblood JE, et al. Testing automated liquid-based cytology samples with a manual liquid-based cytology method using residual cell suspensions from 500 ThinPrep cases. Diagn Cytopathol. 2006;34(6):391–6.
37. DeMay RM. The Pap test. Chicago, IL: ASCP Press; 2005.
38. CLSI. Cervicovaginal cytology based on the Papanicolaou technique; Approved guideline—third edition. CLSI document GP15-A3, Wayne, PA: Clinical and Laboratory Standards Institute; 2008.
39. Ontario Medical Association Laboratory Proficiency Testing Program, Thompson DW. Adequate "pap" smears: a guide for sampling techniques in screening for abnormalities of the uterine cervix. ON, Canada: Laboratory Proficiency Testing Program; 1993.
40. The Gallup Organization. Public Awareness and Behavior Regarding Preventive Health Screenings. College of American Pathologists. Jan 2004.
41. Bigras G, de Marval F. The probability for a Pap test to be abnormal is directly proportional to HPV viral load: results from a Swiss study comparing HPV testing and liquid-based cytology to detect cervical cancer precursors in 13,842 women. Br J Cancer. 2005;93(5):575–81.
42. Feit TD, Mowry DA. Interference potential of personal lubricants and vaginal medications on TPPT. J Am Board Fam Med. 2011;24(2):181–6.
43. Holton T, Smith D, Terry M, et al. The effect of lubricant contamination on TPPT preparations. Cytopathology. 2008;19(4):236–43.
44. Hack LA, O'Brien D, Selvaggi SM. Protocol for the processing of bloody cervical specimens: glacial acetic acid and ThinPrep Pap tests. Diagn Cytopathol. 2006;34(3):210–3.
45. Rowe LR, Bentz JS. A simple method to determine the need for glacial acetic acid treatment of bloody ThinPrep Pap tests before slide processing. Diagn Cytopathol. 2004;31(5):321–5.

Chapter 4
Salt Solutions

Normal saline is neither normal nor physiologic.

Principle No 1

Fresh specimens facilitate specimen processing and cell flattening.

Practice

Use balanced salt solutions to maintain cell viability and morphology.

Historic Milestones

- 1831—Normal saline-like solutions used during cholera pandemic in Europe[1]
- 1882—Ringer identified importance of inorganic salts in maintaining tissue viability[2]
- 1948—Hanks described balanced salt solution for in vitro use[3]

Specimen collection and cytopreparation often use salt solutions. Varieties are available. Their compositions vary, their suitability for use with living cells varies, and they should not be used interchangeably without the user appreciating the differences and

G.W. Gill, *Cytopreparation: Principles & Practice*,
Essentials in Cytopathology 12, DOI 10.1007/978-1-4614-4933-1_4,
© Springer Science+Business Media New York 2013

the possible good and bad consequences. The basic salt solutions include the following: (1) normal saline, (2) balanced electrolyte solutions, and (3) balanced salt solutions.

Normal Saline

Normal saline is 0.9% w/v (i.e., weight/volume) sodium chloride in water. See Tables 4.1 and 4.2.

Don't use normal saline for any purpose (i.e., specimen collection or cytopreparation) unless water is the only available alternative. It bursts cells when its volume is great relative to that of the raw specimen (e.g., a bronchial alveolar lavage collected in normal saline or a cell pellet resuspended in normal saline). See Fig. 4.1.

Nothing is *normal* about normal saline. It is sometimes referred to as isotonic saline or physiologic saline. None of the terms is correct. A physiological saline, for example, is a solution of the necessary ionic concentration and osmotic pressure to allow cells to survive in them without damage.[4] That's certainly not normal saline. The composition of solutions referred to in published articles as physiologic saline should be specified so readers will know exactly what has been used.

"While it is eminently obvious to all who have been exposed to an elementary course in physiology that the term 'normal' saline risks misinterpretation, practical considerations often lead to the abundant use and, perhaps frequently, to the abuse of a 0.9% sodium chloride or other un-physiologic material such as 5% glucose in water for topical wound therapy and for parenteral administration… As might be expected, so-called normal saline is highly damaging to isolated cells and therefore can be shown to be a poor vehicle for substances to be used in biological work."[5]

And: "The use of 0.9% saline is believed to have originated during the cholera pandemic that swept across Europe in 1831. However, an examination of the composition of the fluids used by the pioneering physicians of that era reveals solutions that bear no

TABLE 4.1. Commonly used salt solutions.

	Properties									
Solution	Iso-osmotic	Inorganic ions	pH	Buffered?	Glucose?	Sterile?	Pyrogen-free?	Disrupt cells?	Safe Exposure	Use
Normal saline	Yes	Not Balanced	6.0	No	No	Yes	Yes	Yes	Seconds	Don't
BES[a]	Yes	Balanced	6.0	No	No	Yes	Yes	No	Minutes	In vivo
Hanks' BSS[b]	Yes	Balanced	7.4	Yes	Yes	Yes	No	No	Hours	In vitro

[a]BES = Balanced electrolyte solution (in vivo use [parenteral injection in humans]). Such solutions are sold by companies such as Baxter with a variety of product names (e.g., Normosol-R [replacement], Tis-U-Sol, and PlasmaLyte). These solutions can also be used in vitro.

[b]BSS = Balanced salt solution (in vitro use [irrigation medium in tissue culture]). These should not be used in vivo.

TABLE 4.2. Compositions of commonly used salt solutions.

No.	Ingredient (g/L)	Normal saline	BES	Hanks BSS[9, 10]
1.	Sodium chloride	9.0	4.96	8.0
2.	Sodium acetate	–	7.48	NA
3.	Potassium chloride	–	0.746	0.4
4.	Calcium chloride	–	0.368	0.14
5.	Magnesium chloride · $6H_2O$	–	0.305	0.1
6.	Magnesium sulfate · $7H_2O$	–	–	0.1
7.	Disodium phosphate · $2H_2O$	–	–	0.06
8.	Monosodium phosphate	–	–	0.06
9.	Sodium bicarbonate	–	–	0.35
10.	Dextrose	–	–	1.0

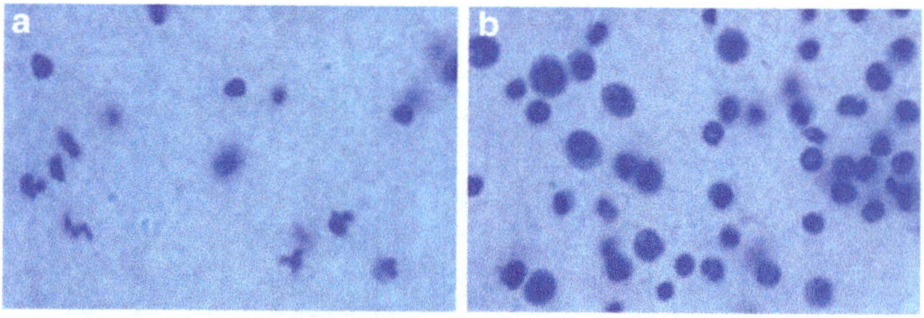

FIG. 4.1. These photomicrographs show white blood cells from a split peripheral blood sample on Millipore filters. Each sample was processed identically, with the exception of the rinse solutions, by a saponin technique, which hemolyzes erythrocytes: (**a**) normal saline, (**b**) Hank's balanced salt solution (Ca- and Mg-free). Norman saline destroys living cells.

resemblance to 0.9% or so-called 'normal' saline which appears to have very little scientific or historical basis for its routine use, except for Hamburger's in vitro studies of red cell lysis. The currently used 0.9% saline solution is without convincing historical basis. Given that the composition of 0.9% sodium chloride is dissimilar to most solutions used in the past, and is in no way 'normal' or 'physiological', our current practice may be based on historical fallacy and misconception."[1]

Balanced Electrolyte Solutions

Balanced electrolyte solutions are also referred to as physiological salt solutions. Physiological salt solutions are based on the ionic composition of plasma. The latter concept is the foundation on which chemically defined media for cell and tissue culture have been built and on which balanced electrolyte solutions are based as well.

English physiologist Sydney Ringer developed physiological salt solutions to study the contractions in isolated amphibian heart and leg muscles.[1, 6–9] Even today commercially available salt solutions for parenteral use bear his name. *Parenteral* means outside the intestine, and refers—by excluding the intestine—to administration by injection. Such nomenclature, in my mind, is similar to the use of nongynecologic to refer to any specimen that does not originate in a gynecologic site.

When I arrived at Hopkins in 1963, two types of salt solutions were used in the Cytopreparatory Laboratory: normal saline and a balanced electrolyte solution known as PolySal. PolySal was used for cerebrospinal fluid specimens and normal saline was used for everything else. We didn't know then about the differences in composition, who introduced their use in our laboratory, or why.

Balanced electrolyte solutions are sterile and pyrogen-free (also referred to as nonpyrogenic) and therefore suitable for both in vitro and in vivo use. Pyrogens are any substance or agent capable of producing fever (i.e., Greek: *pyr,* fire; *genein,* to produce). Cells collected in alcoholic preservative are dead and do not require any salt solution.

Table 4.1 summarizes salient properties of commonly used salt solutions, which properties are reflective of the differences in composition as shown in Table 4.2. This knowledge exemplifies the level of detail needed to ensure quality cytologic preparations.

Balanced Salt Solutions

Balanced salt solutions originated in tissue culture researches. There are two types of balanced salt solutions: transfer and maintenance. Cells grown in culture for months or years (e.g., HeLa) require more exacting chemically defined media (i.e., maintenance) than those media for short-term exposure (i.e., transfer). Cells grown in culture must be resuspended periodically and transferred to another growth vessel. Because the process can be completed in minutes, cells can survive well in transfer media with many fewer ingredients than in maintenance media. Hanks BSS transfer medium is based on ten ingredients; Eagle's maintenance medium, dozens.

Balanced salt solutions have a physiologic pH of 7.4 that is buffered to resist changes in pH, are iso-osmotic, and include dextrose to support continued cell metabolism. BSS are intended primarily for tissue culture. BSS are sterile but not pyrogen-free and therefore are suitable for in vitro use, but not in vivo use. Balanced salt solutions are indicated for clinical specimen collection (e.g., bronchial lavage specimens, not the lavage that is instilled into the patient's bronchi), though balanced electrolyte solutions will suffice. Waymouth has written extensively about balanced salt solutions use in tissue culture.[11–14]

Being more complex than balanced electrolyte solutions and sold in smaller volumes, BSS are much more expensive than BES. Laboratories should choose a mix of the two types of salt solutions based on their particular needs and budgets. In any case, do not buy normal saline to save money. It is false economy that will cost more in wasted specimens than it saves. Particularly, be aware that bacteria flourish in unused salt solutions in opened bottles. Discard the contents; don't use in patients or specimens. Refrigeration doesn't keep bacteria from growing under these circumstances.

Isotonicity and Iso-osmalarity

Isotonicity and iso-osmolarity are terms frequently encountered in the salt solution literature. They may appear to be synonymous but in fact mean different things. All salt solutions, whether intended for long-term cell cultures or brief cell survival, must at least be isotonic, literally "of equal tone," with solutions that maintain human erythrocytes morphologically intact, without shrinking or swelling. Tonicity is commonly described in terms of erythrocytic fragility. An isotonic solution is one that causes no, or minimal, hemolysis.

"Isotonicity is relative to particular situations; osmolarity can be measured in absolute terms. For example, 0.9% NaCl (=155 mM NaCl) prevents change in size and shape of normal human red blood cells, that is, is isotonic with these cells; as well as isosmotic (or isopiestic) for them. A solution of 320 mM urea is isoosmotic with 155 mM NaCl, but not in this case isotonic, because the erythrocyte membrane is permeable to urea, which therefore exerts no osmotic pressure on the cell."[14]

References

1. Awad S, Allison SP, Lobo DN. The history of 0.9% saline. Clin Nutr. 2008;27(2):179–88.
2. Ringer S. Concerning the influence exerted by each of the constituents of the blood on the contraction of the ventricle. J Physiol. 1882;3(5–6):380–93.
3. Hanks JH. The longevity of chick issue cultures without renewal of medium. J Cell Comp Physiol. 1948;31:235–60.
4. Lockwood APM. "Ringer" solutions and some notes on the physiological basis of their ionic composition. Comp Biochem Physiol. 1961;2:241–89.
5. Pomerat CM, Overman RR. Electrolytes and plasma expanders – I. Reaction of human cells in perfusion chambers with phase contrast, time-lapse cine records. Z Zellforsch Mikrosk Anat. 1956;45(1):2–17.
6. Ringer S. A further contribution regarding the influence of the different constituents of the blood on the contraction of the heart. J Physiol. 1883;4(1):29–42.
7. Ringer S. A third contribution regarding the influence of the inorganic constituents of the blood on the ventricular contraction. J Physiol. 1883;4(2–3):222–5.
8. Ringer S. Further experiments regarding the influence of small quantities of lime potassium and other salts on muscular tissue. J Physiol. 1886;7(4):291–308.

9. Ringer S. Further observations regarding the antagonism between calcium salts and sodium potassium and ammonium salts. J Physiol. 1895;18(5–6):425–9.
10. Hanks JH, Wallace RE. Relation of oxygen and temperature in the preservation of tissues by refrigeration. Proc Soc Exp Biol Med. 1949;71(2):196–200.
11. Waymouth C. Construction and use of synthetic media. In: Willmer EN, editor. Cells and tissues in culture, vol. 1. New York, NY: Academic; 1965. p. 99–142.
12. Waymouth C. Simple nutrient solutions for animal cells. Tex Rep Biol Med. 1955;13(3):522–36.
13. Waymouth C. The nutrition of animal cells. Int Rev Cytol. 1954;3(1):1–68.
14. Waymouth C. Osmolality of mammalian blood and of media for culture of mammalian cells. In Vitro. 1970;6(2):109–27.

Chapter 5
Slide Preparation

"Adhesion flattens and retains cells; cohesion fattens and loses cells."

PRINCIPLE NO. 3
Flatten cells to enhance chromatin display.

PRACTICE

Use materials and methods that diminish intracellular cohesive forces and enhance extracellular adhesive forces.

Historic Milestones

- 1691—Bonanni published first illustration of a "slider."[1]
- Early 1800s—Slider transitions to slide.[2]
- 1840—Microscopical Society of London sets 3×1-in. as the standard size for micro slides.[3]
- 1957—Frosted "Dakin" slide patented.[4]
- 1970—ASTM publishes its first Standard Specification for Cover Glasses and Glass Slides for Use in Microscopy.[5]

Slide preparation in the context of cytopreparation simply means transferring cells from suspension in the raw specimen onto a transparent surface for subsequent processing. The transparent support medium is ultimately always glass. However, the cells may have been collected on a cellulosic membrane filter, which is unusual nowadays, or on a polycarbonate filter that is imprinted onto a glass

G.W. Gill, *Cytopreparation: Principles & Practice*,
Essentials in Cytopathology 12, DOI 10.1007/978-1-4614-4933-1_5,
© Springer Science+Business Media New York 2013

slide as is the case with ThinPrep processors. In any case, the details matter. Before going further, however, I want to provide little-known background information about micro slides per se.

Background

Why are slides called slides? In the seventeenth century, slides were known as sliders, which served a similar purpose but in a different format.

> Many, even those who have purchas'd Microscopes, are so little acquainted with their general and extensive Usefulness, and so much at a Loss for Objects to examine by them; that after diverting themselves and their Friends, some few times, with what they find in the Sliders bought with them, or two or three more common Things, the Microscopes are laid aside as of little farther value: and, a Supposition that this must be the Case, prevents many others from buying them…
>
> Most Objects require some Management, in order to bring them properly before the Glasses. — If they are flat and transparent, and such as will not be injured by Pressure, the best Method is to inclose them in Sliders, between two Moscovy Talcs or Isinglass…[6]

The specimen plane was oriented vertically for microscopic examination and was slid back and forth to bring the different specimens into the optical axis of the horizontal microscope. Hence, the specimen holder came to be known as a slider. More than a century later, the optical axis of the microscope became vertical, and sliders became known as slides as the preparations were oriented horizontally, at right angles to the optical axis, as they are today.

By similar reasoning, it may be that color film transparencies are also known as slides. The early projectors held the transparency in the light bath in a 2-transparency holder that the operator slid back and forth, much like a microscope slider.

Perspective: Gyn Versus Non-gyn Specimens

Before the methods are described, it is instructive to compare and contrast various aspects of gyn and non-gyn cytologic specimens, as such an analysis may account in part for the different levels of

Table 5.1. Gyn vs. Non-gyn cytologic specimens.

No.	Attribute	Gyn specimens	Non-gyn specimens
1.	Source	Female genital tract	Both sexes, all body sites
2.	Test status	Screening test	Diagnostic work-up component
3.	Numbers	60,000,000+/year	6,000,000/year (?)
4.	Collection method	Spatula and/or brush	Exfoliation, abrasion, aspiration
5.	Slides per specimen	Usually 1, sometimes 2, rarely 3	1–10+
6.	Preparation	Outside laboratory (pre-LBP)	Inside laboratory
7.	Preparation area	133 mm^2 thru 1,440 mm^2	32 mm^2 thru 1,200 mm^2
8.	Slide prep method	Monolayer device (i.e., LBP) or Pap smear	Smear, cytocentrifuge, monolayer device, membrane filtration
9.	Fixation status	Patented or proprietary preservative followed by fixation or spray-fixed, sometimes wet-fixed	Fresh, preserved, air-dried, or spray-fixed
10.	CT Responsibility	Signs out negatives only	Forward all to pathologist
11.	QC/QA	CLIA-mandated 10% negative review	All reviewed by pathologist
12.	Problematic	False negatives; 90% all negatives/non-high risk seen only once	False positives; pathologists see everything
13.	Reporting system	The Bethesda system	Less standardized reporting system, except for "Thyroid Bethesda"
14.	Patient management	Result influences	Result may impact less if wrong
15.	Accountability	Unrealistic public expectations	Public relatively unaware
16.	Litigation risk	High	Low

attention each receives relative to quality control and quality assessment. See Table 5.1.

Slide Preparation

From the standpoint of specimen collection and slide preparation, it is empowering to appreciate that cytologic specimens are more alike than different. All are suspensions of initially viable cells that require similar handling if they are to be useful diagnostically. Rather than let the body site determine the method of preparation, let the characteristics of the specimen guide the choice of which of several methods are best suited for processing. See Table 5.2.

The descriptive terms small and large, and watery and viscous, are somewhat arbitrary. For lack of a better word, I use "viscous" to describe specimen compositions that will adhere to slides with or without centrifugal enrichment.

- Small volume viscous specimens such as sputum and FNA (Fine needle aspiration) can be spread directly onto a slide.
- Small volume watery specimens such as CSF and urine aren't suitable for direct smears.
- Large volume watery and viscous specimens require centrifugation to reduce the volume and enrich the cell concentration. Depending on the resulting size, the pellets can be spread directly onto a slide or be resuspended in a suitable medium and processed by LBP, cytocentrifugation, or membrane filtration.

Table 5.2. Specimen composition guides slide preparation method.

	Viscosity	
Volume	I. Watery	II. Viscous
A. Small	Cerebrospinal fluids	Sputum
	Urine	FNA
B. Large	Urine	Bloody effusions
	Pleural fluids	Clotted effusions
	Peritoneal fluids	
	Pericardial fluids	

Differences in slide preparation arise primarily in the method of effectively transferring the cells from suspension onto a display medium. The methods menu is straightforward:

1.	Direct smears	**A.II**
2.	Liquid-based preparations, either manual or automated	**A.I, B.I, B.II**
3.	Cytocentrifugation	**A.I, B.I, B.II**
4.	Membrane filtration	**A.I, B.I, B.II**

Of the four basic methods, only devices sold for the purpose of making cervical cytology slides require FDA approval. A laboratory can use any device it wishes (e.g., cytocentrifuge) to make cervical cytology slides without FDA approval. Established in 1848, FDA wasn't known by its present name until 1930. Its modern regulatory functions began with the passage of the 1906 Pure Food and Drugs Act, a law, a quarter-century in the making, that prohibited interstate commerce in adulterated and misbranded food and drugs. In other words, FDA was established before Papanicolaou's test for cervical cancer. Regulatory approval wasn't required then, and it isn't required today, because the conventional Pap test is not marketed as a product with claims of safety and efficacy.

FDA does not require approval for any device that is used to prepare nongynecologic specimens. For example, the ThinPrep processor began to be used to prepare nongynecologic specimens several years before the same processor was FDA-approved for preparing cervical cytology specimens. The difference in regulatory oversight is the fact that the Pap test is considered primarily a screening test, rather than a diagnostic test. Screening tests are applied to asymptomatic individuals among whom a small percentage may harbor an undetected disease (e.g., cervical cancer or precursors). See Table 5.3.

Cytocentrifugation and membrane filtration will be described in Chaps. 6 and 7. For all methods, the goals are identical: make every cell tell.

TABLE 5.3. Summary of possible methods and devices used to prepare gynecologic and nongynecologic specimens and whether FDA approval is required.

	FDA approval required?		
	No		Yes (GYN only)
Specimen	Manual method	Mechanical method	Automated process
Gynecologic	Conventional Pap tests Liquid-based Pap tests	Cytocentrifuge	Automated cervical cytology slide processor: ThinPrep processor (PMA No. P950039) SurePath PrepStain System (PMA No. P970018)
Nongynecologic: body cavity fluids, urines, sputum, CSF, FNA	Direct smears, membrane filtration, membrane filter imprints, cytocentrifugation	BD PrepStain Slide Processor CellSolution 30 system Cytocentrifuge ThinPrep non-gyn	NA

Quality Control and Quality Assessment

In cytopreparation, quality control means using materials and methods that contribute to getting the right answer. Restating what was said in Chap. 2, QC prospectively imparts credibility to the preparation. Quality assessment, on the other hand, retrospectively assesses the impact of the QC. Did we accomplish what we set out to? If not, why not? What can we do the next time to ensure quality every time? The latter goal, while laudable, is impossible to achieve.

For non-gyn specimens, the first step in QC is seeing what's in a specimen before processing it. Don't wait until the specimen has been processed entirely and try to guess why too few cells are on the slide or why the cells are poorly preserved. Use your microscopy expertise up-front.

Examine a drop or two of raw specimen microscopically, while exercising standard precautions. Place the unstained specimen on a slide and cover with a cover glass; lower the substage condenser or close the aperture diaphragm to make the cells visible. Assess the specimen's cellularity, composition, and preservation. An absolute deficiency such as total acellularity cannot be corrected in cytopreparation, whereas a relative deficiency such as low cell concentration may be.

- If the specimen is completely acellular, stop. Document your findings, and don't make any slides. Report the unsatisfactory findings to the clinician.
- If the specimen is hypocellular, centrifuge the specimen to concentrate the cells. Reexamine the button after suspending it in a small volume of salt solution or preservative as appropriate.
- If the specimen is hypercellular, dilute it as needed to avoid overpopulating the preparation.
- If the cells are poorly preserved in a fresh specimen, they may have degenerated because too much time has passed since collection, or the specimen was suspended in normal saline, or the specimen was held for too long at room temperature. I recall receiving a specimen sent from out-of-state that was suspended in a balanced salt solution. Upon receipt, it was processed and reprocessed 5 days later after having sat for 5 days at room temperature. The reprocessed specimen was entirely satisfactory.
- If the cells are poorly preserved in a specimen collected in an alcoholic preservative, the preservative itself may have caused the cells to swell and burst (e.g., low percent concentrations of alcohol).
- If particulates are observed, attempt to dissolve them if possible (e.g., slightly acidifying fresh urine specimens with 0.5 N hydrochloric acid to dissolve phosphate salts) or to separate them by differential sedimentation. Particulates can

be problematical for specimens that are processed automatically in a thin-layer processing device. The particulates can clog Millipore and polycarbonate filter pores and cause cell collection to stop prematurely. If erythrocytes outnumber nucleated cells excessively, they must be removed prior to preparation to prevent their crowding out the nucleated cells and result in a laboratory-based false negative slides and ThinPrep slides. Bloody fresh specimens may be treated with saponin to hemolyze the red blood cells. See Appendix A. A simpler solution is to suspend the cells in 1 part glacial acetic with 9 parts balanced electrolyte solution. Alternatively, bloody preserved specimens may be suspended in one of at least two proprietary hemolytic preservatives. Hemolyzing the RBCs by immersion in a Carnoy-like fixative after the cell spread has been made will not fix the problem.

- If tissue fragments are observed, the preparations should be stained separately to avoid cross-contamination. Also, prepare cell blocks.

Knowing such information in advance shapes expectations and guides actions. Without such information, one cannot know whether the cause of an unsatisfactory preparation originated in the specimen collection or preparation method. Ignoring such information risks wasting time by troubleshooting the wrong problem, needlessly examining unsatisfactory preparations, and incorrectly reporting cases as unsatisfactory or negative.

Someone once told me: “we don’t have time.” Such an attitude defines waste: work that does not add value to the product from the standpoint of the external customer. They had time to do it wrong, but not time to do it right.

Microscopic examination is not needed with every specimen, as to do so can be unacceptably time-consuming. This procedure, however, is exceedingly instructive, especially if one has never done it before. At the very least, it constitutes quality control at the first step of cytopreparation.

Direct Smears

Direct smears are the simplest and most straightforward of the several ways to transfer cells from a raw specimen to a slide. "Direct" means straight to the slide without any intervening processing step. "Smear" is misleading as it diverts attention from the technique needed for successful results. Of all the topics addressed in this volume, for example, only devices that are sold for the purpose of automating the process of making cervical cytology slides require FDA approval.

Smear is rooted in the old English *smerian* "to anoint or rub with grease, oil." The definition that applies here is a thin tissue or blood sample spread on a glass slide and stained for cytologic examination and diagnosis under a microscope. Conventional Pap tests are also known as Pap smears. Indeed, that's the title of Carmichael's 1973 biography: The Pap Smear—Life of George N. Papanicolaou.[7]

The apparent simplicity of direct smears is misleading. One cannot simply slather cells on a slide—like schmear on a bagel—and expect to retain adequate numbers of well-flattened cells. Conventional Pap smears, for example, can be so thick that they can't be coverslipped. Spreading cells as a thin layer on a clean slide that's immediately immersed in alcohol is the key to quality preparations.

Adhesive "Aids"

Watery cytologic specimens such as urine and body cavity fluids are sometimes spread on albuminized slides, frosted slides, or albuminized-frosted slides in an effort to keep cells from falling off when immersed in alcohol. While these special surfaces help cells adhere to slides, they prevent cells from flattening individually, like sunny-side-up eggs, and so the cells cannot display their chromatin informatively. Instead, they remain somewhat rounded up, like hard-cooked eggs. See Fig. 5.1.

Fig. 5.1. Fresh cells in suspension that are spread thinly on a clean glass slide adhere to the surface and flatten—like a sunny-side-up egg. On the other hand, any material or method that promotes intracellular cohesive forces and not extracellular adhesive forces favors cell shrinkage. The end result resembles a hard-cooked egg. The volumes of both eggs are the same, but the contents are distributed differently (i.e., flattened vs. fattened).

Such cells display compacted chromatin that may appear to be stained hyperchromatically and therefore are less useful diagnostically than if flattened. Qualitative display is compromised for quantitative recovery. In 1965, dry fixative slides were described that fixed cells during the slide preparation process.[8] The slides were never commercialized.

Dirty slide surfaces, too much cell suspension, isopropanol fixative, and cells collected in alcoholic preservative are among things that thwart cell flattening. I was taught that collecting cells in alcoholic preservative shrank cells while suspended. Indeed, cells on slides are smaller under those collection circumstances. However, alcoholic preservatives swell cells in suspension (e.g., erythrocytes are hemolyzed by 50% alcohol due to its swelling effect). Alcohol, even dilute alcohol, partially coagulates protein; such protein resists flattening. Smaller cells are the result, but not because they shrank while suspended in alcohol.

Albuminized Slides

To make an albumin solution, we mixed equal parts of freshly separated egg white with glycerin to which we added a thymol crystal as a preservative. The unused solution was refrigerated until needed, at which time slides were coated with a uniform film. Such slides were used to increase recovery of cells in body cavity fluids and urines. Millipore filters were also prepared on the same specimens. In side-by-side comparisons, the cells were consistently more flattened and displayed greater chromatin detail on Millipore filters than cohort cells on the albuminized slides.

The reason for the observable difference was clear. Cells on an albumin-coated slide were unable to flatten. Separated from a glass surface by the albumin, the cells lost their adhesiveness. They adhered to the albumin, which served as natural glue, and in the process rounded-up and became thicker and less useful for interpretation.

Glycerin albumin had been introduced by Mayer in 1883 as an adhesive to attach sectioned tissue to slides.[9] Parenthetically, this is the same Mayer who introduced Mayer hematoxylin in 1903. Glycerin's sole purpose was to keep the adhesive wet during tissue processing. Baker and Jordan have emphasized that under usual circumstances of tissue processing, glycerin is unnecessary and serves no useful purpose.[10] Combine that fact with the inappropriate use of glycerin albumin for cells, one is safe to conclude that glycerin albumin has no place in cytopreparation.

Frosted Slides

These were described by Evelyn Dakin in 1955 as "Improved Adhesion and Visibility of Cytologic Preparations by Use of the Frosted Glass Slide,"[11] and patented in 1957 as "microscope slide."[4] I found it interesting that she worked with J. Ernest Ayre of the Ayre spatula. She wrote:

> Frosted glass comparable to a 150-emery grain has proved most satisfactory for routine cytologic and histologic procedures… . Cells, bacteria, and the inclusions found in body fluids and secretions adhere more readily to a frosted or etched glass surface than to a smooth one. One of the most serious problems in the cytology laboratory today is that of preparing smears to meet the cytologist's requirements for giving an accurate interpretation. Since many specimens encountered are nonmucoid, there is always a problem of a certain loss of important

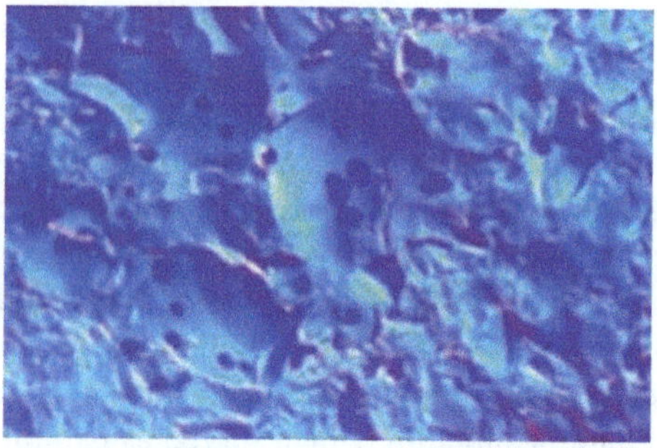

FIG. 5.2. Cells on a frosted slide for size reference. Differential interference contrast photomicrography, original magnification ×100.

> material. Routine staining procedures require much washing, which results in considerable loss of material, regardless of the amount of caution taken during processing... Such substances as egg albumin are now being used for adhesive purposes. However, use of the frosted glass slide eliminates the need for such adhesive substances, which often obscure the important detail, especially in the examination of watery-type specimens.[11]

Dakin slides are more expensive than plain slides with a frosted label end, thinner, more easily broken, and, unless the refractive index of the mounting medium closely matches that of the glass, scatter light. Most important, frosted slides aren't necessary to retain more cells. I believe that more cells "appear" to adhere not because the frosted surface snags the cells, but because the frosted surface is greater in area than 2 square inches, which spreads the suspension medium more thinly, which in turn allows the cells to come into contact with the glass. See Fig. 5.2.

Charged Slides

In 1980,[12] Husain wrote: "After some years of seeking a simple but effective method to lay cells on slides in an evenly dispersed monolayer for automated cell scanners, we managed to utilise what we believe to be the electrical charges on the cell and slide surfaces to achieve this end. The machine scanners required additional techniques to disrupt cell clusters and prevent reaggregation of the cells… These are not necessary for visual screening." An optimized method was described later.[13]

"Adhesion" slides are available commercially today. Whether they're necessary is uncertain, in my view. Cells and tissue drop off slide surfaces because they weren't attached well in the first place.

Why Cells Don't Stick to Glass

Cells do not flatten and stick to glass simply because the glass is dirty. Atmospheric moisture over time accumulates between packaged slides and forms hydrochloric acid that etches the glass. The cells are prevented from touching the slide by a microscopically thin film that reduces wettability and adhesiveness. How do I know that?

Sputum specimens prepared by Saccomanno's homogenization technique[14] usually contained curiously shaped objects that resembled microscopic crèpes. See Fig. 5.3a. I had gotten into the habit of microscopically examining specimens to see what was going on in real time at various steps in a procedure. In the case of sputum, I put a few drops of homogenized specimen on a slide on the microscope stage. Next, I put a second slide on top the first slide, focused with a 10× objective, and saw the specimen at rest. See Fig. 5.3b. Then applying the 2-slide pull technique, first in one direction and then the other, I saw the squamous cells roll up, like crepes, oriented at right angles to the pull direction (Fig. 5.3c) and then reorient themselves at right angles to the next pull direction (Fig. 5.3d).

I applied the same microscopy technique to drops of unfixed mesothelial cells. Being spherical rather than flat, those cells simply rolled along, like microscopic BBs. It was fascinating to watch, but what could it teach me? It taught me that clean and lean are best.

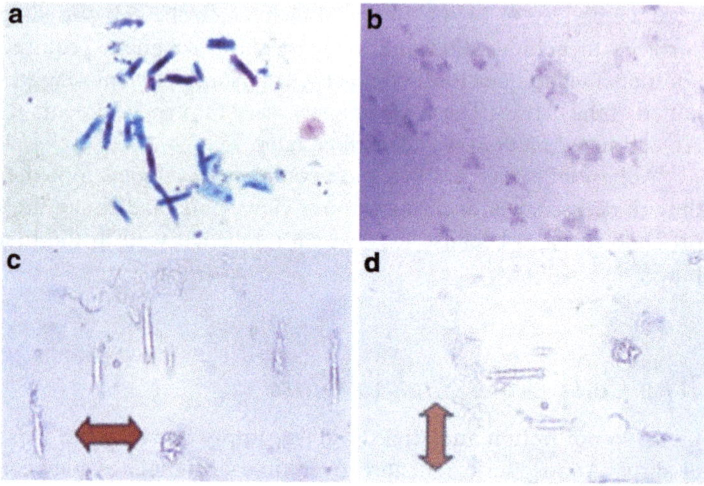

FIG. 5.3. Sputum processed by Saccomanno's homogenization technique and 2-slide pull technique. (**a**) Curious objects of uncertain origin. (**b**) Sputum before 2-slide pull technique. (**c**) Slides pulled face-to-face along long side. (**d**) Slides pulled face-to-face along short side.

Making Glass Wettable

Combining those observations with the knowledge that white blood cells in blood films adhere well to glass and flatten led to the following technique for preparing wettable slides:

- Immediately before use, immerse each slide briefly in alcohol, and using cheesecloth, wipe it squeaky clean to make it wettable. A rack of slides can be kept in alcohol, and the slides withdrawn as needed. It takes but seconds per slide and eliminates the need to purchase expensive charged slides.
- Whether using raw cell suspension or resuspended cell concentrate, transfer a small amount of cell suspension to the slide. Use less than you might think is needed. Only a single layer ("a light dusting") of well-distributed cells is required, not a heavy layer. The specimen should not be able to flow to the edges of the slide when a second slide is applied to spread the cells by

the 2-slide pull technique. You have succeeded when the slides resist sliding. If the cell suspension is expressed beyond the slide boundaries, too much has been added. The suspension medium will keep the cells from touching the glass, which is essential for flattening and retention. See Fig. 5.4.

- Immediately immerse the slide in alcohol. See Fig. 5.5.

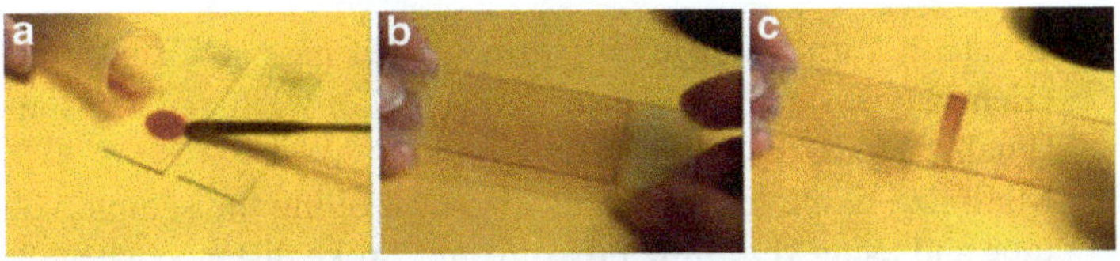

FIG. 5.4. How not to flatten and stick cells to glass using the 2-slide pull technique: (**a**) Too much specimen being placed on the slide. (**b**) The specimen is too much because it spreads spontaneously to the very edges of the slide. (**c**) The slides pull apart easily, without resistance. Under the depicted circumstances, the cells are separated from the glass surface and cannot adhere and flatten. The amount that should be added to *freshly cleaned* slides should be so small that it stops short of the slide edges and resists pulling. Under those circumstances, the suspension medium becomes very thin, which allows the cells to come into contact with, and adhere to, the slide surface.

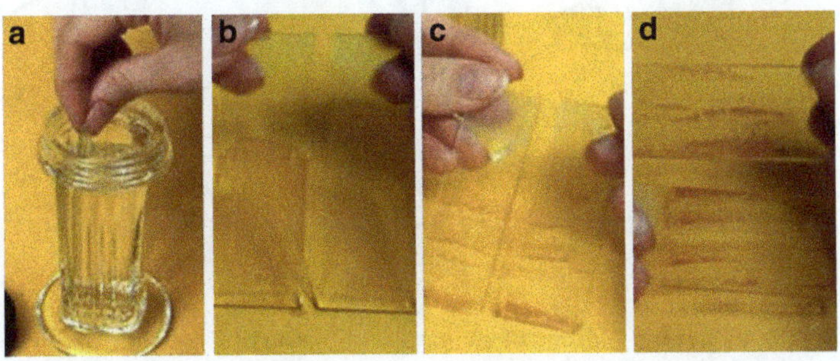

FIG. 5.5. (**a**) Rapidly immerse the cell spread to avoid air-drying and (**b**) retain the original uniform distribution of cells. (**c**) Slow immersion produces a "ribbing" effect of alternating cellular and acellular areas. (**d**) That slow immersion causes the ribbing effect is demonstrated by *slowly* immersing slides long side first in fixative.[15]

Fine Needle Aspiration Smears

The following FNA-related information and figures are excerpted from documentation provided to me by Michael D. Glant, MD. "Mick" is a long-time friend and colleague whom I regard as an FNA expert.

The approach in making FNA smears is unique to my technique. Unlike the Swedish approach of modifying a bone marrow aspirate, I developed this technique (Fig. 5.6) to:

1. Create 2 "mirror-image" type smears per pass: 1 for air-dried Giemsa type stain (rapid analysis) and 1 for wet-fixed Pap stain (final analysis), (see Fig. 5.7).
2. Minimize tissue loss and crush by controlling the sample volume and using very little pressure and stroke distance when the cell suspensions are spread.

FIG. 5.6. Keep 2–4 drops of aspirate in the center of the slide and then sandwich the aspirate between 2 pre-identified slides. As the slides are brought together surface to surface, the aspirate spreads, and the slides are drawn together by aspirate's "natural" adhesiveness. Let the pool spread spontaneously until it slows down and nearly stops. Then pull the slides "sideways," keeping each in contact with the other, which distributes the specimen in an arch-like pattern on each slide. Immerse 1 slide in alcohol and air-dry the other.

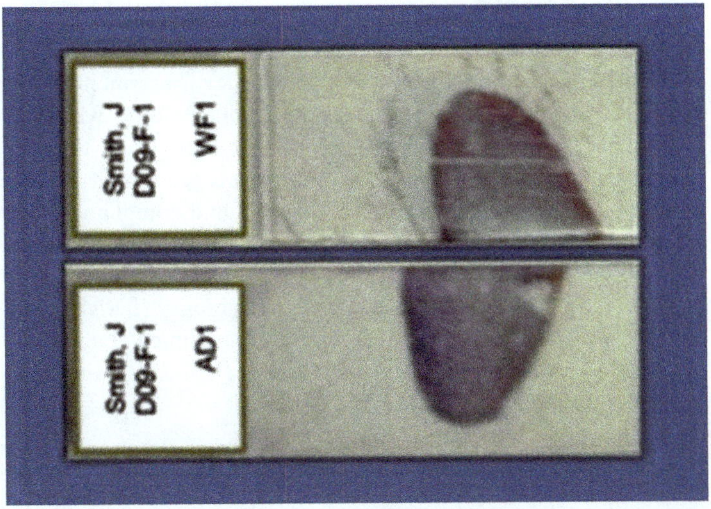

FIG. 5.7. Mirror-image smears per pass: fix one in alcohol and air-dry the other.

3. Make smear creation simple and rapid to reduce clotting.
4. Allow semiquantification of the cellular yield of each of the passes by numbering the slides in advance.

At the end of the procedure, a cell suspension of the needle rinses and, at times additional FNA passes, is created for a cell block or other preparation.

Fixed and stained air-dried slides can be rapidly evaluated on-site to:

- Determine adequacy
- Provide immediate diagnosis
- Proceed with additional samples, if needed
- Triage rinse to RPMI[1] media for additional studies (e.g., flow cytometry if lymphoid)
- Cell block for special stains (solid tumor)

FNA cytology is practiced worldwide; its literature, voluminous. For example, Volumes 2 (609 pages) and 3 (357 pages) of DeMay's 4-volume 2011 edition of The Art & Science of

[1] RPMI is named for Roswell Park Memorial Institute.

Cytopathology are devoted to Superficial Aspiration Cytology and Deep Aspiration Cytology, respectively.[16]

Questions often arise about FNA adequacy assessment billing issues. CPT code 88172 is defined as "cytopathology, evaluation of fine needle aspirate; immediate cytohistologic study to determine adequacy of specimen(s)." For guidance, see Moriarty.[17] The rules are subject to change, so stay current and compliant. Other relevant FNA references include the Joint Commission's National Patient Safety Goal related to time-outs,[18] 13 FNA-related online videos,[19] and NCCLS fine needle aspiration biopsy (FNAB) techniques.[20] NCCLS (National Committee for Clinical Laboratory Standards) changed its name to CLSI (Clinical Laboratory Standards Institute) on January 1, 2005.

Fine Needle Gage

Gage is an arbitrary assignment of numbers to the size of many things, needles among them. Dozens of gage systems are used worldwide. In America, the Stubs system is the only wire gage system recognized by an Act of Congress. Gage as a noun originated in the fifteenth century as a fixed standard of measure.

The Stubs system includes 40 wire gages that range from 0000 to 36. The smallest gage number (i.e., 0000) has the largest outer diameter (i.e., 11.53 mm); largest gage number (i.e., 36), smallest outer diameter (i.e., 0.102 mm). Size steps do not correspond to any mathematical pattern. Fine needles are usually 23–25 gage (i.e., 0.6414–0.5144 mm). See Table 5.4.

TABLE 5.4. The largest gage compared 23-gage fine needle.

Gage 0000 (11.53-mm OD)	Gage 23 (0.6414-mm OD)
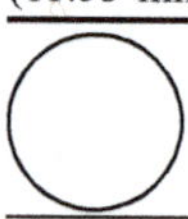	

Manual Liquid-Based Nongynecologic Preparations

Saccomanno Sputum Homogenization[14]

Saccomanno introduced the first liquid-based preparation—33 years before LBP was introduced as a term—when he homogenized preserved sputum in a blender, centrifuged the suspension, transferred a few drops of resuspended cell concentrate to a slide, and air-dried it. The preservative was either 50% ethanol or 50% isopropanol, each with 2% (w/v) Carbowax 1540. Air-drying retained the cells on the slides. If immersed in alcohol without intervening air-drying, the cells wash off the slide. Carbowax precipitates in the cells as the preservative evaporates and protects the cells against the damaging effects of air-drying. Union Carbide later changed the Carbowax numbering from 1540 to 1450.

Saccomanno told me that he had devised the homogenization method to convert the raw sputum specimens to something that resembled a frozen daiquiri. His cytopreparatory technicians didn't like preparing raw sputum. The primary reason may have been that the traditional pick-and-smear method of sampling and preparing sputum specimens is hit-or-miss. Sampling one area of raw sputum means one is sampling against the remaining specimen. One cannot really tell by looking at the specimen whether one area is more or less likely to contain diagnostic cells than another area.

Materials for Saccomanno's Method

• 4-oz specimen collection cup with leak-proof cap	• Semi-micro blender container
• Saccomanno's preservative	• Vortex mixer 30 × 115-mm conical tube 5O ml capacity, with screw cap
• 4 micro slides, plain with one side of one end frosted, 3 x I in., already labeled	• Centrifuge
• Biohazard safety cabinet Blender	

Method of Saccomanno (Modified)

1. The total volume of fixative and specimen is about 100 mL. Pour the mixture into a 250-mL capacity semi-micro blender container.
2. Blend the mixture at high speed for 5–30 s. Grossly, fine threads of material should not be visible. If fine threads are seen, blend the mixture for another 15–20 s until they disappear.
3. Divide the suspension between 2 centrifuge tubes and balance the tubes.
4. Centrifuge the specimen for 5 min at 1500 rpm (410 rcf).
5. Decant the supernatant until 2–3 mL remains.
6. Resuspend the cell concentrate by agitation on a vortex mixer.
7. Aspirate about 1 mL of suspension into a Pasteur pipette. Add two drops to a clean micro slide. For sparsely populated specimens, 3–4 drops or more may be needed.
8. Lay a second slide facedown on the first slide. The frosted ends of the slides should be at opposite ends.
9. After the specimen has spread uniformly between the two slides, separate the slides by smoothly sliding them along their 2-in.-long cell collection area. Do not separate the slides by lifting one from the other.
10. Lay the slides face-up on a level surface and allow the material to dry.
11. After air-drying is complete, align the slides in a staining rack and immerse them in 95% ethyl alcohol for 10 min to complete fixation.

Figure 5.8 shows results obtained by Saccomanno. One laboratory has published the results of its diagnosis of lung cancer experience with paired fresh sputum and Saccomanno's technique over a 21 month period.[21]

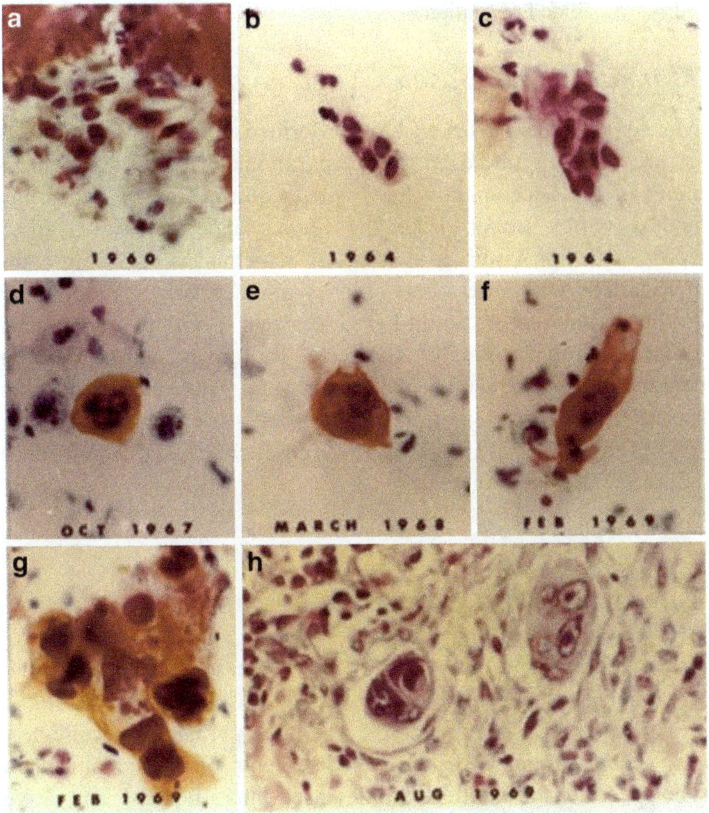

FIG. 5.8 This graphic is a scan of an original composite of photomicrographs given to me by Dr. Saccomanno. It is one of 4 composites in a 4–3/4 × 5–1/8-in. spiral-bound booklet titled Development of Carcinoma of the Lung. The legend he provided for this particular composite.

Case 2

Progression

• 40-year-old male	• Smoked 15 cig/day for 12 yrs.
• Underground uranium miner 12 yrs.	• Quit smoking 1962
• Underground hard rock miner 7 yrs.	• Lobectomy 1969
• Quit mining 1967	• WHO IA

Manual Liquid-Based Gynecologic Preparations

Maksem and his coworkers have described an inexpensive method of preparing liquid-based gyn preparations.[22–28] It is an enhancement of the Saccomanno technique, requiring only a centrifuge and a vortex mixer. Like the Saccomanno technique, FDA approval is not required, as Maksem's method is not being commercialized with safety and efficacy claims. Among noteworthy findings, automated liquid-based cervical cytology instruments possess significant preanalytical error for rare events (i.e., among residual cell suspensions of the parent suspensions of ThinPrep cases, the polymer membrane slides discovered 36 SILs among 200 high risk negatives and 19 additional SILs in a screening cohort of 260 women).[25]

Automated Liquid-Based Preparations

The generic name for a "processor, cervical cytology slide, automated" is cytology slide preparation device, which FDA describes as "a system used to collect and prepare cervical cytology specimens for Pap stain-based screening for cervical cancer. The specimens are collected and rinsed into a liquid preservative fluid from which they are then automatically filtered and deposited in a thin-layer on a glass microscope slide. They are stained and screened by a cytotechnologist and cytopathologist."

Premarket approval (PMA) is the FDA process of scientific and regulatory review to evaluate the safety and effectiveness of class III medical devices. Class III devices are those that support or sustain human life, are of substantial importance in preventing impairment of human health, or which present a potential, unreasonable risk of illness or injury.

Two FDA-approved devices are currently marketed in America: BD PrepStain system (PMA No. P970018, Decision date: 06/17/1999) and Hologic's ThinPrep processor (PMA No. P950039, Decision date: 05/20/1996). The 2 processors likely have greater than 90% market share and are the de facto standard of care for Pap tests in this country. The PrepStain system relies on gravity sedimentation of cells suspended in CytoRich preservative through a density gradient to enrich a specimen's cell content. The ThinPrep processor disperses cells suspended in PreservCyt preservative, collects them on the exterior surface of an 8-μm-pore-diameter polycarbonate filter, and then imprints the cells onto a micro slide. "Surface tension and air pressure cause the cells to adhere to the Slide [*sic*]...."

These same systems are used without FDA approval for nongynecologic specimens. Relative to qualitative and quantitative issues of cytopreparation, the PrepStain system has the potential to produce preparations of higher quality overall than the ThinPrep processor. The PrepStain system processes specimens collected in 24% ethanol. That concentration of alcohol results in cells on slides that are larger than similar cells collected in the 50–55% methanol of PreservCyt. Using gravity sedimentation, the PrepStain truly enriches the cellular content of specimens. Enrichment in this context means concentrating cells of interest and excluding interfering debris.

Beyer-Boon et al. have described the effects of various routine cytopreparatory techniques on normal urothelial cells and their nuclei.[29] The various cytopreparatory techniques were fixation related: 1 was air-dried, 3 were wet fixation related, and the fifth variation studied was a combination of air-drying and wet fixation. None of the methods that were studied included collection in liquid preservative followed by wet fixation, as with the BD PrepStain system or the ThinPrep processor: not the particular alcohols or concentrations. Beyer-Boon's results, however, are consistent with expectations based on other published measurements.[30] Nuclei in air-dried cells had the largest mean nuclear areas, while those that had been wet fixed in 96% ethanol had the smallest mean nuclear areas. In my view, Beyer-Boon's findings support my observation that cells and their nuclei collected in 24% ethanol are noticeably larger than those collected in 50–55% methanol.

Yang has provided the mathematical basis that accounts for the difference in nuclear areas between air-dried and wet-fixed cells.[31] While air-dried cells aren't encountered in liquid-based preparations, the point is that there's often a quantitative explanation for much of what we encounter in cytology. Whether air-dried, wet fixed, or preserved in suspension, cells expand or shrink in response to many forces, but in no case is their size in life maintained. Thus, Yang's contribution:

It is known that the area of a circle equals πr^2 and that the volume of a sphere equals 4/3 πr^3. Assume that surface tension on a glass slide causes flattening of a fresh, unfixed, elastic, spherical nucleus with a radius of r into a flat discoid nucleus with a radius of R and a thickness of T. As long as the nuclear membrane remains intact, the nuclear volume will remain constant. Therefore, the nuclear volume of a discoid nucleus is equivalent to the nuclear volume of a spherical nucleus:

$$\pi R^2 \times T = 4/3\pi r^3$$

thus

$$\pi R^2 = (4/3\pi r^3)/T$$

Observed nuclear area after air during (NA).

r	$\text{na} = (\pi r^2)$	$\text{NA} = (3/4\pi r^3)/T$
2	13	17
2.5	20	33
3	28	57
3.5	38	90
4	50	134
4.5	64	191
5	79	261
5.5	95	348
6	113	451

In the interest of scientific accuracy, surface tension is a property of liquids, and not solids such as a glass slide. Surface tension is the elastic-like force existing in the surface of a body, especially a liquid, tending to minimize the area of the surface, caused by asymmetries in the intermolecular forces between surface molecules. Cell surfaces have surface tension but not glass surfaces.

The ThinPrep processor uses upside-down filtration. In right-side-up filtration, everything added from above the filter is deposited—the good, the bad, and the ugly—on the surface. In upside-down filtration, that's not the case. Cells on the bottom surface are pulled up to it by vacuum. For well-behaved "clean" specimens that contain relatively little competing debris, upside-down filtration produces satisfactory results. Cells of likely interest are deposited on the filter and then transferred to a micro slide by imprinting. The filter is discarded.

On the other hand, specimens with competing "noise" fare less well. When abundant, erythrocytes, leukocytes, and mucus will be drawn first to the filter, crowd out cells-of-interest, and cover and/or occupy the pores, which the system's sensor interprets as stop. Work-arounds have been published.[32–36] The system's features are patented but so are its limitations.[37, 38]

Conclusion

With the benefit of hindsight, I am convinced that the technique by which cells are applied to glass slides contributes to interpretable cell display more than any other step in the cytopreparatory chain. Reliably preparing slides that retain and flatten cells is essential to displaying interpretable cytomorphology. Well-behaved specimens prepared in the laboratory are less problematical than those that are not. FNA are a prime example of the latter. Instructing those who procure and prepare FNA in the fundamentals may help them overcome some—but not all—of the inherent technical challenges posed.

Apppendix A. Saponin Technique for Bloody Fresh Cell Suspensions

When erythrocytes outnumber nucleated cells in specimens from any body site, they exclude such cells—including cancer cells—from micro preparations. These preparations are technically satisfactory but functionally unsatisfactory, meaning that the preparations represent the true mix of cells in the raw specimens and have been well prepared but are useless for cancer detection. Specifically the preparation method has invalidated the value of the raw specimen. The results are reported as within normal limits, when in fact the preparation method is unsatisfactory and the reported result is a false negative.

This common problem can be remedied by eliminating erythrocytes from cytologic specimens *before* the cell concentrate is collected on slides or filters. Such an approach to specimen enrichment is entirely different than hemolyzing RBCs after a cell spread is prepared by immersing it in a Carnoy's-type hemolytic fixative or in 2 M urea. The latter techniques merely increase the visibility of the remaining cells but leave their numbers unchanged.

The saponin method that follows hemolyzes erythrocytes while in suspension, thus proportionally increasing the number of nucleated cells available for microscopic examination and permitting these cells to occupy the additional collection/display area that is now available. This enrichment technique was used in a research project on circulating cancer cells in peripheral blood.

The first time it was used for a clinical application, the control preparations exhibited countless erythrocytes but no cancer cells. The experimental preparations that had been processed with the saponin method described below, on the other hand, exhibited the exact opposite results: no erythrocytes and an abundance of cancer cells. If the specimen had not been processed with saponin, it would have been reported as negative and satisfactory. In other words, it would have been a laboratory technique-based false negative.

This technique should be applied to all cytologic specimens in which erythrocytes are visible in the cell concentrate—no matter how small—following the initial centrifugation.

A.1 Materials

- Hemolytic agent: 1% (w/v) saponin in distilled water with 0.2% sodium p-hydroxybenzoate as a preservative (optional). Filter through a 5 μm pore size cellulosic filter after preparation (also optional). Keep refrigerated

- Antihemolytic agent: 3% (w/v) calcium gluconate in distilled water with 0.2% sodium p-hydroxybenzoate (optional). Filter as above. Keep refrigerated
- Balanced electrolyte solution (not normal saline)
- 50-mL plastic centrifuge tubes
- Transfer pipettes
- Vortex mixer

A.2 Method

1. Centrifuge the specimen, up to 50 mL, for 10 min at 3,000 rpm.
2. Discard the supernatant.
3. Add 25 mL balanced electrolyte solution.
4. Resuspend the cell concentrate by repeatedly inverting the centrifuge tube, or better, by agitating the contents on a vortex mixer.
5. Add balanced electrolyte solution to the 45-mL level and mix.
6. Add 2 mL saponin and invert several times to mix.
7. After 1 min, add 3 mL calcium gluconate.
8. Centrifuge 10 min at 3,000 rpm.
9. Decant the supernatant.
10. Prepare cell spreads if volume permits. Otherwise, resuspend the button in 5 mL balanced salt solution for collection by cytocentrifugation or membrane filtration.

A.3 Results

After step 8, the supernatant will be colored red (the depth of color is a function of amount of hemoglobin released by the RBCs); the cell concentrate, white. RBC ghosts remain suspended in the supernatant and thus cannot contaminate the cell concentrate. The increased number of nucleated cells is remarkable. Cancer cells are often present in the specimen that otherwise would be absent in the final preparation.

A.4 Discussion

The value of the results is self-evident. The relative amount of RBC to nucleated cells determines whether the specimen should be treated with saponin. In other words, even small total numbers qualify for saponin enrichment. The saponin and calcium gluconate solutions may grow microbes. It is uncertain how long saponin solutions remain effective. Historically, we discarded saponin solutions after 1 week. Saponin is a plant derivative that varies in activity. Some lots may require higher concentrations and/or longer exposure times. Saponin hemolyzes RBCs by etching holes in the membranes, thus allowing hemoglobin to escape. Nucleated cells will begin to show cytoplasmic damage if exposed to saponin for too long. Saponin does not work in alcohol-preserved specimens.

References

1. Bonanni F. Observationes circa viventia, quae in rebus non viventibus reperiuntur. Cum micrographia curiosa. Rome: DA Hercules; 1691.
2. Gould C. The companion to the microscope. London: W Cary; 1827.
3. Baker JR. Charter centenary of The Royal Microscopical Society. Nature. 1966;210(5036):564–5.
4. Dakin ES, inventor. Microscope slide. US Patent 2801568. 6 Aug 1957.
5. American Society for Testing and Materials. Standard specification for cover glasses and glass slides for use in microscopy. E211-70.
6. Baker H. The microscope made easy. London: R Dodsley; 1743.
7. Carmichael DE. The Pap smear: life of George N. Papanicolaou. Springfield, IL: Charles C. Thomas; 1973.
8. Kohn J, Earle JHO. Use of dry fixative slides for exfoliative cytology. J Clin Pathol. 1965;18:479.
9. Mayer P. Mitt zool Stat Neapel. 1883;4:521.
10. Baker JR, Jordan BM. Miscellaneous contributions to microtechnique. Q J Microsc Sci. 1953;94(3):237–42.
11. Dakin ES. Improved adhesion and visibility of cytologic preparations by use of the frosted glass slide. Science. 1955;121(3144):424–5.
12. Husain OA, Millett JA, Grainger JM. Use of polylysine-coated slides in preparation of cell samples for diagnostic cytology with special reference to urine sample. J Clin Pathol. 1980;33(3):309–11.

13. Watts KC, Husain OAN. Optimal use of cationic polyelectrolyte poly-L-lysine in preparation of cell monolayers for diagnostic cytopathology. J Clin Pathol. 1984;37(7):829–31.
14. Saccomanno G, Saunders RP, Ellis H, Archer VE, Wood BG, Beckler PA. Concentration of carcinoma or atypical cells in sputum. Acta Cytol. 1963;7(5):305–10.
15. Naylor B. The elimination of a ribbing effect observed in cytologic smears. Am J Clin Pathol. 1958;30(2):143–4.
16. DeMay RM. The art & science of cytopathology. 2nd ed. Chicago, IL: ASCP Press; 2011.
17. Moriarty AT. When and how to use CPT code 88172. CAP Today 2006;20(9). Available at http://www.cap.org/apps/portlets/contentViewer/show.do?printFriendly=true&contentReference=cap_today%2Fpap_ngc%2F0906NGC_CPT.html. Accessed 6 May 2012.
18. The Joint Commission. NPSG UP.01.03.01: A time-out is performed before the procedure. National Patient Safety Goals Effective January 1, 2012. Available at http://www.jointcommission.org/assets/1/6/NPSG_Chapter_Jan2012_HAP.pdf. Accessed 6 May 2012.
19. Optimal FNA Techniques…. A series of educational videos from the DVD by Britt-Marie Liung MD, UCSF. Available at http://papsociety.org/fna.html. Accessed 14 May 2012.
20. NCCLS. Fine Needle Aspiration Biopsy (FNAB) techniques; Approved guideline—second edition. NCCLS document GP20-A2 (ISBN 1-56238-509-7). NCCLS, 940 West Valley Road, Suite 1400, Wayne, PA 19087-1898 USA; 2003.
21. Perlman EJ, Erozan YS, Howdon A. The role of the Saccomanno technique in sputum cytopathologic diagnosis of lung cancer. Am J Clin Pathol. 1989;91(1):57–60.
22. Maksem JA, Dhanwada V, Trueblood JE, Weidmann J, Kane B, Bolick DR, et al. Testing automated liquid-based cytology samples with a manual liquid-based cytology method using residual cell suspensions from 500 ThinPrep cases. Diagn Cytopathol. 2006;34(6):391–6.
23. Lee JM, Kelly D, Gravitt PE, Fansler Z, Maksem JA, Clark DP. Validation of a low-cost, liquid-based screening method for cervical intraepithelial neoplasia. Am J Obstet Gynecol. 2006;195(4):965–70.
24. Maksem JA, Bedrossian CW, Kurtycz D, Sewall S, Shalkham J, Dhanwada V, et al. Resolving ASCUS without recourse to HPV testing: manual reprocessing of residual automated liquid-based cytology (ALBC) material using manual liquid-based cytology (MLBC). Diagn Cytopathol. 2005;33(6):434–40.

25. Maksem JA, Dhanwada V, Trueblood JE, Weidmann J, Kane B, Bolick DR, Bedrossian CWM. Incorporating suspended articles into insoluble thin membranes reveals significant preanalytical error in the automated liquid-based Pap test. Cancer Cytopathol. 2004;48(5):664–5. Supplement Abstract 17.
26. Maksem JA, Finnemore M, Belsheim BL, Roose EB, Makkapati SR, Eatwell L, Weidmann J. Manual method for liquid-based cytology: a demonstration using 1,000 gynecological cytologies collected directly to vial and prepared by a smear-slide technique. Diagn Cytopathol. 2001;25(5):334–8.
27. Maksem JA, Weidmann J. Specialized preparative devices are not needed for liquid-based, thin-layer cytology: an alternate manual method using a metastable alcoholic gel. Diagn Cytopathol. 2001;25(4):262–4.
28. Johnson T, Maksem JA, Belsheim BL, Roose EB, Klock LA, Eatwell L. Liquid-based cervical-cell collection with brushes and wooden spatulas: a comparison of 100 conventional smears from high-risk women to liquid-fixed cytocentrifuge slides, demonstrating a cost-effective, alternative monolayer slide preparation method. Diagn Cytopathol. 2000;22(2):86–91.
29. Beyer-Boon ME, van der Voorn-Den Hollander MJ, Arentz PW, Cornelisse CJ, Schaberg A, Fox CH. Effect of various routine cyto-preparatory techniques on normal urothelial cells and their nuclei. Acta Pathol Microbiol Scand A. 1979;87(1):63–9.
30. Gill GW. Air-dried/rehydrated CV smears are different. Diagn Cytopathol. 1998;18(5):381–2.
31. Yang GCH. The mathematical basis for the increased sensitivity in cancer detection in air-dried cytopreparations. Mod Pathol. 1994;7(6):681–4.
32. Pang Y, Smola B, Pu RT, Michael CW. Restoring satisfactory status in ThinPrep Pap test specimens with too few squamous cells and containing microscopic red blood cells. Diagn Cytopathol. 2008;36(10):696–700.
33. Haack LA, O'Brien D, Selvaggi SM. Protocol for the processing of bloody cervical specimens: glacial acetic acid and the ThinPrep Pap Test. Diagn Cytopathol. 2006;34(3):210–3.
34. Dalton P, MacDonald S, Boerner S. Acetic acid recovery of gynecologic liquid-based samples of apparent low squamous cellularity. Acta Cytol. 2006;50(2):136–40.
35. Islam S, West AM, Saboorian MH, Ashfaq R. Reprocessing unsatisfactory ThinPrep Papanicolaou test specimens increases sample ade-

quacy and detection of significant cervicovaginal lesions. Cancer Cytopathol. 2004;102(2):67–73.

36. Rowe LR, Bentz JS. A simple method to determine the need for glacial acetic acid treatment of bloody ThinPrep Pap tests before slide processing. Diagn Cytopathol. 2004;31(5):321–5.
37. Bentz JS, Rowe LR, Gopez EV, Marshall CJ. The unsatisfactory ThinPrep Pap Test: missed opportunity for disease detection? Am J Clin Pathol. 2002;117(3):457–63.
38. Song LH, Goh ES, Phang LC, Poh WT, Tay SK. Technical aspect of ThinPrep. Singapore Med J. 2000;41(12):575–8.

Chapter 6
Cytocentrifugation

Principle No. 2

Make specimen-representative preparations.

Principle No. 3

Flatten cells to enhance chromatin display.

Practice

Examine a drop of resuspended cell concentrate. Hemolyze the erythrocytes if necessary. Determine the number of drops that will produce useful cell population density. Add to the specimen chamber and cytocentrifuge.

Historic Milestones

- 1685—Newton in his *Principia* coins "centrifuge," meaning to "flee from the center"[1]
- 1965—Doré and Balfour describe "a device for preparing cell spreads"[2]
- 1966—Watson describes "a slide centrifuge: an apparatus for concentrating cells in suspension onto a microscope slide"[3]
- 1972—"Cytospin" as a product name first used in commerce
- 1972—Mikat and Mikat flatten tissue using centrifugal force[4]
- 1981—Leif patents centrifugal cytology buckets[5]

G.W. Gill, *Cytopreparation: Principles & Practice*,
Essentials in Cytopathology 12, DOI 10.1007/978-1-4614-4933-1_6,
© Springer Science+Business Media New York 2013

- 1982—Shandon Southern Instruments Limited registers Cytospin as trademark
- 1982—Gill writes first user manual for Shandon Cytospin 2[6]
- 1983—Cytocentrifuge patented[7]
- 2000—Cytocentrifuge used to prepare thin-layer cervical cytology[8]
- 2005—Thermo Electron renews registration of Cytospin as trademark

Until the mid-twentieth century, sparsely populated cell suspensions challenged conventional slide preparation techniques. Usually, such suspensions were centrifuged first. Assuming cells were present at all, they then had to be transferred from the centrifuge tube to a slide. Often in such cases, the pellet was invisible. So then what?

In 1956, Seal introduced the use of Millipore filters to meet the challenge. As the next chapter illustrates, Millipore filters are capable of producing excellent quantitative recovery and qualitative cytomorphological display in the right hands. Achieving useful results, however, requires special techniques and more work than most labs can afford. Today, Millipore filters are a footnote in the history of cytopreparation.

Ten years after 1956, Watson described Type A and Type B apparatuses for concentrating cells in suspension onto a microscope slide. An O ring projected from the primitive specimen chamber, thus "compressing the periphery of the hole in the filter paper and reducing cell loss by absorption into the paper." A footnote on page 495 stated: "A commercial version of the Type A apparatus is to be marketed by Shandon Scientific Co. Ltd., Pound Lane, Willesden, London, N.W. 10, England."[3] Shandon's address at the time was about 26 miles west of Watson's address in Taplow, Maidenhead, Berkshire, England.

The cytocentrifuge uses centrifugal force to concentrate and flatten cells directly onto a glass slide without the additional work required by membrane filters. The FDA identifies cytocentrifuge as "a centrifuge used to concentrate cells from biological cell suspensions (e.g., cerebrospinal fluid) and to deposit these cells on a glass microscope slide for cytological examination and classifies them as Class I (general controls). The device is exempt from

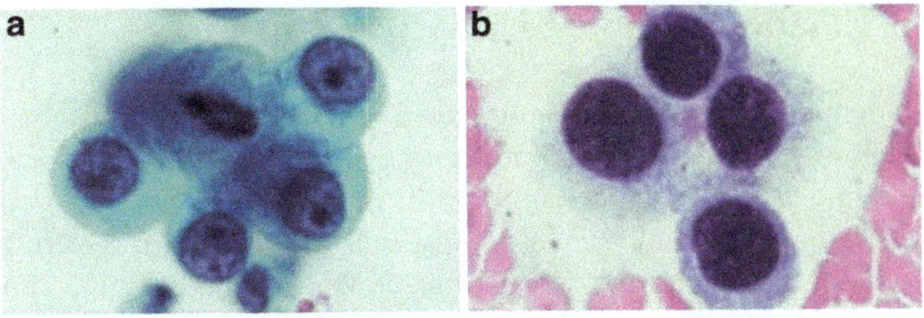

FIG. 6.1. Centrifugal force flattens cells constructively, as is plainly evident in these two photomicrographs. At 1,500 rpm, a Cytospin 4 generates relative centrifugal force of 231 × g. The rotor radius—from the center to the vertically oriented slide surface during centrifugation—is 9.2 cm.

premarket notification procedures."[9] However, it's not exempt if a manufacturer markets its device for use with gynecological cytology specimens. While manufacturers cannot promote their cytocentrifuges for such an off-label use, users can use cytocentrifuges to prepare SpinThins,[8] PapSpins,[10] and the like.[11,12]

Cytocentrifugation is another way to retrieve few cells from suspension and transfer them onto a slide surface. Watson described the first cytocentrifuge in 1966. It was soon brought into commercial production by Shandon and by others years later. Today's models are engineered to prevent many of the early limitations (e.g., cells being absorbed in the blotter). A major benefit of cytocentrifugation is the constructive flattening of cells by centrifugal force. See Fig. 6.1.

Shandon Scientific Company was the first to market a cytocentrifuge. Shandon as a boy's name in English means God is gracious. Shandon Scientific is now Thermo Scientific. More than 15,000 units have been sold worldwide.

Numerous cytocentrifuges are now on the market (e.g., Hettich's Cyto-System, Sakura Finetek's Cyto-Tek, StatSpin's Cytofuge2, and Wescor's Cytopro). I've not used these cytocentrifuges and therefore cannot comment on their relative merits. I imagine they are more alike than different. Assuming that is the case, common challenges include (1) estimating sample size, (2) not adding a volume of specimen greater than the maximum volume of the sample chamber, and (3) retaining cells on slide after cytocentrifugation.

Estimating Sample Size

In manual (as opposed to automated) cytopreparation, it is common practice to transfer a portion of cell suspension to the processing device (e.g., membrane filter, Cytospin) without first assessing its cellularity and adjusting the volume to assure a satisfactorily cellular preparation. Most of the time, this approach apparently yields satisfactory results as it continues to be used widely, or it's not occurred to laboratorians to look. However, problematical specimens point out the shortcomings of proceeding blindly. Clear urine specimens, for example, appear to be acellular but in fact often contain hundreds, if not thousands, of cells.

> If you don't examine the raw specimen microscopically before cytocentrifugation, how will you know whether the cause of an acellular or sparse preparation is specimen-based or preparation-based? You might troubleshoot a problem that doesn't exist or ignore a technical problem that does exist. Neither is acceptable laboratory practice.

Therefore, concentrate all specimens first by conventional centrifugation. The sole exception may be CSF. Whether these precious specimens require centrifugation before cytocentrifugation depends on cell concentration. Examine a drop of raw suspension microscopically. If few cells are seen, centrifuge to concentrate them into a pellet that can be resuspended and added in toto to a single cytofunnel. If many cells are present and the overall volume exceeds the 0.5 mL capacity of a single cytofunnel chamber, add 0.5 mL volumes to several cytofunnels. Alternatively, one can use a larger volume specimen chamber (e.g., Megafunnel).

For specimens that are centrifuged, discard the supernatant, and resuspend the pellet in few mL of residual supernatant or added balanced salt solution. Put a drop of unstained resuspension on a slide and cover with a cover glass. Examine the temporary preparation microscopically, lowering the substage condenser slightly to make the unstained cells visible. Quickly review the cell distribution using a 4× or 10× objective and 10× eyepieces,

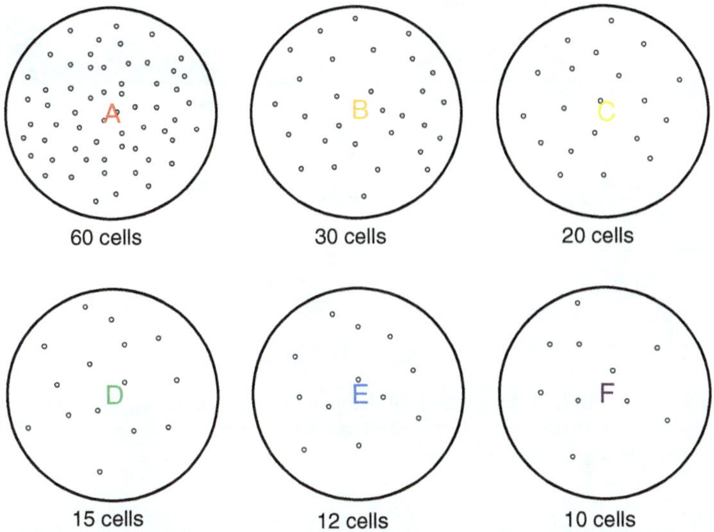

FIG. 6.2. Match approximately the cells seen in a representative ×40 field of view with the closest approximation in this figure. Then, match the letter of that field-of-view with that in Fig. 6.3. The number in the cell below that intersects the row and size of the collection device in use is the recommended volume. Alternatively, count the cells and divide into 60 to equal the number of drops required for a single cytofunnel.

and select a representative field. Then, using a 40× objective, compare the observed cellularity with that depicted in the six graphical fields of view in Fig. 6.2. Then, match the letter of that field-of-view with that in Fig. 6.3. The number in the cell below that intersects the row and size of the collection device in use is the recommended volume.

	Drops per FOV population density					
	A	B	C	D	E	F
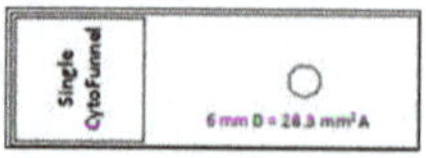	1	2	3	4	5	6
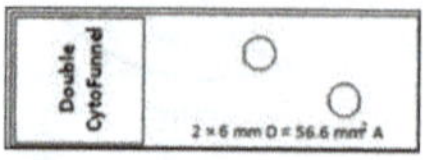	1 × 2	2 × 2	3 × 2	4 × 2	5 × 2	6 × 2
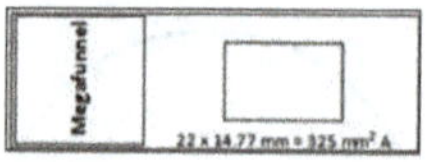	12	24	36	48	60	72

FIG. 6.3. Adding the number of drops of suspension as indicated to the various Cytospin specimen chambers will produce preparations with satisfactory population densities.

25-mm membrane filter with 18-mm collection area ($250\ mm^2/28.3\ mm^2$)	× 9
47-mm membrane filter with 35-mm collection area ($960\ mm^2/28.3\ mm^2$)	× 34
19 × 42-mm membrane filter with 16.5 × 36.5-mm collection area ($600\ mm^2/28.3\ mm^2$)	× 21
20-mm TransCyt filter ($314\ mm^2/28.3\ mm^2$)	× 11
13-mm SurePath Prep ($133\ mm^2/28.3\ mm^2$)	× 5

The numbers in Fig. 6.3 can be scaled up for other types of preparations with larger cell collection areas by multiplying the drops per FOV in Row 1 for a single cytofunnel × the correction factor as follows:

Body cavity fluids can be so hypercellular that even a single drop of resuspended cell concentrate can overpopulate the 28.3-mm^2 area of a single cytofunnel. In such cases, dilute the cell concentrate tenfold with balanced salt solution, and examine a drop as before.

Don't Add More Specimen than Sample Chamber Can Hold

Shandon's original non-disposable single cytofunnel has a vertical conical portion and a horizontal cylindrical chamber. The combined volumes are 3.3 mL: 2.8 mL for the former (85% of the total); 0.5 mL (15% of the total), the latter. The latter, and only he latter, is the sample chamber. If the specimen volume exceeds 0.5 mL, regardless of Cytospin model number, that excessive volume cannot "turn the corner" to the horizontal chamber. Centrifugal force plasters any specimen in the cone against its outer wall and holds it there. See Fig. 6.4. The total specimen volume, therefore, should not exceed 0.5 mL.

If the total specimen volume is greater than 0.5 mL when added to the resting rotor, and if the chamber is horizontal at the time, the specimen will be wicked into the "filter card" and substantially reduce cell recovery. See Fig. 6.5. Filter card is a misnomer. It

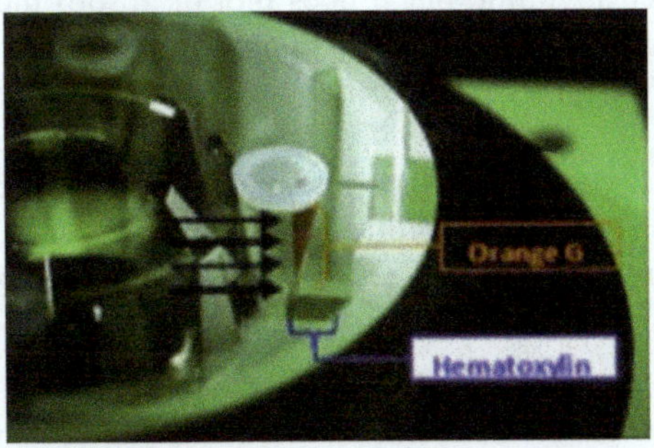

FIG. 6.4. Stroboscopic illumination that was synchronized to flash once per revolution made the spinning rotor appear to "stand still" when photographed. I added hematoxylin to the horizontal specimen chamber and, for contrast, orange G to the vertical cone-shaped chamber. The 4 right-pointing arrows indicate the direction of centrifugal force, which accounts for the reoriented orange G in the chamber.

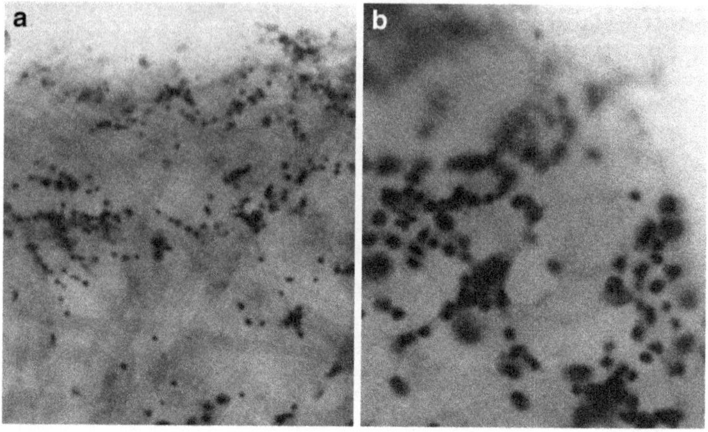

FIG. 6.5. (**a**) Cells radiating from hole in absorbent cytocentrifuge pad (×40 original magnification). (**b**) Mesothelial cells and leukocytes present at edge of hole in the cytocentrifuge pad (×100 original magnification).[13]

doesn't filter anything, but it does blot or absorb the suspension fluid during cytocentrifugation. Filter card is holdover terminology from Watson's 1966 paper in which he described using filter paper.

Cell loss by wicking is not possible in rotors that hold the cytofunnel at a downward incline away from the slide and blotter. Gravity works to advantage. Upon start-up, the cytofunnel is reoriented horizontally.

Retaining Cells on Slides

If the slide surface is wet when the slide's being unloaded from the cytofunnel, the cells aren't adhering well to the glass and will likely wash-off when immersed in alcohol. If you hadn't examined a drop of resuspended cell concentrate up-front to see the starting cellularity, you couldn't know whether an acellular slide resulted from an acellular or sparsely populated specimen or a cellular specimen that ran into unidentified technical difficulties.

Let's assume that the starting specimen is cellular but the finished slide is not. What can be done proactively that will ensure cell recovery?

- Use fresh specimens that haven't been collected in alcoholic preservative.
- Use freshly squeaky-cleaned slides.
- Centrifuge each specimen, and resuspend the cell concentrate in about 5 mL balanced electrolyte solution.
- Examine microscopically and determine the number of drops that will produce satisfactory cellularity on the slide.
- Fill each cytofunnel chamber to its 0.5 mL capacity, which is about 20 drops from a narrow tip glass pipette. To equalize differences in the number of drops of different specimens that will be cytocentrifuged during a given run, subtract from 20 the number of drops of specimen that will be added, and divide the difference by 2. Half the difference should be balanced electrolyte solution (BES), the other half, 95% ethanol. For example, a 2-drop hypercellular specimen will be added to the cytofunnel, followed by 9 drops BES and then 9 drops alcohol.
- Filling each chamber to capacity as described ensures that all cytofunnels will "run dry" at about the same time. Otherwise, some may finish before others and leave the cells on the slide "high and dry." The BES separates the specimen from the alcohol; the alcohol fixes the flattened cells. Optionally, 95% ethanol with 2% Carbowax can be used instead of alcohol alone.
- Upon removing each cytofunnel from the rotor, orient it so the slide is cell-side up. Remove the cytofunnel. If the slide surface is still wet, lay the slide flat, and wait until the remaining liquid evaporates. When the slide is no longer wet, immerse it in alcohol.

Conclusion

Cytocentrifugation requires attention to details for satisfactory qualitative and quantitative results. I wrote the first user's manual for Shandon in 1982. Most of that guidance is included in the manufacturer's current user's manual, so it would be redundant to include it in this chapter. I've not used other cytocentrifuges or seen the respective user's manuals but assume the information presented in this chapter applies. In all cases, details count. After all, that's where the devil lives.

References

1. Online Etymology Dictionary. Centrifugal. Available at http://www.etymonline.com/index.php?term=centrifugal. Accessed 16 May 2012.
2. Doré CF, Balfour BM. A device for preparing cell spreads. Immunology. 1965;9(4):403–5.
3. Watson P. A slide centrifuge: an apparatus for concentrating cells in suspension onto a microscope slide. J Lab Clin Med. 1966;68(3):494–501.
4. Mikat KW, Mikat DM. Tissue flattening: a method of enhancing histologic detail. Lab Med. 1972;3(7):41–2.
5. Leif RC, inventor. Swinging buckets. US patent 4250830. 17 Feb 1981.
6. Gill GW. The Shandon Cytospin 2 in diagnostic cytology – tips, techniques, and troubleshooting. Sewickley, PA: Shandon Southern Instruments; 1982 (73 pages).
7. Gordon AJ, inventor; Shandon Southern Products Limited, assignee. Cytocentrifuge. US patent 4391710. 5 Jul 1983.
8. Khalbuss WE, Rudomina D, Kauff ND, Chuang L, Melamed MR. SpinThin, a simple, inexpensive technique for preparation of thin-layer cervical cytology from liquid-based specimens: data on 791 cases. Cancer Cytopathol. 2000;90(3):135–42.
9. 21 CFR Ch. I (4–1–11 Edition) § 864.3300 Cytocentrifuge. Subpart D—Pathology Instrumentation and Accessories, p. 232–3.
10. Rosenthal DL, Geddes S, Trimble CL, Carson KA, Alli PM. The PapSpin: a reasonable alternative to other, more expensive liquid-based Papanicolaou tests. Cancer Cytopathol. 2006;108(3):137–43.

11. Garbar C, Mascaux C, Fontaine V. Efficiency of an inexpensive liquid-based cytology performed by cytocentrifugations: a comparative study using the histology as reference standard. Cytojournal. 2005;2:15.
12. Rimiene J, Petronyte J, Gudleviciene Z, Smailyte G, Krasauskaite I, Laurinavicius A. A Shandon PapSpin liquid-based gynecological test: a split-sample and direct-to-vial test with histology follow-up study. Cytojournal. 2010;7:2.
13. Barrett DL. Sources of cell loss using membrane filter and cytocentrifuge preparatory techniques. Cytotechnol Bull. 1975;12(5):7–8.

Chapter 7
Membrane Filtration

PRINCIPLE NO. 3

Flatten cells to enhance chromatin display.

PRACTICE

Use fresh specimens suspended in balanced salt solution, and apply 100-mm Hg negative pressure. Keep the cells wet during filtration and all subsequent steps until coverslipped.

Historic Milestones

- 1956—Seal introduces Millipore filters "for concentrating cancer cells suspended in large quantities of fluid."[1]
- 1957—Del Vecchio et al. extend Millipore filtration to "all specimens, regardless of volume or cell concentration."[2]
- 1964—Seal introduces Nuclepore filters as "a sieve for the isolation of cancer cells and other large cells from the blood."[3]
- 1967—Reynaud and King describe the use of Nuclepore filters for diagnostic cytology.[4]
- 1969—Gill publishes comprehensive method for processing Millipore filters for exfoliative cytology.[5]

Specimens with few cells do not lend themselves to direct smears. As a practical matter, how does one retrieve the cells and transfer them onto a slide surface for cytopreparation? For example,

G.W. Gill, *Cytopreparation: Principles & Practice*,
Essentials in Cytopathology 12, DOI 10.1007/978-1-4614-4933-1_7,
© Springer Science+Business Media New York 2013

following centrifugation of a fluid, a barely visible cell pellet is obtained. Now what do you do?

Doing something with nothing was problematical before the introduction of cytocentrifugation and liquid-based cytology processors. Membrane filtration was introduced in the mid-1950s to concentrate cancer cells from large quantities of fluid. About that time, there was great interest in detecting cancer cells circulating in the blood of cancer patients. My first research project was part of the Circulating Cancer Cell Cooperative. It hadn't occurred to us initially that a cancer cell had to pass by the tip of an inserted hypodermic needle at the exact moment to be captured, processed, and survive intact to be recognized. If a cancer patient had enough cancer cells circulating in the bloodstream to satisfy that unlikely sequence of events, a fatal outcome would likely be a foregone conclusion.

The most substantive and practical accomplishment from that work was elevating membrane filtration from a hit-or-miss art to a science.[6] Membrane filtration was extended to collecting cells suspended in fluids from all body sites, not only in our own cytopreparatory laboratory, but in other laboratories as well. When performed properly, membrane filtration exhibits great quantitative cell recovery and so is particularly useful for retrieving cells of all kinds, regardless of numbers, from specimens of any volume from any body site.

"When performed properly" means work that few today want to take the time to do. It means adapting the materials and methods of specimen collection, staining, and mounting to the chemical, physical, and optical properties of the filters. Filtration captures the cells on the filter surface, while the suspension fluid passes through. The filter and cells are stained together and mounted for microscopic examination. When processed by a filtration-staining-mounting technique that's been modified to accommodate the filter's physical and chemical properties, the results are superb. Membrane filtration not only provides a quantitative means to capture cells that might otherwise be impossible to recover under certain circumstances, but it utilizes negative pressure that constructively flattens cells. See Fig. 7.1.

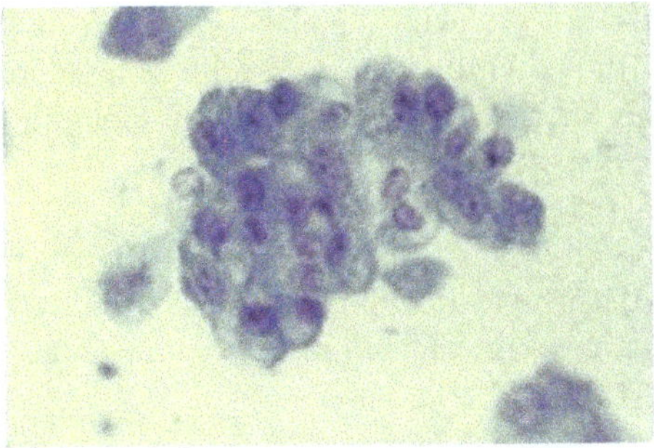

Fig. 7.1. Bladder cancer cells on Millipore filter. Papanicolaou stain. Original magnification ×400.

Relatively few laboratories today use membrane filters. Indeed, their use is almost a footnote in cytopreparatory history. Nonetheless, they continue to be used in a few cytology laboratories—especially those whose directors were trained at Hopkins and know firsthand how good they can be.

Materials and Methods

Cerebrospinal fluids can be filtered directly. All other specimens, especially clear urines, should be centrifuged for 10 min at 3,000 rpm, the supernatant discarded, and the cell concentrate resuspended in balanced electrolyte solution before filtration. When a cell pellet is not obvious, aspirate the supernatant until 5 mL remains. Then, resuspend the "invisible" pellet in balanced electrolyte solution and proceed.

Materials

- Cellulosic membrane filters, 5-μm pore size: discs (SMWP02500 [25 mm] or SMWP04700 [47 mm]) or rectangles (SMWP0190R [19 × 42 mm])

- Glass filter holders with stainless steel screen filter support (e.g., XX10 047 00 for 47-mm discs) (see footnote 1)
- Flat-tip forceps (i.e., XX6200006P) (see footnote 1)
- Indelible ink marking pen
- Petri dish (100 × 15 mm) with 95% ethanol
- Vacuum flask (1 L), vacuum tubing, and 3-way stopcock
- Balanced electrolyte solution
- Vacuum source with vacuum regulator and gauge

Methods (Performed Within a Biological Safety Cabinet)

1. Write the specimen's accession number along the filter's border in indelible ink. Write the number twice, side-by-side, for the 47-mm discs, which will be cut in half before being mounted.
2. Use fresh specimens, not ones collected in preservative.
3. Centrifuge in balanced capped centrifuge tubes for 10 min at 3,000 rpm.
4. Discard the supernatant and resuspend the pellet in balanced electrolyte solution.
5. Immerse the Millipore filter in a Petri dish of 95% ethanol for 10 s to "pre-expand" it (Fig. 7.2a).
6. Using flat-tip filter forceps, remove the filter by the edge opposite the numbers and lay it flat across the filter-holder base (Fig. 7.2b, c). Position the numbers immediately adjacent to, but not outside, the margin of the support medium (i.e., glass frit or metal screen).
7. Place the filter-holder funnel on top of the filter (Fig. 7.2d).
8. Add balanced electrolyte solution to a depth of about 5 cm in the funnel. See Fig. 7.2e, f. Wet the hydrophobic filter's pores by momentarily applying negative pressure to pull some solution through the filter. This step makes filtration proceed smoothly.
9. Add conservative amounts of cell suspension as needed to the center of the dilution reservoir of salt solution (Fig. 7.2g).

10. Apply 100-mm Hg negative pressure for cellulosic filters (Fig. 7.2h).
11. As filtration proceeds, wash the interior wall of the funnel with salt solution squirted from a squeeze bottle. Don't let the filter dry, not even for a nanosecond.
12. When a shallow pool of clear salt solution covers the filter (Figs. 7.2i and 7.2i-before), add 20 to 30 mL of filtered 95% ethanol along the interior wall of the funnel (Figs. 7.2j and 7.2j-after). This key step, "fixation in situ," anchors the cells to the filter surface and prevents them from being washed off in Step 14.
13. When a shallow pool remains, disconnect the vacuum and remove the funnel.
14. Using forceps, quickly remove the filter (Fig. 7.2k); smoothly slip it into the Petri dish of alcohol for 15 min.

Results

When subsequently stained and mounted properly, the preparations exhibit usefully flattened cells with morphology that is a boost, and not a barrier, to interpretation. See Fig. 7.3a–d.

Filters must be kept wet and not be allowed to air-dry. Vacuum-assisted air-drying rapidly "sucks the life" out of cells and ruins morphology. See Fig. 7.4.

Discussion

Given my conviction that membrane filters are a valuable tool, I have described in-depth the technique of membrane filtration. Few readers are likely to use it, but it's too useful to be abandoned entirely.

Other membrane filters may be used. Gelman (Pall) membrane filters are one such product. Like Millipore filters, it too is a cellulosic product, but its particular composition is less reactive than Millipore's

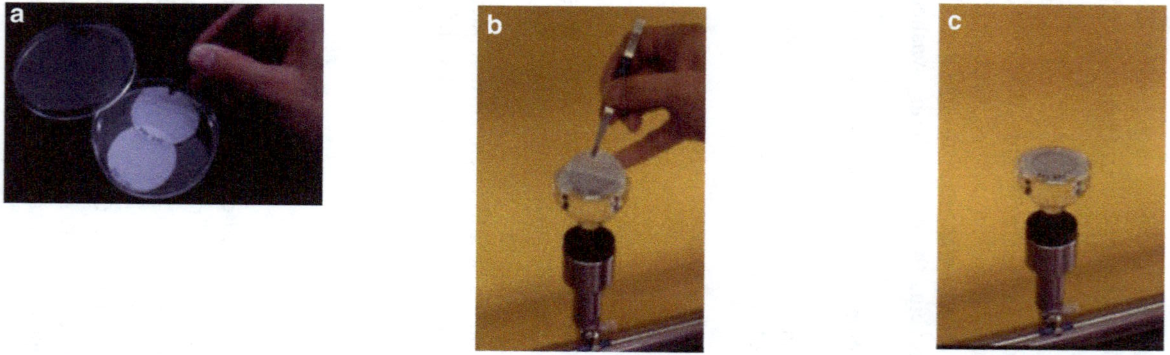

FIG. 7.2. (**a**) Type MF Millipore filters are made of mixed esters of cellulose that will expand slightly, but noticeably, when alcohol is added to the filter preparation in Step 12. Unless "pre-expanded" by immersion in alcohol before the filter is secured on the filter-holder base by the filter-holder funnel, the filter will wrinkle. Wrinkled filters don't coverslip flat.

(**b**) Carefully lay the filter onto the filter-holder base.

(**c**) Lay it flat. Position the numbers immediately adjacent to, but not outside, the margin of the support medium.

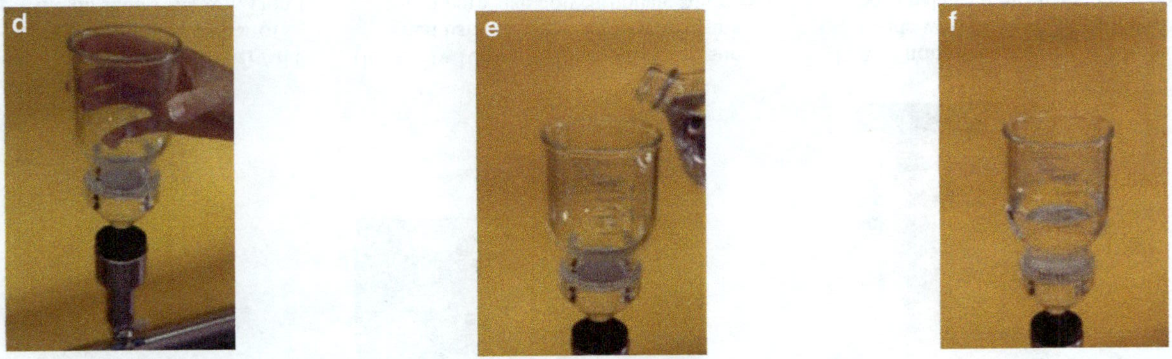

FIG. 7.2. (continued) (**d**) Center the filter-holder funnel on top of the filter-holder base. (**e**) Add balanced electrolyte solution first. Don't use normal saline. It destroys cells. (**f**) Add enough to dilute the specimen and promote uniform cell distribution.

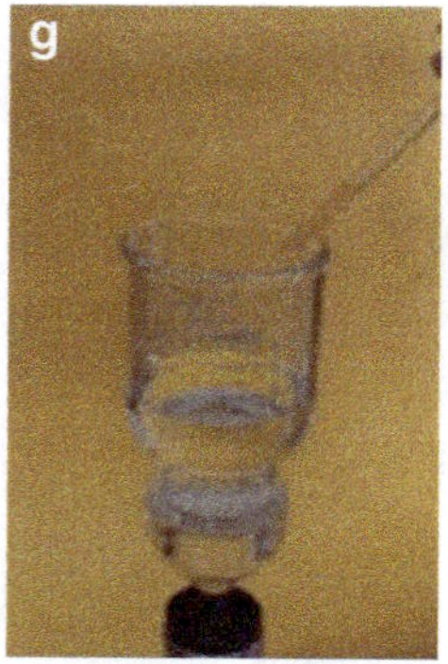

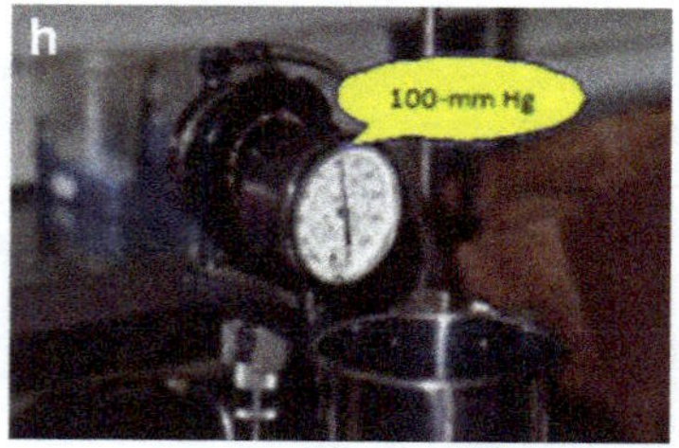

FIG. 7.2. (continued) (**g**) Add specimen to the center of the dilution reservoir. Don't overload the filter.

(**h**) Apply 100-mm Hg negative pressure when using Millipore filters. If using 5-µm pore size polycarbonate filters (e.g., Nuclepore), use 20 mm.

(**i**) As filtration continues, wash the inside wall of the filter-holder funnel with balanced electrolyte solution. See (**i-before**).

FIG. 7.2. (continued) (**j**) Never let the filter dry out. Add filtered alcohol to the inside wall of the filterholder funnel. This anchors the cells to the filter. See (**j-after**).

(**k**) Quickly remove the filter-holder. Using flat-tip forceps, remove the filter and immerse it in alcohol as in (a).

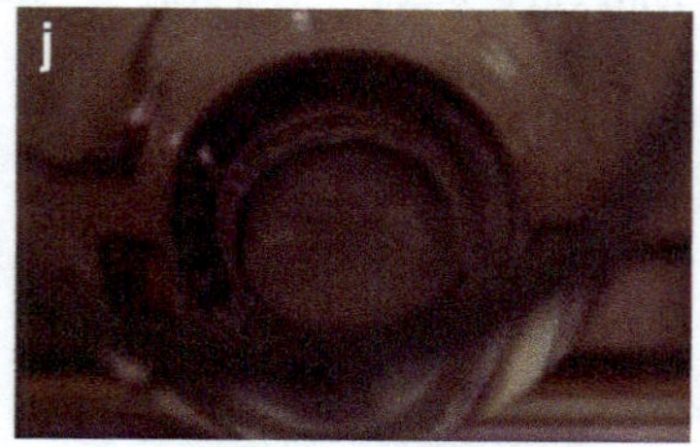

FIG. 7.2. (continued) **(i-before)** Before alcohol is added to the dilution reservoir, blood is obvious when present. **(j-after)** After alcohol is added to the dilution reservoir, it is diluted and hemolyzes blood if present. Red cell ghosts are seen microscopically.

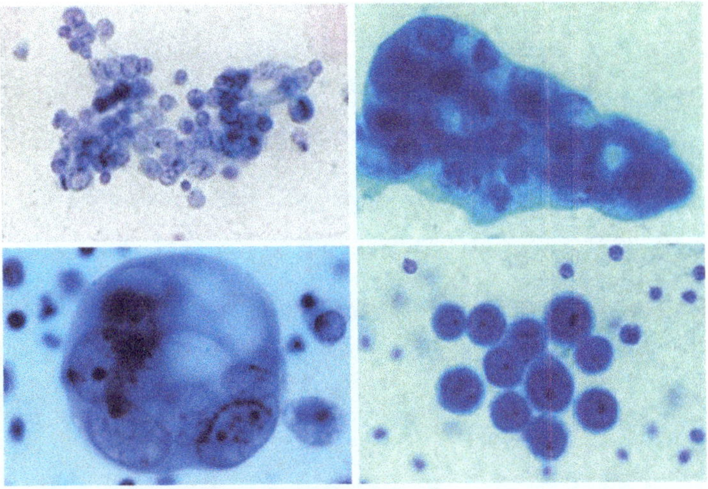

Fig. 7.3. (**a–d**) Malignant cells on Millipore filters. (**a**) shows oat cell carcinoma in a sputum specimen. When slides were also prepared on any specimen, the cells on the Millipore filters were always larger and displayed superior chromatin details. That observation contributed to a better understanding of slide preparation techniques and to slight modifications that improved the quality of slide preparations overall.

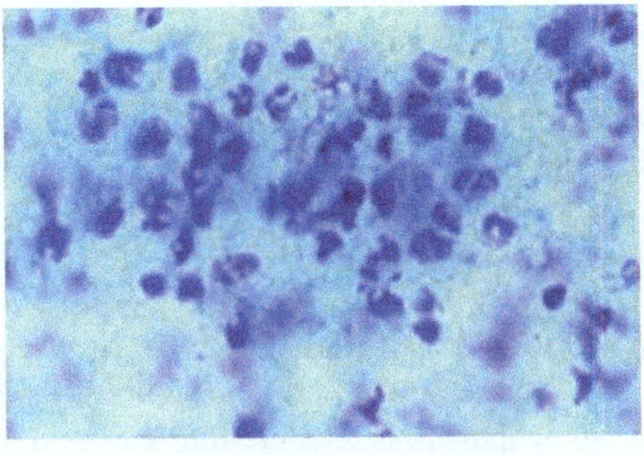

Fig. 7.4. Vacuum-assisted air-drying of cells on Millipore filters rapidly ruins cytomorphology.

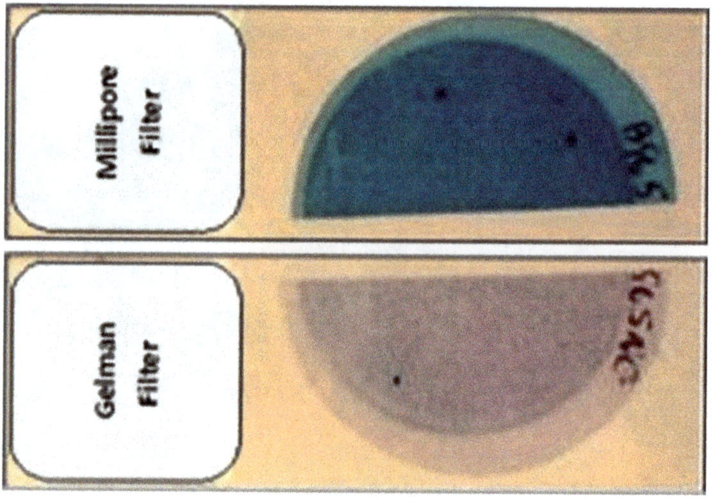

FIG. 7.5. The particular cellulosic composition of Gelman filters adsorbs less stain than the mixed cellulosic esters of Millipore filters. The difference in color as illustrated here is consistent when the filters are stained as described in Chap. 10. Both outcomes, however, depend entirely on the technique of filtration and staining. Using preserved specimens, inadequate rinsing to remove trapped protein, and staining with inadequate rinsing to remove passively trapped stain, can turn the filters as dark as the midnight sky. The colored tint of the unoccupied area of the slides in this figure is an artifact of photography.

mixed esters of cellulose. It does not need “pre-expansion” in alcohol, and it stains more lightly than Millipore filters. See Fig. 7.5. The green background color of the Millipore filter doesn’t diminish its diagnostic utility.

Figure 7.6 illustrates the increase in contrast between Pap-stained cells and the lightly stained Gelman filter.

The second basic type of membrane filter that was to be used in cytology is made of polycarbonate. The first polycarbonate filters available commercially were Nuclepore filters (without an “o” [i.e., not Nucleopore]). Nuclepore filters were developed by a team of engineers at General Electric. The product did not fit GE’s

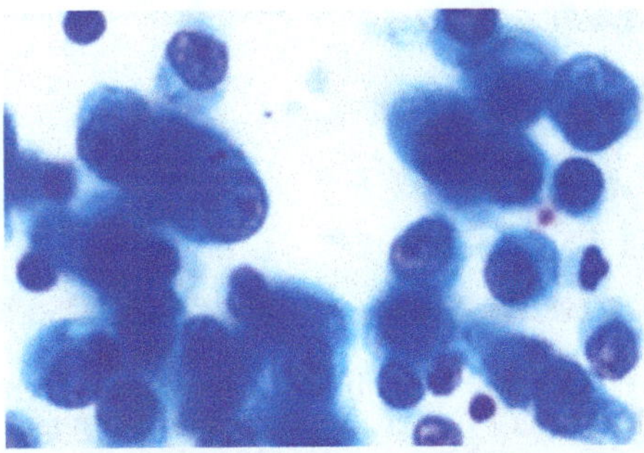

FIG. 7.6. Malignant cells from a pleural fluid. Papanicolaou stain. Original magnification ×400. Not all Pap stains are created equal. This one had been modified in a variety of ways to enhance performance. See Chap. 10.

family of products, so with GE's blessing the engineers left and formed the Nuclepore Corporation.

"Thin plastic sieves with precisely controlled hole size and density can be made by irradiating plastic films with fission fragments and etching out the material traversed by the fragments. These filters may be used for the nondestructive separation of cells of closely similar sizes."[7] Hence, the origin of Nuclepore. Controlling the radiation intensity and immersing the irradiated films in 6 N NaOH at 75° for a prescribed time results in filters with the desired pore diameter and density.[7]

Nuclepore filters are about 1/16th as thick as Millipore filters (i.e., 10 μm vs. 160 μm). See Fig. 7.7. Their properties differ in several ways.

Type SM means Separatory Medium (i.e., 5 μm), and MF means Millipore Filter comprised of mixed esters of cellulose acetate and cellulose nitrate. These filters are easily torn when dry, 160-μm thick, 84% porous, and have a refractive index of 1.495. The thickness is the same as that of a thick No. 1 cover glass. Every pore follows a tortuous path through the filter and emerges as a corresponding pore opening. The pore openings on the upper filter surface are slightly smaller than those on the bottom. See Fig. 7.8.

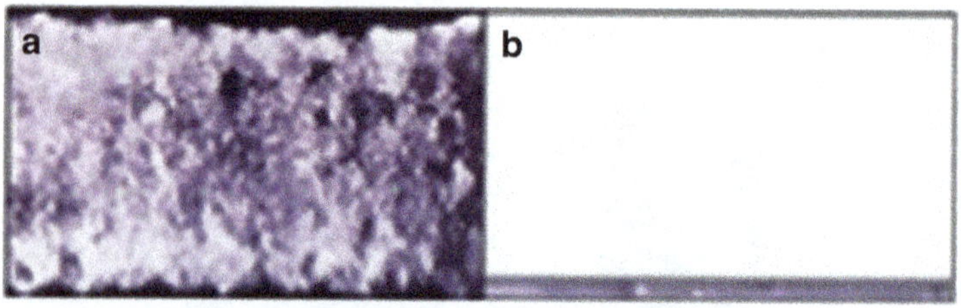

FIG. 7.7. (**a**) Type SM MF Millipore filters are recommended for specimen collection in cytology. (**b**) Polycarbonate filters were used in cytology for a few years following their 1964 introduction to the marketplace. Polycarbonate filters are the TransCyt filters used in the ThinPrep Processor for gyn and non-gyn specimens (i.e., 8-μm and 5-μm pore sizes, respectively).

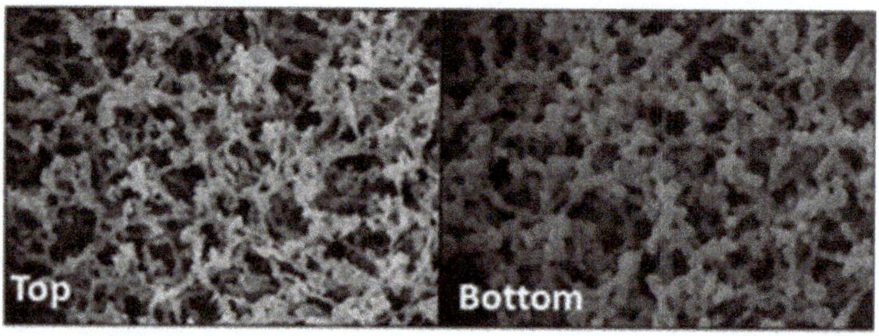

FIG. 7.8. Type SM MF pores have 5-μm mean diameters that are wider on the upper filter surface than on the lower surface. The pore-size distribution is determined by a mercury-infusion method.[8] Scanning electron microscope photomicrograph, magnification ×2,500.

The 1.495 refractive index is lower than that of most mounting media for cytological applications. Millipore filters are white when dry and transparent when mounted in a medium of similar refractive index. Unless absolute isopropanol is substituted for absolute ethanol in the Pap stain final dehydration series, the filter may semi-dissolve and become unmanageable to handle.

Nuclepore filters, on the other hand, are tough (i.e., think Lexan), 10-μm thick, 5–7% porous, naturally transparent macroscopically, and birefringent. Birefringent means two refractive indexes. Practically speaking, the boundary between the pore

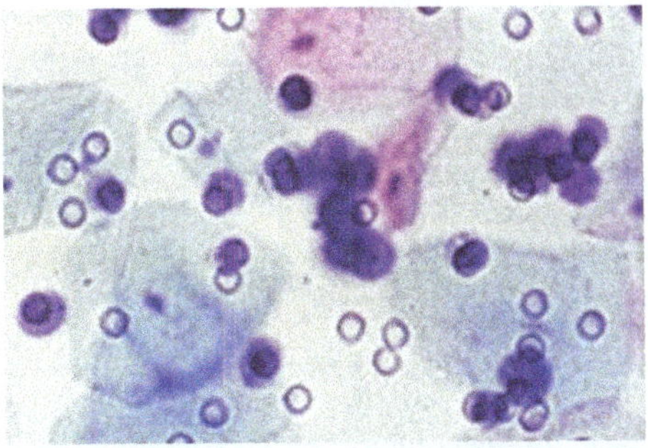

FIG. 7.9. The plane of best focus in this photomicrograph is deliberately selected to exaggerate the outline of the pores. Focusing up and down on cells-of-interest images the pores unacceptably.

(i.e., the hole itself) and the surrounding filter could not be made to "disappear" by mounting the filter in a medium of matching refractive index, unlike cellulosic filters. See Fig. 7.9.

Some laboratories eliminated the microscopically distracting pores by dissolving the Nuclepore filter in chloroform while the filter and its cells lay flat on a horizontal microscope slide. The rapid evaporation of the volatile chloroform decreased the surface temperature below the dew point, the temperature at which atmospheric water condenses to cause blushing. Blushing is the milky cloudy deposit that is left behind. In addition, the air-drying damaged the cells.

This technique presented potential health risks to laboratory personnel and diminished the quality of the preparations. For these and other reasons, Nuclepore filters aren't used today in cytology.

References

1. Seal SH. A method for concentrating cancer cells suspended in large quantities of fluid. Cancer. 1956;9(5):866–8.
2. Del Vecchio PR, De Witt SH, Borelli JI, et al. Application of Millipore filtration technique to cytologic material. J Natl Cancer Inst. 1959;22(2):427–31.

3. Seal SH. A sieve for the isolation of cancer cells and other large cells from the blood. Cancer. 1964;17(5):637–42.
4. Reynaud AJ, King EB. A new filter for diagnostic cytology. Acta Cytol. 1967;11(4):289–94.
5. Gill GW. Exfoliative cytology – Application Report AR-24. Bedford, MA: Millipore Corporation; 1969.
6. Frost JK, Gill GW, Hankins AG, LaCorte FJ, Miller RA, Hollander DH. Cytology filter preparations: factors affecting their quality for study of circulating cancer cells in the blood. Acta Cytol. 1967;11(5):363–73.
7. Fleischer RL, Price PB, Symes EM. Novel filter for biological materials. Science. 1964;143(3603):249–50.
8. Honold E, Skau EL. Application of mercury-intrusion method for determination of pore-size distribution to membrane filters. Science. 1954;120(3124):805–6.1Filtration materials available from EMD Millipore: http://www.millipore.com.

Chapter 8
Fixation

The microscopic interpretation of any type of smear depends largely on the excellence of the preparations.

Ruth M. Graham

PRINCIPLE NO. 4

Fix preparations immediately to maintain morphology.

PRACTICE

There is a hierarchy of best fixation practices that is a function of who's preparing what kinds of specimens, where, and available resources. The best practice uses fresh cells that are spread thinly on a clean slide and immediately immersed in alcohol. Everything else lessens cell display and cell recovery to various degrees.

Historic Milestones

- 1851—Clarke mixes alcohol with acetic acid as a fixative.[1]
- 1875—"Preservative" as a chemical added to foods to keep them from rotting.[2]
- 1880s—"Fixation" as a term comes into general use.[3]
- 1887—Carnoy adds chloroform to Clarke's fluid.[4]
- 1917—Alcohol–ether introduced as fixative.[5]

G.W. Gill, *Cytopreparation: Principles & Practice*, Essentials in Cytopathology 12, DOI 10.1007/978-1-4614-4933-1_8,
© Springer Science+Business Media New York 2013

- 1952—Sills introduces Carbowax in alcohol–formalin mixture as a preservative.[6]
- 1954—*"If ether is not available, ethyl alcohol (95%) can be used alone."*[7]
- 1957—International write-in symposium considers whether air-drying causes cellular changes in Pap smears.[8]
- 1958—Union Carbide trademarks polyethylene glycol as Carbowax.[9]
- 1963—Saccomanno homogenizes sputum preserved in 50% alcohol with 2% Carbowax 1540.[10]
- 1966—Bonime rehydrates air-dried Pap smears in 50% glycerin.[11]
- 1968—Ehrenreich and Zadeikis patent spray fixative with alcohol and polyethylene glycol.[12]
- 1993—"Cell preservative solution" (i.e., PreservCyt) patented.[13]

The foregoing milestones span more than a century. Relative to cytology as we know it, they relate to the preservation and fixation of gyn and non-gyn specimens, whether they are air-dried or not, whether air-drying alters cell morphology or not, and if so, how to protect cells from the effects of air-drying. Before describing what works and what doesn't for cytological applications, I want to give some historical background.

Early pioneers in microtechnique had no experience on which to build. They had no expectations for the outcomes of their experimentation. The early microscopes imaged with extreme curvature, added artificial background color, and demonstrated poor resolution. Living cells quickly self-digest and become unusable. Needless to say, they're ephemeral, not permanent. Preservation and fixation were born out of necessity.

The ideal preservative neither shrinks nor swells immersed cells or tissues, does not dissolve or distort its constituent parts, kills bacteria and molds, and prevents autolysis. The ideal fixative is just that an ideal. It does not exist. Fixatives take preservatives at least one step further: they modify the biological materials so they retain their form when exposed to subsequent processing. Historically, subsequent processing meant tissue sliced thinly by a razor blade. To remain intact and not crumble, early fixatives hardened tissue, and were sometimes called hardening fluids. When embedding made it possible to stabilize tissues for sectioning, hardening became less important. Fixation displaced hardening as a term.[14]

All fixatives are preservatives, therefore, but no preservatives are fixatives. Both are forward-looking solutions that are used in anticipation of the intended use, whether the specimen is cytologic or histologic. In cytology, the intended use is cytomorphology—especially nuclear morphology and chromatin patterns—on which interpretations of cellular health and disease are based.

The major difference between the same chemical used as a fixative or as a preservative is its concentration. Ethanol at 95% concentration is a fixative and at 50% a preservative. There is a gray zone between these 2 percentages that is indeterminate. The final concentration depends on the proportion of alcohol to specimen. Nongynecological large-volume specimens such as effusions dilute the starting concentration the most.

Fixation in the context of diagnostic cytopathology today traces its roots to Dr. Papanicolaou's use of equal parts of diethyl ether and 95% ethyl alcohol. This fixative became the standard in diagnostic cytopathology after Papanicolaou recorded its use in the 1942 monograph that he coauthored with Herbert Traut.

According to Papanicolaou, he first described this fixative in a 1917 paper[5] about the existence of a typical estrous cycle in the guinea pig: "Smears are fixed immediately (before drying) in equal arts of 95 per cent alcohol and ether (original method of Stockard and Papanicolaou)."[15] Fixative, fixation, ether, and alcohol are not mentioned in the cited paper.

However, a footnote on page 69 of a 1973 biography is illuminating: "Papanicolaou initially used Carnoy's fixative consisting of absolute alcohol, chloroform and acetic acid for wet fixation but later deleted acetic acid, exchanged ether for chloroform because of the latter's distasteful odor and used 95% alcohol instead of absolute alcohol, which was difficult to procure."[16] (I am indebted to Paul Elgert, CT(ASCP), CMIAC for calling my attention to this information.)

Ether has no fixative properties. Since it is an effective fat solvent, ether may serve as an adjuvant, which means it speeds alcohol's

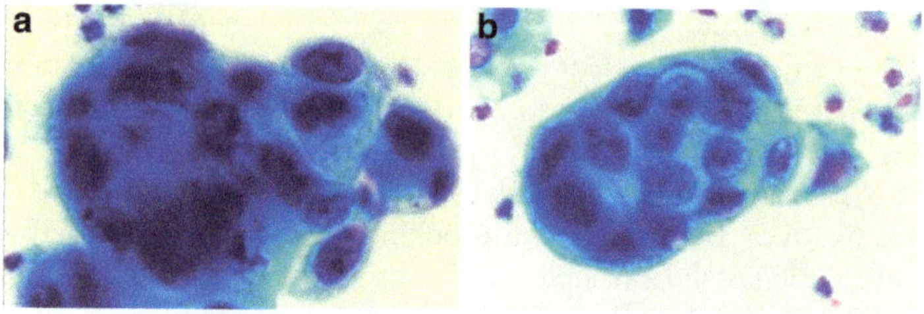

FIG. 8.1. Adenocarcinoma tissue fragments from a fresh pleural effusion that were fixed in (**a**) alcohol-ether or (**b**) 95% ethyl alcohol are cytomorphologically indistinguishable.

penetration into cells without fixing the cells per se. At some time in the 1970s, an unnamed pathologist told me at a workshop that Papanicolaou told him that he had included ether to clean the dirty glass slides that were the norm when he began using it.

Ether, of course, is highly volatile and explosive and should not be handled by amateurs. More than a few non-explosion-proof refrigerators have exploded when storing ether that was ignited by a spark. Explosion-proof suggests an armored refrigerator that can withstand an explosive blast, but that is not what explosion-proof means. Explosion-proof refrigerators are designed *not* to produce sparks with sufficient energy to trigger an explosion.

Consequently, ether's use in alcohol as a fixative was generally discontinued by the late 1950s, leaving 95% ethanol alone as the standard fixative for Pap smears. Indeed, that was the conclusion of 12 international participants in a write-in symposium published in the very first issue of Acta Cytologica in 1957.[8] See Fig. 8.1.

As Papanicolaou's cytological method became more widely used, it was applied in more settings and to nongynecologic cytologic specimens. The standard method of immediate wet fixation of fresh cells could not be readily adapted, which led to many variations of fixatives and fixation methods, not all of which produced cytomorphologically equivalent results.

The goal of fixation is make every cell reveal its health or disease status by its cytomorphology. The goal is *not* to preserve the lifelike appearance of cells, unless one seeks to study the lifelike appearance of cells per se. See Fig. 8.2.

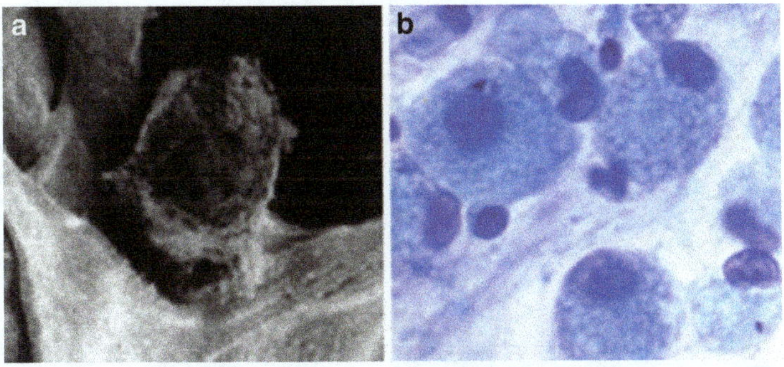

FIG. 8.2. (**a**) Highly magnified scanning electron micrograph of an unstained pulmonary alveolar macrophage in a hamster lung. (**b**) Histiocytes in fresh sputum that had been immediately wet fixed. Papanicolaou stain (original magnification ×1,000).

Hierarchy of Fixation Materials and Methods

There are four basic alternatives to immediate wet fixation in 95% ethanol:[14]

1. Substitute alcohols
2. Air-drying of protected fixed cells
3. Air-drying and rehydration of unprotected cells
4. Preservation

Substitute Alcohols

Substitute alcohols are those alcohols that can be used in the same manner as 95% ethanol without noticeable morphological differences. "Same manner" means immediate wet fixation of fresh specimens. These substitute alcohols include (1) reagent alcohol, (2) absolute methanol, (3) proprietary grade alcohol, (4) 80% isopropanol, and (5) 90% acetone. Acetone is not an alcohol, but it is included in this category for convenience. *Caution: All are flammable and must be handled with the usual fire safety precautions.* See Fig. 8.3.

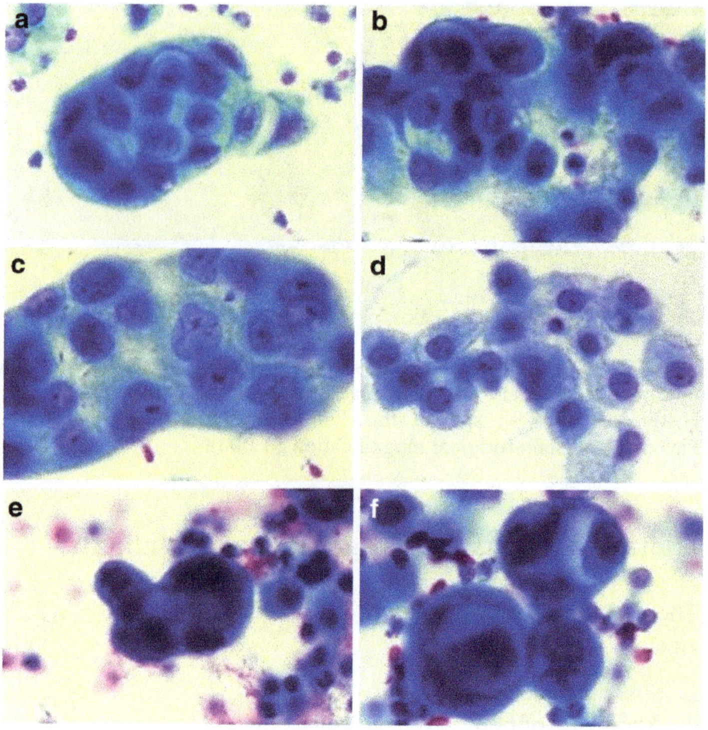

FIG. 8.3. With the exception of the cells fixed in proprietary grade alcohol (**d**), all these cells are from the same pleural fluid: (**a**) 95% ethanol, (**b**) reagent grade alcohol, (**c**) absolute methanol, (**d**) proprietary grade alcohol, (**e**) absolute isopropanol (IPA), and (**f**) 90% acetone. Proprietary grade alcohol is a mixture of 100 parts alcohol and 1 part each of ethyl acetate, methyl isobutyl ketone, and aviation gasoline. A thru D morphologically similar results, IPA and acetone (**e** and **f**) did not. The cells shown in (**e**) were fixed in absolute IPA, which shrinks cells excessively. Diluting IPA to 80% concentration shrinks cell less, thus making 80% IPA a suitable alternative. Acetone has been described as an alternative fixative, but its volatility and odor make it less attractive as a practical matter. All six solutions are flammable and should be handled safely.

Absolute Methanol. Methanol is the first of three possible substitute alcohols. Methanol shrinks cells less relative to the shrinkage observed with 95% ethanol. For this reason, methanol can be used as absolute (i.e., 100%) strength.

Reagent Alcohol. Reagent alcohol is comprised of 9 parts ethanol and ½ parts each of methanol and isopropanol. The latter 2 alcohols denature the ethanol, which makes it unfit for consumption as a beverage, and thereby makes it untaxable. According to federal law, "reagent alcohol shall be packaged by the manufacturer in containers not exceeding four liters." The latter restriction increases the cost of reagent alcohol.

There is no need to dilute reagent alcohol to 95% concentration, as it makes no visibly discernible difference in cytomorphology when used full-strength. Indeed, ethanol can be used full-strength as well. Using ethanol at 95% concentration saves money, however, as it can be purchased at that concentration and is less expensive than if purchased at absolute concentration. "Absolute" means anhydrous (i.e., no water). Removing the remaining few percent of water from alcohol requires more expensive chemical processing.

Reagent alcohol is 1 of 50 specially denatured alcohol formulations approved by the federal government for manufacture for assorted applications.[17] Some of the formulations that are not denatured by denaturants likely to impact cytomorphology and also are not restricted to sale in containers not to exceed 4 L capacity. Special denatured alcohol Formula No. 3-C, for example, is 100 gallons of alcohol and 5 gallons of isopropyl alcohol.

80% Isopropanol. Absolute isopropanol shrinks cells excessively. Diluting isopropanol to 80% concentration diminishes the shrinkage to that comparable to ethanol.

90% Acetone. Historically, acetone has been used full-strength, but it hardens cells and tissue unacceptably. Diluting acetone to 90% concentration produces acceptable results. Its odor, however, makes it unattractive for routine use.

Air-Drying of Protected Fixed Cells

Air-drying of protected fixed cells is a process by which cells on a slide are fixed in alcohol with polyethylene glycol (PEG) and subsequently air-dried. At least two obvious approaches are possible:

- *Spray fixation.* Fresh cells can be spread onto a slide–avoiding air-drying—and covered with alcohol with PEG, which may be dropped on the slide or delivered as a spray (i.e., spray fixation) and then allowed to air-dry.
- *Immerse first, air-dry second.* Fresh cells can be spread onto a slide and immediately immersed in alcohol (i.e., wet fixed) with PEG, removed after several minutes, and allowed to air-dry.

Ten to 12 inches is reported to be the optimal range of distance from which a fluorocarbon propellant powered spray fixative should be delivered to a cell spread.[18] Nearer distances result in nuclear shrinkage, while farther distances result in air-drying. Although the causes of these limitations were not suggested, it is possible that at near distances, the propellant-driven blast of spray fixative tears cells from their moorings and shrinks them as they float in the pool of fixative. And at farther distances, it is possible that the inverse square law diminishes the density of spray particles and so retards the rate of deposition on the cells, thus allowing sufficient time for partial air-drying. Therefore, the optimal distance may vary with the force of delivery and the density of spray droplets of a given aerosol fixative. Short distance sensitivity has not been reported for pump spray fixatives, which deliver the fixative with less force.

Hair spray (i.e., Aqua Net) was recommended as a less expensive alternative to commercial spray fixatives (e.g., Spray-Cyte).[19] As time passed, its use fell into disfavor. The ingredients were considered proprietary, so one could never be certain what was being sprayed onto cells. That consideration would be moot if the spray-fixed cells were cytomorphologically useful, but they were not.

Air-Drying and Rehydration of Unprotected Cells

Cells that are spread on a slide and simply air-dried are useless for interpretation cytomorphologically. Indeed, air-drying is one of

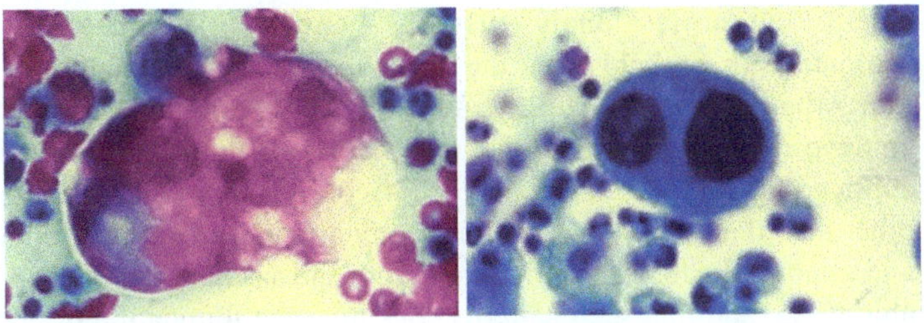

FIG. 8.4. Pap-stained cells from the same pleural fluid: (**a**) air-dried and (**b**) air-dried followed by rehydration in 50% glycerin ×3 min, followed by 95% ethanol.[11]

the major causes of unsatisfactory conventional Pap tests, which helped spur the development of liquid-based cytology.

Air-drying alters cells physically in at least three observable ways. It *irreversibly* (1) degrades chromatin structure and detail, (2) enlarges or ensmalls cells, and *reversibly* (3) slows dye uptake.

Changes in chromatin display and cell diameter are *irreversible* by all rehydration methods I have tried. The effects of the more tightly woven protein texture that selectively blocks stain uptake, however, are *reversible* upon immersion in glycerin (Fig. 8.4). Glycerin is a trihydric alcohol, which means it has three OH groups instead of one like isopropanol, which is the same carbon chain length. It penetrates and "wets" protein molecules, thereby allowing all dyes of the Pap stain to penetrate and color the cells normochromatically. The quality of the restoration is dependent on the quality of the Pap stain in use at the time in any given laboratory; Koss's fleeting reference to the lack of glandular cell detail may reflect this fact. Parenthetically, glycerin is a humectant used in the manufacture of chocolate candies to keep the exterior surfaces looking fresh and edibly attractive.

While morphologically more interpretable than if not rehydrated and diagnostic outcomes may be comparable to those of wet-fixed cohorts, rehydrated air-dried cells are not identical with alcohol wet-fixed cells in diameter or chromatin patterns. Given the choice between examining alcohol wet-fixed cells and rehydrated air-dried cells, cytoprofessionals usually choose the standard presentation—probably on the basis of subjective familiarity rather than on objective utility. After all, wet fixation in alcohol is an empirical method used in no other comparable biomedical application. It is what Papanicolaou used for no particular reason other than that he could. Alcohol precipitates chromatin in normal and abnormal cells in visibly distinctive and visually distinguishable ways, and that such patterns constitute the foundation of our collective memory banks of cytopathological images.

Air-drying of Pap smears[20] and subsequent rehydration in normal saline[21] have sometimes been recommended, especially in low resource settings. Rehydration in normal saline has been applied to fine needle aspirations as well.[22] Apart from the doubtful efficacy of such practices, using normal saline would make sense only if the cells were alive. Dried cells are dead, however, and incapable of responding to a salt solution of any kind. Plain water will suffice.

Preservation

Preservation is collecting cells in preservative, either gynecologic cytologic samples for liquid-based cytology or nongynecologic cytologic cell suspensions such as sputum, urine, and body cavity fluids. Many years ago, preservation was sometimes referred to as prefixation.

In general, collecting cell suspensions in an equal volume of 50% ethanol, or comparable substitute alcohol, is recommended. If cells will be air-dried after being spread on a slide, include Carbowax 1450 at a 2% (w/v) concentration. Otherwise, do not include Carbowax, as Carbowax does not preserve or fix cells. It's OK to use Carbowax-based preservatives, but using Carbowax when it is noncontributory is wasteful.

Putting the Pieces of the Puzzle Together

In diagnostic cytopathology, visually interpretable chromatin detail is everything. Such detail is exhibited best when every material and method in the cytopreparatory chain is selected for its contribution to the final result. Sound biological, chemical, physical, and optical principles underpin the foundation of the practice. When the links in the chain are strong, useful cytomorphology results; when weak, cytomorphology is compromised.

From the standpoint of cytopreparation, cells are more alike than different—regardless of the originating body site. The goal is to transfer the cells from suspension onto a surface for fixing, staining, mounting, and microscopy. It matters little whether the cells are prepared as a cell spread, a cytocentrifuged preparation, a cellulosic or polycarbonate filter preparation, a liquid-based preparation, or cell blocks.

Of all the reactions to the quality of my preparations I've encountered over the years, 2 are particularly memorable because they are polar opposites. An education coordinator of a cytotechnology program stated: "If my preparations looked this good, I'd call histiocytes cancer." A pathologist remarked: "If my preparations looked this good, I think I could do cytology!" So, high quality cytologic preparations can be either a help or a hindrance, depending on what one's accustomed to. These reactions underscore the artifactual nature of cytomorphology and the need to control it tightly.

A cytologic preparation has two basic components when viewed microscopically: the object itself and the image derived from the object. Thus, there are 4 possible combinations of satisfactory and unsatisfactory quality. See Fig. 8.5. One skilled in the art and science of cytopreparation should be able to discern microscopically the reasons for deviations from optimal results and implement stable corrections as needed.

Cell flattening is good; cell shrinkage is bad. Extremes are undesirable. Standard fixation method strikes a balance. See Fig. 8.6.

Standard fixation methods in cytology strike a balance in specified applications. Since it is impossible to foresee every conceivable fixation scheme readers may employ or encounter, I have elected to explain the chemical and physical mechanisms that influence fixation outcomes. Readers may want to use this information to optimize their fixation protocols.

The degree to which shrinkage occurs, or doesn't, depends on (1) cellular water content, (2) cell location when preserved or

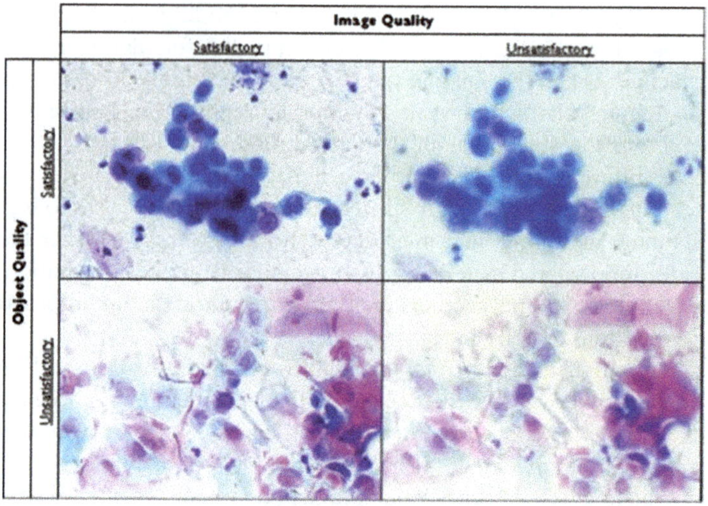

FIG. 8.5. Sitting at one's microscope, one should be able to discern the object and the image derived therefrom, evaluate the quality of the two components, and troubleshoot the likely cause(s) of any limitations. For example, in both the unsatisfactory images, the loss of contrast can be due to glare, flare, and/or spherical aberration. Fixes include (1) minimal thickness of mounting medium, (2) No. 1 thickness cover glass, (3) clean microscope, and (4) Köhler illumination.

fixed (i.e., in suspension or on a surface), (3) alcohol chain length (i.e., methyl, ethyl, isopropyl [1, 2, 3 carbons, respectively]), (4) alcohol concentration, (5) whether maintained wet or allowed to air-dry, (6) location if and when air-dried (i.e., separate from, or in contact with, the slide surface), and (7) whether Carbowax is present in the preservative or fixative when air-drying takes place.

Cellular Water Content

The water content of materials in the human body ranges from practically nothing in tooth enamel up to approximately 85% in neurons of the brain's gray matter. This leads to the notion of cells

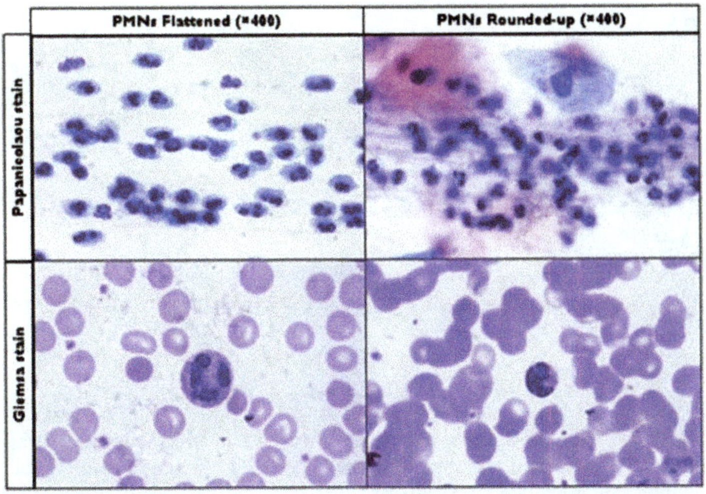

FIG. 8.6. The PMNs in the Pap smear and the air-dried blood film adhered to the clean glass surface, which caused them to flatten. Subsequent fixation maintained the flattening. In other areas of the same slides, respectively, the same cell types are distinctly smaller because they couldn't flatten. They were suspended above the glass surface in a stream of proteinaceous material in the Pap smear and by the thickness of the crowded surrounding erythrocytes in the blood film. Each pair of photomicrographs was taken at the same magnification.

with low and high water content in considerations of cellular responses to alternative fixatives and fixation methods used in diagnostic cytopathology. See Fig. 8.7.

Low water content cells include normal and abnormal intermediate and superficial squamous cells, while high water content cells are everything else. Water content is important because alcohol and air-drying independently extract water and will change cell diameters—with associated changes in nuclear area, thickness, and chromatin display—to different degrees.

As shown in Fig. 8.8, low water content cells collected in preservative are essentially unchanged in diameter relative to alcohol wet-fixed controls. When air-dried on glass, however, such cells increase in diameter nearly 1–1⁄2 times.

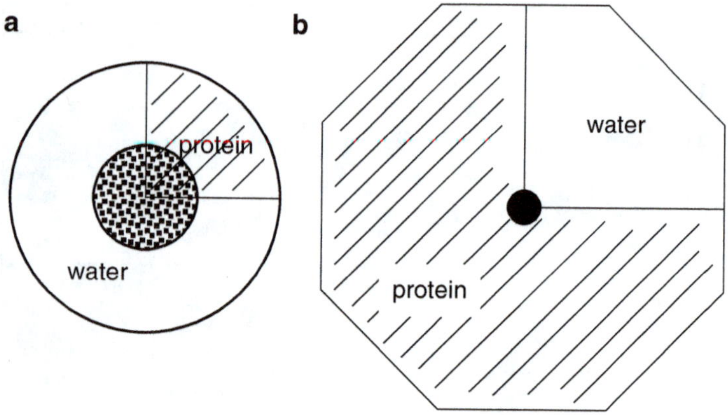

FIG. 8.7. Cellular water content limits the extent to which cells will shrink or swell under various conditions of fixation. Low water content cells include intermediate and superficial squamous cells and keratinizing squamous carcinoma cells. High water content cells are everything else. The latter are more sensitive indicators of tolerance to alternative materials and methods of fixation.

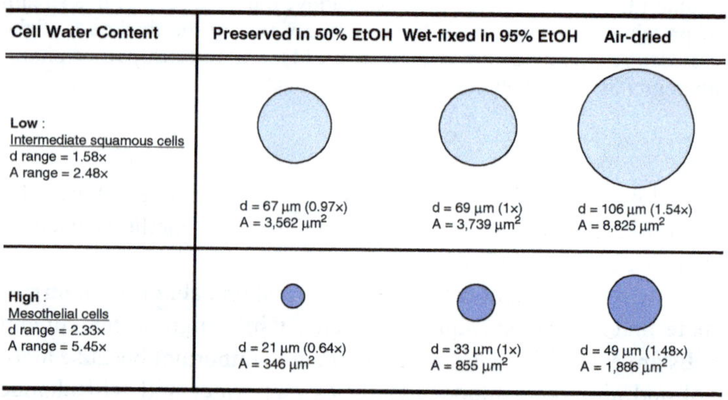

FIG. 8.8. Wet fixation vs. air-drying. Fresh (i.e., not preserved) wet-fixed cells are the standard against which alternative fixation materials and methods are judged. Deviations from that fixation protocol degrade chromatin display, which means less information content.[23] Modified from Kirby.[24]

High water content cells collected in preservative shrink to two-thirds the diameter of control cells, becoming thicker and more optically dense when stained. Preservatives swell cells in suspension initially (e.g., erythrocytes leak hemoglobin) and increases cohesive forces, which promotes subsequent shrinkage after the cells are spread on a slide and wet fixed. When air-dried on glass, high water content cells display diameters enlarged also by nearly 1–1⁄2 times—approximately the same amount as by air-dried low water content cells. Air-dried high water content cells are up to 5–1⁄2 times greater in area than their preserved counterparts. Such enlargement is a plus—indeed essential—for Romanovsky-stained preparations.

Cell Location When Preserved or Fixed

In the standard fixation method—whether for gyn or non-gyn specimens—fresh cells are in contact with the slide surface when fixed. Such contact is essential to flattening cells, as the contact favors the adhesive forces between the cells and the glass, which are generally stronger than the cohesive forces within fresh cells that resist flattening. See Fig. 8.9.

Clean slide surfaces facilitate cell flattening; dirty slide surfaces thwart cell flattening.

Adhesion may be defined as a property that causes unlike substances to stick together; cohesion causes like substances to cling

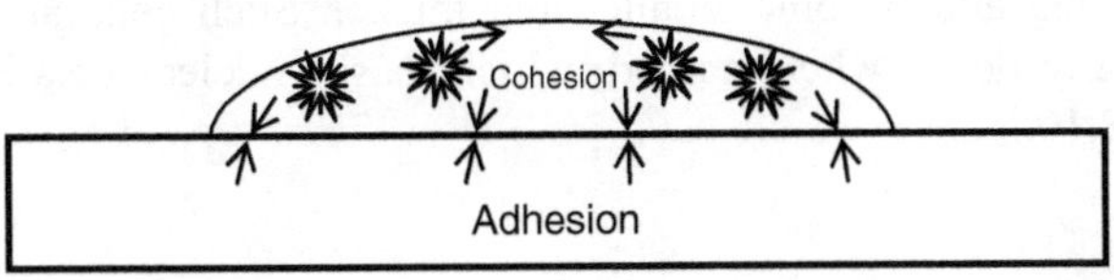

FIG. 8.9. Cell flattening facilitates visual access to nuclear details. In terms of useful outcomes, flattening is to cells as sectioning is to tissue. The degree to which cells flatten depends on the balance of cohesive forces within the cells vs. the adhesive forces between the cell and the slide surface. When cohesion exceeds adhesion, cells round up. When adhesion exceeds cohesion, cells flatten.

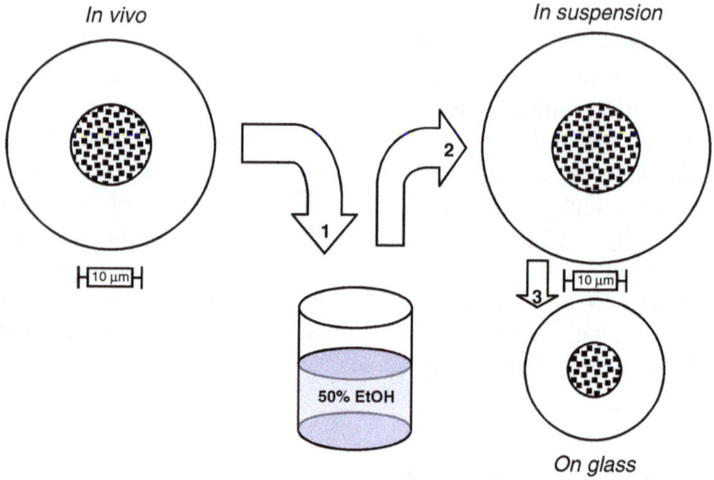

FIG. 8.10. (1) Collecting cells in alcoholic preservative (2) swells high water content cells while in suspension and partially coagulates the protein, thereby increasing the cohesive forces within the cells. When subsequently put into contact with the surface of a glass slide, (3) the increased cohesive forces make cells resist flattening. The end result is that such cells shrink to about two-thirds of the diameter of similar initially fresh cells when wet fixed.

together. Collecting cells as preserved suspensions, however, have the opposite effect. It increases the cohesive forces within cells. While cells in preservatives first swell, they ultimately shrink when immersed in fixative because the increased cohesive forces are greater than the adhesive forces. Thus, preserved cells resist flattening and become smaller and thicker. Such cells are more optically dense when stained, which masks nuclear details. See Fig. 8.10.

Alcohol Chain Length

All alcohols are organic derivatives of water. The chemical formula for water, H_2O, can be written as H-OH, which shows a single hydrogen atom and a hydroxyl group OH, which is the basis for alcohols. Methanol, ethanol, and isopropanol consist of 1, 2, and 3 carbon atoms and a single hydroxyl group: CH_3OH, C_2H_5OH, and C_3H_7OH. As the carbon chain length increases, changes in fixative properties occur: the longer the chain length, the greater the cellular shrinkage that results.

The relationship between alcohol chain length and cellular shrinkage may be due to several factors: (1) greater solubilities of lipids in the higher alcohols; (2) the polarity of the alcohol, the greater the polarity, the faster the penetration of the cell and the faster the solvation of cellular structure; and (3) cellular water leaves the cell by diffusion, and the cellular contents become progressively more dehydrated. I suspect the dehydration is the result of the coagulation of the proteins, which brings the molecules closer together, thereby in effect squeezing the water out of the cells.

Solvation, commonly called dissolution, is the process of attraction and association of molecules of a solvent with molecules or ions of a solute. As ions dissolve in a solvent, they spread out and become surrounded by solvent molecules. The bigger the ion, the more solvent molecules are able to surround it and the more it becomes solvated.

Alcohol Concentration

Alcohols are coagulating, nonadditive fixatives. "Coagulating" means transforming protoplasm into a microscopical spongework. "Nonadditive" means there is no obvious permanent addition of atoms to some part of the protein. As might be expected, the degree of coagulation, or hardening, is greatest for each alcohol at its maximum concentration. At lower concentrations, alcohols lose their fixative properties—those making cells and tissues capable of resisting subsequent treatments—and gain preservative properties. Collecting a small volume of cells in a large volume of concentrated alcohol, for example, coagulates dissolved protein and shrinks cells unacceptably, which complicate subsequent cytopreparation.

An alternative expression of shrinkage in alcohol can be seen in Fig. 8.11.

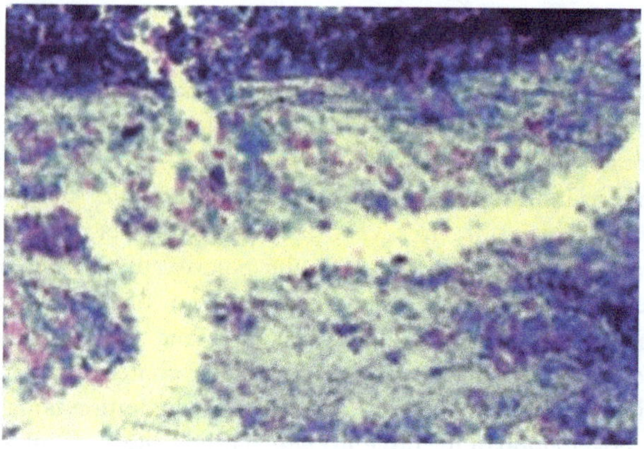

FIG. 8.11. These fissures express another presentation of alcohol-induced shrinkage. As indicated by the *arrows*, the protein blanket in which the cells are distributed was intact prior to wet fixation in 95% ethanol. Alcohol caused the protein to shrink and pull apart (Conventional Pap smear, original magnification ×100).

Whether Maintained Wet or Allowed to Air-Dry

As a cell air-dries, its moisture escapes at the air–water interface. Passing through the cell, it exerts tremendous surface tension forces that denature and disrupt proteins—forever altering chromatin display relative to its alcohol wet-fixed appearance. This force has been calculated to be 320 tons/in^2, which—less dramatically—equals 450 mg/μm^2.[25]

Location If and When Air-Dried

When in contact with a clean glass surface during air-drying, cells increase in diameter as the forces of extracellular adhesion exceed those of intracellular cohesion. The opposite electrostatic charges of cell and glass surfaces are attractive, as are the inward cohesive forces. In other words, the balance of forces favors cells flattening like sunny-side-up cooked eggs instead of rounding-up like hard-cooked eggs.

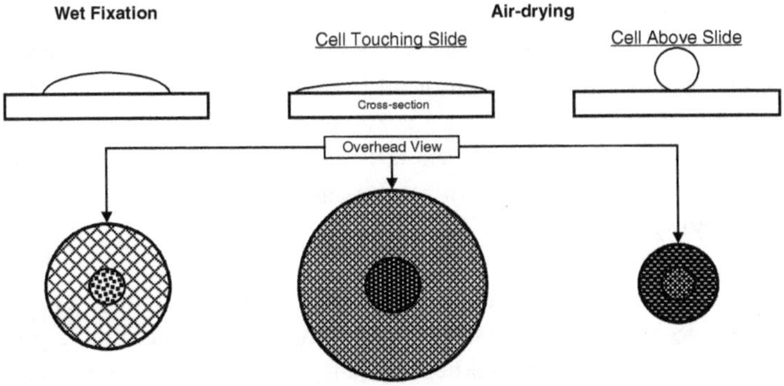

FIG. 8.12. Air-dried cells increase in area and decrease in thickness when touching the slide and do the opposite when not touching the slide. In this illustration, 2 high water content cells have been air-dried: 1 is touching the slide, the other, raised above it. In the "touching" case, the cellular diameter is enlarged approximately 1–1/2 times compared to its wet-fixed counterpart, the cell is thinner than if it had been wet fixed, and the cellular material is more closely textured, having collapsed upon itself.[26] As a consequence, biological dyes penetrate more slowly, leading overall to diffuse eosinophilia that masks chromatin edge boundaries. The adhesive forces exceed the cohesive forces. In the "raised" case, the cellular diameter is ensmalled to 2/3 the diameter of the wet-fixed cell since the cohesive forces within the cell have greatly exceeded the adhesive forces. This imbalance is observed in cells with a tight radius of curvature (e.g., PMNs) and in cells that are not making substantial contact with the slide (e.g., cells in tissue fragments, cells suspended above the slide in mucus, etc.).

On the other hand, a cell suspended above the glass surface (e.g., on an albumin adhesive film or frosted slide surface, in a mucus stream, or tissue fragment) becomes smaller in diameter as the intracellular cohesive forces are now relatively greater. See Fig. 8.12.

So air-drying may enlarge or ensmall cells diametrically, depending on which force is stronger under the circumstances—with the cells becoming thinner or thicker in the process, respectively. See Fig. 8.12. Cells that are air-dried and then fixed flatten/shrink more than if fixed first and then air-dried.

Air-drying compresses protein and reduces the intermolecular spaces through which dyes must subsequently pass to reach bonding sites. Of the Papanicolaou stain dyes, eosin penetrates tight spaces best. Consequently, air-dried cells appear eosinophilic throughout.[27] Diffuse eosinophilia eliminates color and optical density differences among cellular components, makes cells visually impenetrable, blurs particle boundaries, and exaggerates the apparent visible effects of air-drying.

A potential practical consequence is that normal squamous cells with nuclei as small as 8 μm may be misinterpreted as ASC-US because air-drying artifactually enlarges them to 12 μm, while reducing their thickness to 3.4 μm.[28] Rehydrated air-dried cells in groups may appear more optically dense. Some automated coverslippers allow xylene to evaporate from slides before applying the cover glasses.

Direct microscopic examination of cells in groups reveals that they shrink suddenly as the xylene evaporates completely—though not in single cells anchored to the slide surface. Such an observation raises the question of whether such coverslippers should be used at all and underscores the basis for coverslipping slides one at a time. Once fixed, cells should always be kept wet until mounted.

Overall, air-drying may make it difficult to compare quantitative cellular features with those of wet-fixed cells. Any quantitative study of cellular features should control the materials and methods of fixation and so should specify the details of fixation in reporting findings. Since rehydration is not considered to be standard practice, it might be considered unfavorably in any false negative Pap smear litigation. If using any alternative fixative and fixation method, evaluate high water content cells as they are more sensitive quality indicators.

Whether Carbowax Is Present When Air-Drying Takes Place

In no instance should cells be allowed to air-dry before being fixed unless intended (e.g., FNA, blood films). After being applied to a slide, the alcohol evaporates. As it does, an air–

alcohol interface recedes—or advances, depending on one's perspective— through the cells. Absent PEG, substantial denaturation forces are exerted—as already described. When part of the fixative, PEG precipitates as the alcohol evaporates and preempts the forces of distortion, in effect embedding and protecting the fixed cells in situ.

Carbowax is The Dow Chemical Company's trademarked name for its line of polyethylene glycols of varying molecular weights, of which there are approximately 11—depending on the application—that range from 300 to 8,000. Numbers below 600 are liquids, above 900, solids. The melting points of the solids become progressively higher with increasing number. Carbowax 1450 is a water-soluble solid wax with a melting temperature range of 43–46 °C, which makes it suitable for cytological applications. Carbowax 1450 was known as Carbowax 1540 at one time (e.g., in 1963, when Saccomanno included it in 50% ethanol as part of his preservative for sputum). The inclusion of Carbowax in a fixative in cytology can be traced back to 1946.

Carbowax is solid at room temperature. It is available in flaked form, which simplifies weighing for those who prepare their own alcohol with Carbowax. Before it was made available as flakes, Carbowax was sold solid in gallon flat-sided cans with a small opening. To simplify handling, I used to melt it in a hot oven and subsequently mix it with an equal volume of water. At 50% (w/v) concentration, 2 mL contains 1 g Carbowax. One pair of inventors patented "Carbowax sticks for preparation of the Carbowax fixative use in cancer cytology" to avoid "the laborious task of digging out small chunks, weighing this out on scales that might not be available…."[29] Selling Carbowax as a 50% solution hadn't occurred to them.

In practice, a fresh cellular sample is spread onto a glass slide and immediately fixed by plunging it into alcohol (i.e., wet fixation). Air-drying should not be allowed to occur either before, during, or after fixation. "Fresh" means the cell sample had just been collected and not been suspended in preservative prior to being applied to the slide surface. After the standard fixation method, the hierarchy of potential alternatives, best first, *for preparations that are to be Pap stained* is as follows:

GYN and FNA

1. Immediate wet fixation in 1 of the following: absolute reagent alcohol, absolute methanol, proprietary grade alcohol, 80% isopropanol, or 90% acetone.
2. Immediate wet fixation in any of these reagents with 2% Carbowax 1450 (w/v), followed by air-drying.[10]
3. Spray fixation with ethanol-based fixative that contains Carbowax. Spray from a distance of 10–12 in.
4. Air-drying followed by rehydration in 50% glycerin × 3 min, followed by rinsing in water and immersion in 95% alcohol.[11]

NON-GYN

1. Collection in equal parts of 50% alcohol.
2. Collection in equal parts of 50% alcohol with 2% Carbowax, followed by concentration, spreading, and air-drying.
3. Collection in commercial preservatives as instructed by the manufacturer (e.g., BD CytoRich Preservative, BD CytoRich Red Preservative, BD CytoRich Blue Preservative, Cytyc CytoLyt, Cytyc PreservCyt). PreservCyt's composition fails the "non-obvious to someone skilled in the art" criterion for patentability, in my view.

Global Observations and Considerations

- Fixatives "fix" cells as they are immediately before being fixed. Cells that are flattened well by proper slide preparation techniques *when fixed* will display chromatin so it's visually accessible and useful. *In other words, the specimen collection and slide preparation techniques are part and parcel of the fixation outcome.*

- Alcohol means ethyl alcohol, by convention. Such short-hand works only if the speaker/writer and the listener/reader are aware of the convention. Ethyl alcohol is also known as ethanol and is sometimes written as EtOH.
- Alcohol is sometimes referred to as a dehydrating fixative, meaning it pulls water out of cells. More accurately, alcohol "pushes" water out of cells by shrinking proteins, and in effect, squeezing the water out.
- Preservation was referred to as prefixation at one time. Prefixation is a misnomer. Anything that precedes fixation is prefixation.
- Air-drying precipitates cellular proteins but does not fix them. For example, a blood film air-dried *without* subsequent fixation in methanol will be stained lightly by a Giemsa stain. The unfixed erythrocytes will be hemolyzed.
- The method of fixation is coupled with the methods of slide preparation *and* staining. Blood stains perform best with cells that have been air-dried, flattened, and fixed in methyl alcohol. If fixed in ethanol, the cells may shrink enough to unattractively alter the uptake of the blood stain. On the other hand, high water content cells wet fixed in methanol, instead of ethanol, shrink less and are well stained by the Pap stain.
- While morphologically more interpretable than if not rehydrated, rehydrated air-dried cells are not identical with alcohol wet-fixed cells in diameter or chromatin patterns. Given the choice between examining alcohol wet-fixed cells and rehydrated air-dried cells, cytoprofessionals usually choose the standard presentation—probably on the basis of subjective familiarity rather than on objective utility. After all, wet fixation in alcohol is an empirical method used in no other comparable biomedical application. It just so happens that it is what Papanicolaou used for no particular reason other than that he could, that it precipitates chromatin in normal and abnormal cells in visibly distinctive and visually distinguishable ways and that such patterns constitute the foundation of our collective memory banks of cytopathological images.

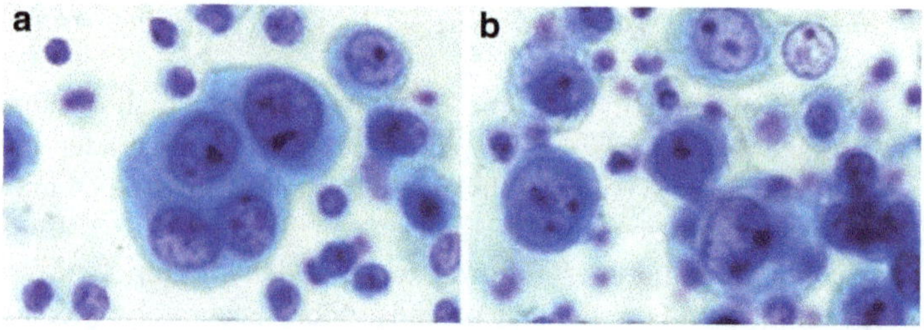

FIG. 8.13. Nongynecologic cytology specimens do not always need to be collected in preservative. These two photomicrographs are of malignant cells in the same pleural fluid that was collected fresh in a heparinized container and prepared by Millipore filtration. (**a**) Processed on a Monday, when received. (**b**) Refrigerated for 4 days and processed on a Friday. The latter shows some nuclear clearing consistent with degeneration, and more specks were evident grossly on the filter. However, the preparation is still morphologically interpretable. An extreme delay was allowed simply to demonstrate that fresh specimens can remain fresh for at least hours, especially when refrigerated, and that routine suspension in preservative is unnecessary. Body cavity fluids normally contain protein in solution that protects cells, which is a well-known phenomenon in tissue culture.

- Avoid unprotected air-drying (i.e., air-drying without benefit of Carbowax) unless the preparation will be stained by a blood stain (e.g., Giemsa, Diff-Quik).
- Suspensions of fresh non-gyn cytologic specimens can tolerate delays of at least 4 h from collection to cytopreparation and fixation. If longer delays are anticipated, the specimens can be safely refrigerated for up to several days.[30] See Fig. 8.13.
- The absolute best way to learn what's happening to specimens during cytopreparation is to concentrate them by conventional centrifugation, resuspend them in BES if fresh or in same preservative if preserved, and examine an unstained drop microscopically. Partially closing the substage condenser iris diaphragm will create a poor man's phase contrast microscope that will let you see all you need to see. Examine preparations at subsequent steps as needed (e.g., steps in the staining process).
- Although seldom used today, Carnoy's fixative will hemolyze erythrocytes in bloody cell spreads. Such spreads should remain

TABLE 8.1. Clarke introduced alcohol and glacial fixative in 1851.

Formulation	Parts by volume 95% Alcohol	Absolute alcohol	Chloroform	Glacial acetic acid
Clarke's fixative (1851)	–	3	–	1
Carnoy's fixative (1887)	–	6	3	1
Modified Carnoy's fixative (1957)	7	–	2.5	0.5

Carnoy added chloroform in 1887. DeWitt and coworkers modified it slightly in 1957.

immersed in Carnoy's for no longer than approximately 30 min. Longer exposure shrinks cells excessively, causing them to round up and become smaller in diameter and thicker. On a historical note, there are 3 variants of Carnoy's, but only 1 is the "real McC(arn)oy." See Table 8.1.

Clarke, an Englishman, introduced his hemolytic fixative in 1851. Carnoy, a Frenchman, added chloroform to his variant in 1887. DeWitt et al., who were Americans, modified Carnoy's formulation slightly in 1957.[31] Of the 3 variants, the 1957 variant is preferred as it does not shrink cells as much. Glacial acetic acid does not coagulate protein, which means it does not harden it. That lack of hardening accounts for the shrinkage induced by alcohol. A common *caveat* is "don't fix cells in Carnoy's fixative for more than 30 minutes." Chloroform may have been added for a reason similar to that for including ether in alcohol, it is a fat solvent. Chloroform, of course, is a liver toxin—among its various potentially dangerous properties—and should be handled with care.

TABLE 8.2. FACTORS THAT CONTRIBUTE TO CELL FLATTENING BEFORE, DURING, OR AFTER CELLS CONTACT A GLASS SURFACE.

Contact Timing	Promote Cohesion (Ensmall)	Promote "Balance" (Standard)	Promote Adhesion (Enlarge)
Pre-	High water content cells	Low or high water content cells	Low or high water content cells
	Preservative suspension	Fresh cells	Fresh cells
	Higher ROH concentrations		
During	Dirty slide surface	Clean slide surface	Clean slide surface
	Above slide	On slide	On slide
	Glycerin : albumin coating	Force: centrifugal, negative pressure	Force: centrifugal, negative pressure
	Slow immersion	Immediate immersion	Immediate immersion
	Isopropanol	95% ethanol, methanol, reagent alcohol, 80% isopropanol, 90% acetone	Methanol
Post-	Air-drying	Keep wet	Keep wet
		Use Carbowax ad hoc	Use Carbowax ad hoc

Summary

Table 8.2 summarizes the various materials and methods that impact adhesion and cohesion before, during, and after contact between a cell and a clean glass slide.

It is unknown why fixation in alcohol elicits the microscopically visible differences in nuclear morphology among benign and malignant cells at a submicroscopic, molecular level. Some have published musings on the topic, but without offering definitive explanation or understanding.[32] Truth be told, in my view, it makes no difference whether we understand the molecular basis in the context of morphology-based diagnostic cytopathology. According to Baker, alcohol does not fix nucleoprotein, it precipitates it.[33] If that is indeed the case, then differences in chromatin appearance may be due to larger pieces of stained nucleoprotein in abnormal cells being distributed in nuclear space more or less usefully in response to the details of the fixation method.

The criteria of cellular health and disease are nothing more than descriptions of cytomorphology that have been artifactually modified by fixation. As a result, fixation's materials, methods, and results are mutually dependent and consistent: the results we expect determine our materials and methods, and the materials and methods we use determine our results. Although such a relationship is obvious and inevitable, so stating it emphasizes the fact that the cytomorphologic changes on which we rely to signal cellular health and disease are not fundamental, dependable features of cells in vivo. Nature provides the starting materials, and man provides the finishing touches.

References

1. Carnoy JB. Quelques observations sur la division cellulaire chez les animaux. Cellule. 1887;3:225–324.
2. Online Etymology Dictionary. Available at http://www.etymonline.com/index.php?allowed_in_frame=0&search=preservative&searchmode=none. Accessed 6 Apr 2012.
3. Bracegirdle B. A history of microtechnique. Ithaca, NY: Cornell University Press; 1978.

4. Clarke JL. Researches into the structure of the spinal cord. Philos Trans R Soc. 1851;141:601–22.
5. Stockard CR, Papanicolaou GN. The existence of a typical oestrous cycle in the guinea pig with a study of its histological and physiological changes. Am J Anat. 1917;22:225–83.
6. Sills B. Use of polyethylene glycols in dry preservation of anatomic and pathologic specimens. Lab Invest. 1952;1(3):378–81.
7. Papanicolaou GN. Atlas of exfoliative cytology. Cambridge, MA: Harvard University Press; 1954.
8. Pundel JP, Ferriera CDA, Peters H, deBrux J, Nieburgs HE, Schueller E, et al. II. Experiences with various methods of fixation of smears. Do cellular changes occur as a result of air drying of smears? Acta Cytol. 1957;1(1):62–9.
9. Hajdu SI. A note on the history of Carbowax in cytology. Acta Cytol. 1983;27(2):204–5.
10. Saccomanno G, Saunders RP, Ellis H, Archer VE, Wood BG, Beckler PA. Concentration of carcinoma or atypical cells in sputum. Acta Cytol. 1963;7(5):305–10.
11. Bonime RG. Air-dried smear for cytologic studies. Obstet Gynecol. 1966;27(6):783–90.
12. Ehrenreich T, Zadeikis SK, inventors. Fixing and drying cytological smears. US patent 3389052. 18 Jun 1968.
13. Hurley AA, Lapen DC, Oud PS, inventors; Cytyc Corporation, assignee. Cell preservative solution. US patent 5256571. 26 Oct 1993.
14. Gill GW, Miller KA, Frost JK. Cytomorphological consequences of fixation methods other than ethanolic wet fixation. Platform presentation delivered at American Society of Cytopathology's 21st Annual Scientific Meeting, Salt Lake City, UT, 8 Nov 1973.
15. Papanicolaou GN. Some improved methods for staining vaginal smears. J Lab Clin Med. 1941;26:1200–5.
16. Carmichael DE. The Pap smear: life of George N. Papanicolaou. Springfield, IL: Charles C. Thomas; 1973. p. 69.
17. 27 CFR. Alcohol, Tobacco and Firearms. Part 21–Formulas for Denatured Alcohol and Rum. Available at http://ecfr.gpoaccess.gov/cgi/t/text/text-idx?c=ecfr&rgn=div5&view=text&node=27:1.0.1.1.17&idno=27. Accessed 15 Apr 2012.
18. Holmquist MD. The effect of distance in aerosol fixation of cytologic specimens. Cytotechnol Bull. 1978;15(2):25–7.
19. Freeman JA. Hair spray: an inexpensive aerosol fixative for cytodiagnosis. Acta Cytol. 1969;13(7):416–9.

20. Randall B, van Amerongen L. Commercial lab practice evaluation of air-dried/rehydrated cervicovaginal smears vs. traditionally-fixed smears. Diagn Cytopathol. 1997;16(2):174–6.
21. Sivaraman G, Iyengar KR. Rehydrated air-dried Pap smears as an alternative to wet-fixed smears. Acta Cytol. 2002;46(4):713–7.
22. Chan JK, Kung IT. Rehydration of air-dried smears with normal saline. Application in fine-needle aspiration cytologic examination. Am J Clin Pathol. 1988;89(1):30–4.
23. Gill GW. Air-dried/rehydrated CV smears are different. Diagn Cytopathol. 1998;18(5):381–2.
24. Kirby JP. The effects of different fixatives and fixation methods on cell areas. Gill GW, Erozan YS, advisors. Baltimore MD: The Johns Hopkins Hospital, School of Cytotechnology, 29 Jun 1976.
25. Anderson TF. Electron microscopy of microorganisms. In: Oster G, Pollister AW, editors. Cells and tissues. Chapter 5 in Physical techniques in biological research. New York, NY; Academic; 1955. p. 177–240
26. Beyer-Boon ME, van der Voorn-Den Hollander MJ, Arentz PW, Cornelisse CJ, Schaberg A. Fox CH Effect of various routine cytopreparatory techniques on normal urothelial cells and their nuclei. Acta Pathol Microbiol Scand A. 1979;87(1):63–9.
27. Schulte E. Air drying as a preparatory factor in cytology: investigation of its influence on dye uptake and dye binding. Diagn Cytopathol. 1986;2(2):160–7. Air-drying.
28. Yang GCH. The mathematical basis for the increased sensitivity in cancer detection in air-dried cytopreparations. Mod Pathol. 1994;7(6):681–4.
29. Sills K, Sills B, inventors. Carbowax sticks for preparation of the Carbowax fixative use in cancer cytology. US patent 4565721. 21 Jan 1986.
30. Manosca F. Diagnostic effects of prolonged storage on fresh effusion samples. Diagn Cytopathol. 2007;35(1):6–11.
31. Dewitt SH, Del Vecchio PR, Borelli JI, Hilberg AW. A method for preparing wound washings and bloody fluids for cytologic evaluation. J Natl Cancer Inst. 1957;19(1):115–21.
32. Dey P. Chromatin pattern alteration in malignant cells: an enigma. Diagn Cytopathol. 2005;32(1):25–30.
33. Baker JR. Principles of biological microtechnique – a study of fixation and dyeing. Bungay, Suffolk, England: Methuen & Co. Ltd.; 1958.

Chapter 9
Cell Block Preparation

Principle No. 2

Make specimen-representative preparations.

Practice

Check for cellularity, centrifuge, concentrate, consolidate, cut, and color.

Historic Milestones

- 1882—First cytologic cancer diagnosis in body cavity fluid.[1]
- 1896—Cell block prepared in celloidin embedding medium and sectioned using microtome.[2]
- 1901—Centrifugation introduced to increase cellularity of cell blocks.[3]
- 1917—Malignant tumors diagnosed in paraffin sections of centrifuged exudates.[4]
- 1947—Cell block enhances cellularity of serous effusions.[5]
- 1959—Bacterial agar introduced for preparing cell blocks.[6]
- 1973—Plasma–thrombin clot introduced for preparing cell blocks.[7]
- 2005—"Rapid cell block embedding method and apparatus" patented.[8]
- 2007—Automated cell block system (i.e., Cellient) introduced into commerce.

G.W. Gill, *Cytopreparation: Principles & Practice*,
Essentials in Cytopathology 12, DOI 10.1007/978-1-4614-4933-1_9,
© Springer Science+Business Media New York 2013

Cells blocks for effusions and other nongynecologic specimens began to be used about the same time the Pap test was introduced, and more than a decade before, cell concentration preparation methods, such as membrane filtration and cytocentrifugation, were introduced. Remember, too, it was the preelectronic age. News of anything new was disseminated, shall I say, less instantaneously than it is today.

All cell block techniques are more alike than different:

1. Cells in suspension are centrifuged to form a cell concentrate or pellet.
2. Cells are fixed as needed for subsequent applications.
3. The cells are embedded in situ so they can be removed *en bloc* from the centrifuge tube.
4. Processed as though tissue.

If the specimen concentration preparation methods were as good then as they later became, cell blocks might not have been needed at all. However, the development of immunohistochemical methods that can answer questions that cannot be answered by morphological interpretation would have caused the cell block to be revisited. If you don't already have a satisfactory cell block technique, I recommend—based on first-hand experience—the thrombin-clot technique.

Thrombin-Clot Technique

A cell block can be prepared from the pellet of any centrifuged cell suspension by embedding it in a clot formed by added plasma and thrombin. These definitions are in this protocol:

- BSS = balanced salt solution
- FFP = fresh frozen plasma
- FNA = fine needle aspirate
- IHC = immunohistochemistry
- NBF = neutral buffered formalin
- RPMI = Roswell Park Memorial Institute (place of origin of one kind of BSS)

This protocol is suitable for major body cavity fluids, washing specimens (e.g., bronchial, gastric), urines, FNA rinses, and any cell suspension that might benefit from a cell block:

Materials

- Balanced salt solution for FNA rinses (e.g., RPMI 1640, Hanks' BSS)
- FFP (fresh frozen plasma; OK if outdated)
- Thrombin, Topical USP (bovine origin)—store in refrigerator. Thrombin, Topical U.S.P., 5,000 U.S. unit vial with 5 mL diluent, NDC 052604-7102-1 (Jones Pharma)
- 50-mL centrifuge tube
- 10% neutral buffered formalin
- Disposable transfer pipette

Method

Use standard precautions when handling any cytology specimen:

Thaw the FFP Plasma and Prepare Ten 1-mL Aliquots

Thaw FFP by immersing bag in cold water bath for 30 min. Aliquot 10 mL thawed plasma into each of 20 labeled 50-mL centrifuge tubes as follows:

- 10 mL FFP.
- Store at –20°C.
- Write the date prepared.
- Write the expiration date on the tube (add 3 years to date prepared).
- Reserve 1 aliquot tube for daily use. Refrigerate at 4–8°C when not in use.
- Freeze the remaining labeled aliquot tubes of FFP in –20°C non-defrosting freezer for future use.

- Defrost a single frozen aliquot tube as needed in a refrigerator for several days before anticipated use. If necessary, defrost a single frozen aliquot tube quickly in a cold water bath for 30 min.

Prepare Thrombin

Dissolve thrombin in the sterile saline solution provided. Use the transfer double-hypodermic needle provided. If transfer is incomplete, complete using a standard hypodermic needle and syringe.

Prepare Cell Block Using Plasma/Thrombin Clot

1. Balance the 50-mL specimen centrifuge tube, labeled with patient name and cytology number, and centrifuge for 10 min @ 3,000 rpm.
2. Following centrifugation, discard the supernatant into biohazardous waste container. **Note**: Remove as much supernatant as possible from the sediment prior to making cell blocks.
3. Using a clean disposable transfer pipette, place four drops of plasma on top of the sediment. **Note**: Do not contaminate pipette with the specimen.
4. Mix the plasma and sediment by gently swirling the contents. **Note**: Do not vortex. Vortexing creates air bubbles that disrupt the button.
5. Using the transfer pipette, add 1–2 drops of thrombin to the plasma/cell suspension. Discard the transfer pipette.
6. Allow approximately 1–5 min for the clot to form.
7. Tilt the centrifuge tube to check for clot formation. If a clot does not form, wait 5 more minutes and check again.
8. If a clot still does not form, centrifuge again.
9. Add enough 10% NBF gently down the inside wall of the tube to float the clot. **Note**: Use a micro spatula to loosen the clot from the bottom of the centrifuge tube. Process as though tissue.

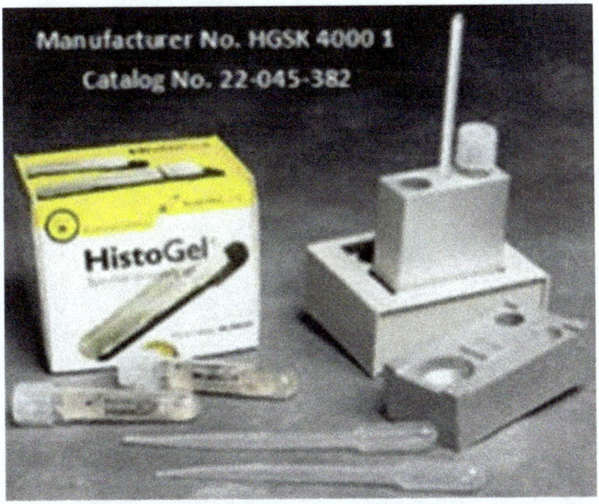

FIG. 9.1. The starter kit includes twelve 10-mL HistoGel tubes, two Wonderblock aluminum cooling blocks to rapidly cool the gel drops, a dry bath incubator, and 500 disposable dispensing pipettes.

Alternative Cell Block Methods

Alternative manual cell block methods include HistoGel[9] and the Cytoblock technique.[10] Both products are sold by Thermo Fisher Scientific. See Figs. 9.1 and 9.2. According to the manufacturer's website, the HistoGel starter kit:

- Is the solution for small or viscous histological/cytological specimens
- Completely encapsulates and retains the entire specimen during processing
- Will not retain histological stains, eliminating the unwanted discoloration around specimens on slides
- Is virtually unnoticeable during sectioning and will not "pop out" of the paraffin block during sectioning
- Increases ability to process scanty or unseen material for cell block and reduces loss of critical material

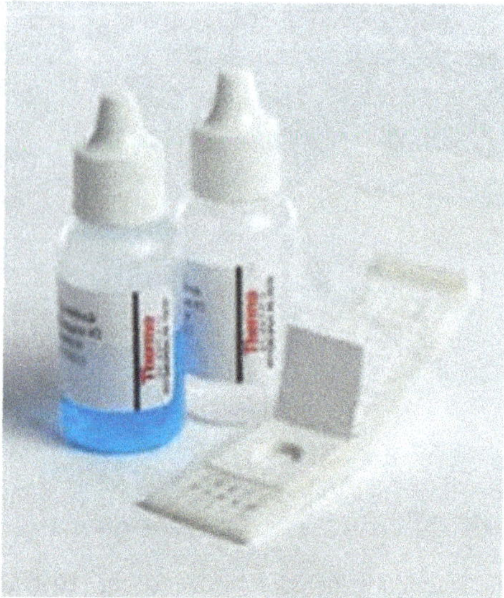

FIG. 9.2. The Cytoblock kit includes 50 Cytoblock cassettes with backing papers and board-inserts, one bottle each of Reagent 1 and 2. See the manufacturer's instructions for details.[9]

See the HistoGel instructions for use for details.[9] An online instructional video is available.[10]

The Shandon Cytoblock cell block preparation system is the second alternative manual cell block preparation product.[11] By design, it concentrates cells by cytocentrifugation in a Thermo Shandon Cytospin. The Shandon Cytoblock system is designed to facilitate the preparation of paraffin-embedded cell suspensions, cell aggregates, and tissue fragments. The system simplifies the production of paraffin blocks from cellular material and increases the yield of useful blocks from cellular suspensions and cell aggregates.

The Cytoblock system can also be used to process tissue biopsies and fragments that are difficult or impossible to process using other techniques. Cytoblock can be used to produce

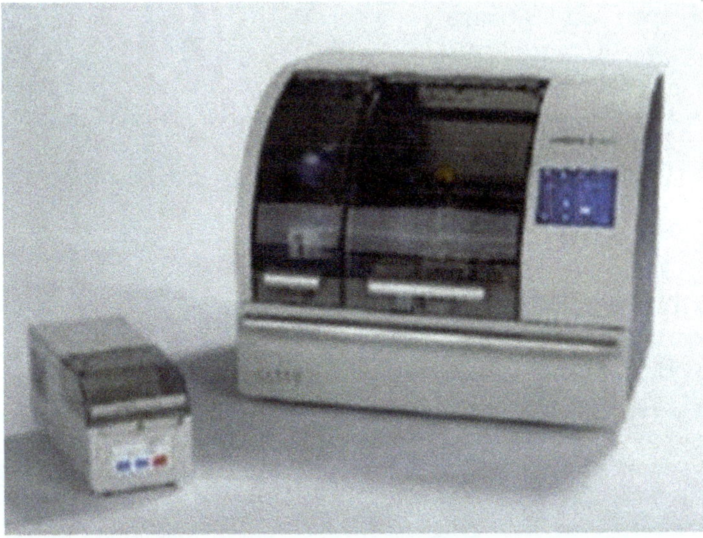

FIG. 9.3. The Cellient automated cell block system processor and finishing station.[12]

paraffin-embedded blocks from fine needle aspirates, cutting needle cores, body fluids, and residual sediment from other cytological preparations. Cytoblock is also an ideal method for processing tissue fragments such as small biopsies, curettings, and other specimens that are too small to be processed in standard cassettes. The use of the Cytoblock system eliminates the need for tea bags, tissue wrapping, and the potential loss of tiny fragments. Cytoblock preparations are suitable for immunohistochemical techniques.[11]

In contrast to manual methods of preparing cell blocks, Hologic's Cellient is the only fully automated cell block preparation system on the market at this time. See Fig. 9.3.

According to the manufacturer's website: "With the Cellient system, expect improvements over your current cell block technology—improvements in capture, presentation, and consistency.

Improved Capture

- Vacuum-assisted filtration
- Captures available cells, maximizing cellularity even from small/scanty samples
- Built on ThinPrep technology

Improved Presentation

- Helps maintain crisp, clear, and cellular architecture
- Creates concentrations of cells within the block
- Reviews of cytology and cell block simultaneously
- Supports easier and more productive pathology review

Improved Consistency

- High-quality blocks
- Fully automated with minimal operator dependency
- Less cross-contamination risk
- Consistently rapid processing time (45 min or less)"[13]

An online morphology atlas is available.[14] The Cellient is priced at US $49000.

Cell Blocks and Immunohistochemistry

Cell block sections often include cancer cells that aren't in the companion cytologic samples.[15–19] Cell block sections are especially useful for immunohistochemical (IHC) stains and special stains.[20, 21] Entering the following search terms in PubMed (http://www.ncbi.nlm.nih.gov/pubmed) produces these results:

Search term	Number of citations
"Cell block" and cytology	722
"Cell block" and cytology and immunocytochemistry	314
"Cell block" and cytology and immunohistochemistry	292
"Cell block" and fluids	100

IHC techniques require stringent quality control measures. The following are excerpted from the College of American Pathologists Anatomic Pathology Checklist:[22]

Specimen Modification: If the laboratory performs immunohistochemical staining on specimens other than formalin-fixed, paraffin-embedded tissue, the procedure manual includes appropriate modifications to address such specimens.

Buffer pH: The pH of the buffers used in immunohistochemistry is routinely monitored. *Note: pH must be tested when a new batch is prepared or received.*

QC: Antibodies: Appropriate negative controls and positive tissue controls are used for each antibody.

Endogenous Biotin: If the laboratory uses an avidin–biotin complex (ABC) detection system (or a related system such as streptavidin–biotin or neutravidin–biotin), there is a policy that addresses nonspecific false-positive staining from endogenous biotin.

Control Slide Review: When batch controls are run, the laboratory director or designee reviews all control slides each day of patient testing.

Antibody Validation: The laboratory has documented validation of new antibodies, prior to use in patient diagnosis.

New Reagent Lot Verification: The performance of new lots of antibody and detection system reagents is compared with old lots before or concurrently with being placed into service.

Slide Quality: The immunohistochemical stains produced are of acceptable technical quality.

Discussion

Use whatever works best in your laboratory's circumstances. Regardless of the cell block method, *always* examine a drop of specimen microscopically to see whether cells are present or not. If cells are few or absent, stop. Cell blocks can't salvage acellular specimens. Residual PreservCyt samples in unsatisfactory ThinPrep Pap test vials don't benefit from cell blocks.[23] In a 2010 published comparison of Hologic's automated cell block preparations (i.e., Cellient) with traditional cell block preparations, for example, inadequate cellularity prevented full evaluation in IHC preparations of 23% (4/17) of patients. Adequate cellularity was demonstrated by all traditional cell block preparations. The authors did not attempt to account for the difference.[20]

References

1. Quincke H. Uber die geformten Bestandtheile von Transsudaten. Deutsches Arch klin Med. 1881–1882;30:580–8.
2. Bahrenberg L. On the diagnostic results of the microscopical examination of the ascitic fluid in two cases of carcinoma involving the peritoneum. Cleveland Med Gaz. 1896;11:274–8.
3. Josefson A. Primur lungkancer med svulstceller i pleuraexsudat och sputum. Hygiea. 1901;63:435–45.
4. Mandlebaum F. Diagnosis of malignant tumors by paraffin sections of centrifuged exudates. J Lab Clin Med. 1917;2:580.
5. Chapman CB, Whalen EJ. The examination of serous fluids by the cell block technic. N Engl J Med. 1947;237(7):215–20.
6. Shackelford RI, Jones JL. An embedding medium for permanent sections of exudative material and fragments of tissue removed for biopsy. Tech Bull Regist Med Technol. 1959;29:155–6.
7. Kaltenbach FJ, Hillemanns HG, Fetttig O, Hilgarth M. Thrombin cell block technic in gynecologic cytodiagnosis. Acta Cytol. 1973;17(2):128–30.
8. Fischer AH, inventor; University of Massachusetts, assignee. Rapid cell block embedding method and apparatus. US patent 6913921. 5 Jul 2005.

9. Thermo Scientific Richard-Allan Scientific. Histogel: instructions for use. Five-page brochure 92957066 R11/08. Available at http://www.thermo.com/eThermo/CMA/PDFs/Various/File_9759.pdf. Accessed 3 Apr 2012.
10. Varsegi GM, Shidham V. Cell block preparation [using Histogel] from cytology specimen with predominance of individually scattered cells. J Vis Exp. 2009. Link to online video: http://www.jove.com/details.php?id=1316. Accessed 3 Apr 2012.
11. Thermo Electron Corporation. Shandon Cytoblock cell block preparation system. Brochure, Rev. 4, 09/03, P/N 231803. Available at https://www.thermo.com/eThermo/CMA/PDFs/.../File_23523.pdf. Accessed 3 Apr 2012.
12. Hologic, Inc. Hologic Product Catalog 2010. Available at http://www.hologic.com/data/File/pdf/Hologic_PrdctCatlg.pdf. Accessed 4 Apr 2012.
13. Hologic Inc. Cellient automated cell block system benefits. Available at http://www.cellientsystem.com/benefits.htm. Accessed 5 Apr 2012.
14. Hologic Inc. Cellient Cell Block Morphology Atlas. Available at http://cytologystuff.com/indexcellblock.htm. Accessed 4 Apr 2012.
15. Harris MJ. Cell block preparations: 3% bacterial agar and plasma-thrombin clot methods. Cytotechnol Bull. 1974;11(3):6–7.
16. Nigro K, Tynski Z, Wasman J, Abdul-Karim F, Wang N. Comparison of cell block preparation methods for nongynecologic ThinPrep specimens. Diagn Cytopathol. 2007;35(10):640–3.
17. Kulkarni MB, Desai SB, Ajit D, Chinoy RF. Utility of the thromboplastin-plasma cell-block technique for fine-needle aspiration and serous effusions. Diagn Cytopathol. 2009;37(2):86–90.
18. Thapar M, Mishra RK, Sharma A, Goyal V, Goyal V. Critical analysis of cell block versus smear examination in effusions. J Cytol. 2009;26(2):60–4. Free text available at http://www.ncbi.nlm.nih.gov/pmc/articles/PMC3168020/?report=printable. Accessed 3 Apr 2012.
19. Shivakumarswamy U, Arakeri SU, Karigowdar MH, Yelikar B. Diagnostic utility of the cell block method versus the conventional smear study in pleural fluid cytology. J Cytol. 2012;29(1):11–5. Free text available at http://www.ncbi.nlm.nih.gov/pmc/articles/PMC3307444/?report=printable. Accessed 3 Apr 2012.
20. Wagner DG, Russell DK, Benson JM, Schneider AE, Hoda RS, Bonfiglio TA. Cellient™ automated cell block versus traditional cell block preparation: a comparison of morphologic features and immunohistochemical staining. Diagn Cytopathol. 2011;39(10):730–6.

21. Shidham VB, Epple J. Appendix I. Collection and processing of effusion fluids for cytopathologic evaluation. In: Shidham VB, Atkinson BF, editors. Cytopathologic diagnosis of serous fluids. Philadelphia, PA: Saunders; 2007. p. 207–37.
22. CAP Accreditation Program. Anatomic Pathology Checklist. Northfield, IL: College of American Pathologists; 07.11.2011 revision.
23. Kabbani W, Raisanen J, Thomas S, Saboorian MH, Ashfaq R. Cell block findings from residual PreservCyt samples in unsatisfactory ThinPrep Paps: no additional benefit. Diagn Cytopathol. 2002;27(4):238–43.

Chapter 10
Papanicolaou Stain

Papanicolaou's original stain and its subsequent modifications are not standardized.

PRINCIPLE NO. 5

Stain preparations to facilitate cell visibility, detection, and interpretation.

PRACTICE

Apply chemically competent stains in the prescribed sequence for the right times and rinse sufficiently in clean solutions.

Historic Milestones

- 1865—Böhmer adds mordant to hematoxylin to make first successful hematoxylin stain[1]
- 1875—Fischer first to use eosin[2]
- 1876—Wissowzky first to use hematoxylin and eosin (H&E)[3]
- 1878—Weigert first to use Bismarck brown Y[4]
- 1879—Ehrlich first to use orange G[4]

G.W. Gill, *Cytopreparation: Principles & Practice*, Essentials in Cytopathology 12, DOI 10.1007/978-1-4614-4933-1_10,
© Springer Science+Business Media New York 2013

- 1881—Flemming introduces differentiation of hematoxylin in dilute hydrochloric acid[5]
- 1886—Griesbach first to use light green SF yellowish[4]
- 1900—Harris introduces his eponymous hematoxylin[6]
- 1912—Scott introduces his tap water substitute to blue hematoxylin[7]
- 1931—Trotman and Frearson demonstrate that phosphotungstic acid precipitates basic dyes[8]
- 1942—Papanicolaou publishes Pap stain with 5 dyes[9]
- 1954—Papanicolaou modifies Pap stain[10]
- 1960—Papanicolaou again modifies Pap stain[11]
- 1974—Gill introduces progressive "half-oxidized hematoxylin"[12]
- 1975—Gill recommends that Bismarck brown be omitted from EA formulations[13]
- 1998—Gill publishes environmentally friendly Enviro-Pap[14]

Hematoxylin—the first Pap stain component dye—was introduced 1865. By 1942, the other four dyes had been put into use, and differentiation and bluing had been incorporated routinely into cell and tissue stain protocols. The Pap stain is basically H&E on steroids. Papanicolaou modified his 1942 "Pap" stain twice over the next 18 years.

Papanicolaou described three chief objectives for "staining of vaginal, cervical, and endometrial smears:

1. Definition of nuclear details. Because of the widespread nuclear abnormalities of cancer cells and their diagnostic significance, good staining of the nucleus is of primary importance.
2. Transparency. This is of particular importance because of the varying thickness and the frequent overlapping of cells.
3. Differentiation of cells. Differences in the staining reaction such as that between acidophilic and basophilic cells help greatly in the identification of certain cell types found in smears."[10]

Fig. 10.1. The five dyes that comprise the Papanicolaou stain are in three solutions. Hematoxylin and orange G are each in separate solutions; the third solution EA contains Bismarck brown Y, light green SF yellowish, and eosin Y. I used to prepare aqueous stock solutions in volumetric flasks, as shown, to control the molar concentration of the counterstain dyes.

Classically, five dyes—used in conjunction with numerous nonstaining solutions—comprise the Pap stain. The dyes are hematoxylin, orange G, Bismarck brown Y, light green SF yellowish, and eosin Y. See Fig. 10.1.

Objective No. 1: Definition of Nuclear Details. Harris hematoxylin is the nuclear stain. Papanicolaou used it in a variety of ways. See Table 10.1.

Since Harris hematoxylin won't be discussed again in this chapter, this is the best place to discuss the contribution of its relatively high concentration of hematein and aluminum to the surface

TABLE 10.1. Papanicolaou's use of Harris hematoxylin.

Material	1942	1954	1960
Harris hematoxylin	Full-strength, no acetic acid, × 5–10 min	Half-strength, no acetic acid, × 6 min	Full-strength with 4% acetic acid, × 45 s
Mode	Regressive	Regressive	Progressive
Differentiation	0.5% HCl × 3–4 dips	0.25% HCl × 6 dips	None
Bluing	Li_2Co_3 × 1 min	Tap water × 6 min	1.5% NH_4OH in 70% alcohol × 1 min

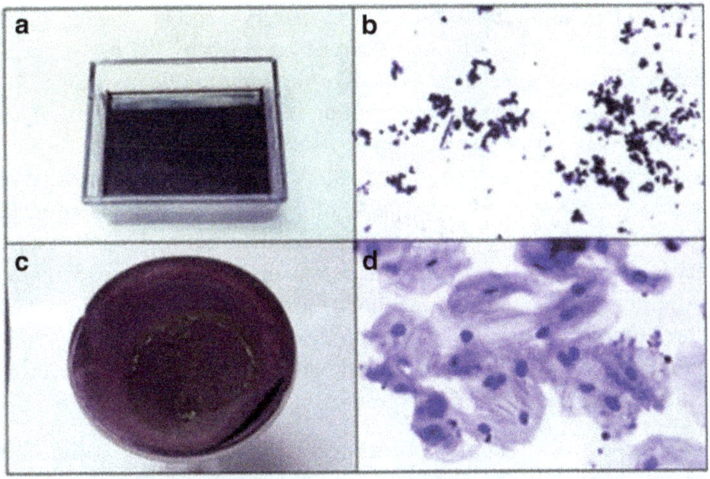

Fig. 10.2. The metallic appearing scum on the surface of Harris hematoxylin and other high-concentration hematoxylin formulations (e.g., Gill hematoxylin No.3) precipitates because the amount of hematein–aluminum complex exceeds its solubility limit in water. Contrary to popular belief, it is not an overoxidation product.

precipitate that forms: Why it forms, what it is, how to identify it, and how to avoid its forming in the first place.

See Fig. 10.2.

(a) Harris hematoxylin with greenish-golden surface precipitate	(b) If not removed by filtration before staining, the precipitate is deposited on microslides
(c) The precipitate can be captured by filtration	(d) When the precipitate is dissolved in 25% ethylene glycol, its identity as aluminum–hematein is confirmed by its ability to stain cells

Objective No. 2: Transparency. Papanicolaou wrote: "Methods which proved successful in other applications were found not to be entirely satisfactory for this particular work because of a common disadvantage. The staining of the cells was too deep to permit a sharp definition of their outlines in smears that were relatively thick or contained much blood… After long experimentation it was found that a much greater transparency and an equally good color differentiate of the cells can be obtained by the use of solutions of stains in 95%. [*sic*] alcohol instead of aqueous stains."[9]

Cells are naturally transparent. Coloring them with dyes reduces this transparency. Dyes dissolved in water are rapidly taken up. Dyes dissolved in alcohol have a slower rate of dye uptake, resulting in less reduction of cellular transparency. Therefore, alcohol as a solvent for dyes does not make cells transparent; it simply helps them to retain their natural transparency. Another benefit of using alcohol as a solvent is that it minimizes the solubility of salts found in dyes, thereby mitigating their potential positive or negative impact on staining performance. Such variables compromise predictable staining results.

Objective No. 3: Differentiation of Cells. Orange G is the first counterstain dye in OG-6. OG is an abbreviation of Orange G, in which G is the first letter of the German word *gelb,* meaning yellow. The *6* indicates that this is the sixth variation in a series of experimental OG solutions that contained phosphotungstic acid (PTA). In footnote 8 of his 1942 paper, Papanicolaou wrote: "The addition of phosphotungstic acid to the Orange G solution intensifies the orange color. For normal slides a slight acidification of 0.010 g per 100 cc (OG 8) or 0.015 g per 100 cc (OG 6) is suggested. For cancer diagnosis a higher acidification of 0.025 g per 100 cc (OG 5) is often preferable, as it gives a sharper contrast of the abnormal cell types."[9]

A counterstain colors cytoplasm a color that contrasts with the nuclear stain. Counterstains make the cells more visible overall. Different color counterstains help the microscopist differentiate one cell type from another. Cytologic preparations can be examined without the benefit of counterstains, but the examination would be tiresome and perhaps less effective. Or one could examine cells stained with hematoxylin and eosin, which some do.

Papanicolaou varied his use of OG-6 as well, but not nearly as much as he did Harris hematoxylin. See Table 10.2.

Light green, eosin Y, and Bismarck brown Y comprise the second counterstain EA. EA was Papanicolaou's laboratory record

Table 10.2. Papanicolaou's use of OG-6.

Materials and time	1942	1954	1960
Orange G	5 g	5 g	5 g
Phosphotungstic acid	0.15 g	0.15 g	0.15 g
Alcohol "95 percent"	1,000 mL	1,000 mL	1,000 mL
Time	1 min	1.5 min	1.25 min

Table 10.3. Papanicolaou's use of EA.

	EA-36 (g/L)[a]		EA-65 (g/L)[b]	
Materials and time	1942	1954	1954	1960
Light green SF yellowish	2.25	**0.45**	0.225	0.225
Eosin Y	2.25	2.25	2.25	2.25
Bismarck brown Y[c]	0.5	0.5	0.5	0.5
Phosphotungstic acid	2.0	2.0	2.0	**6.0**
Lithium carbonate	10 drops	10 drops	10 drops	Omitted
Time	2 min	1.5 min	1.5 min	3 min

[a]EA-36 not modified in 1960.

[b]EA-65 not described in 1942.

[c]Named after Otto von Bismarck, Chancellor of Germany, 1867–1871.

keeping code for alcoholic solutions of those three dyes plus phosphotungstic acid and lithium carbonate. EA is not an abbreviation for eosin alcoholic or eosin–azure. The -36 and -65 designations indicate different proportions of the ingredients. Papanicolaou described other EA stains (e.g., EA-25, EA-31). In his 1954 Atlas, Papanicolaou wrote: "EA 50, prepared and marketed by the Ortho Pharmaceutical Corporation, Raritan, NJ, is a stain comparable to EA 36."[10] Eosin is derived from the Greek word *eos,* meaning dawn. The *Y* following eosin and Bismarck brown means *yellow*. The *SF* following light green is derived from the German word *Saurefärbstoff,* meaning acid dye.

The EA formulations are particularly problematic. EA stains are not quantitatively reproducible, as Papanicolaou did not specify total dye content. EA contains two chemically incompatible ingredients, phosphotungstic acid and Bismarck brown Y, that precipitate one another from solution—thereby severely compromising the differential staining of eosin Y and light green SF yellowish and shortening the useful working life of EA. All staining times were changed in each of his three relevant publications. See Table 10.3.

More than half a century since its 1942 introduction, however, the Pap stain is still far from standardized in day-to-day practice—not the materials, methods, nor results. In this regard, it's just like the 1876 H&E stain. Staining results often differ among laboratories, as well as within the same laboratory daily. Indeed, the color plates in Papanicolaou's Atlas of Exfoliative Cytology show the wide variations that Papanicolaou himself experienced. Manufacturers of imaging devices for liquid-based gynecologic preparations require that their proprietary Papanicolaou stains and protocols be used.

Materials and Methods

The following are the materials and methods with which I am most familiar. Other hematoxylin formulations, OG-6, and EAs can be used, but with the *caveat* that the while the stain names are identical, the compositions are not. For example, the composition of one vendor's EA-36 may be quite different from that of another vendor. Expect variation in performance. Unfortunately, not all vendors label their products thoroughly, so how can one tell what is good and what is not?

Most laboratories use readymade stains, and so it is unlikely that any will prepare the stains described here. Nonetheless, readers should be aware that the staining results described include my fixes to the limitations in Papanicolaou stain. Preparing stains from scratch constitutes true quality control—assuming one has the requisite knowledge provided in this chapter. Since the Pap stain is not standardized, one cannot be certain of the composition or likely performance.

Gill Hematoxylins

Gill hematoxylin was originally described in 2 strengths: No. 1 and No. 2.[12] Gill hematoxylin No. 3 was introduced by the late Irwin "Win" Lerner to satisfy some of his customers who wanted a Harris hematoxylin-like strength stain. Gill hematoxylin No. 1 was originally used for cytology only; No. 2, thin sections of

TABLE 10.4. Composition of Gill hematoxylins.

No.	Component Mix in order at room temperature:	Gill hematoxylin No. 1	 No. 2	 No. 3
1.	Distilled water	730 mL	710 mL	690 mL
2.	Ethylene glycol	250 mL	250 mL	250 mL
3.	Hematoxylin, anhydrous	2.0 g	4.0 g	6.0 g
4.	Sodium iodate	0.2 g	0.4 g	0.6 g
5.	Aluminum sulfate	17.6 g	35 g	54 g
6.	Glacial acetic acid	20 mL	40 mL	60 mL

paraffin-embedded cell, concentrates. I now recommend using Gill hematoxylin No. 2 for cytology and histology applications.[15]

Gill hematoxylins are prepared as shown in Table 10.4.

Gill Modified OG

This modified OG replaces phosphotungstic acid with glacial acetic acid. Absent PTA, therefore, this modified OG is unnumbered (i.e., it's not OG-5, OG-6, or OG-8). PTA undoubtedly was intended to acidify OG to enhance cellular uptake of orange G. Its concentration is too low to reliably boost performance.

To prepare an alcoholic solution of OG, begin by preparing a liter of aqueous stock concentrate of 10% total dye content (TDC) orange G (certified, C.I. No. 16230).

Certified means certified by the Biological Stain Commission, which is headquartered in the Pathology Department in the University of Rochester Medical Center in Rochester, New York. Sixty-four stains are on a certification basis with the Biological Stain Commission. All but two, hematoxylin and orcein, are synthetic dyes. Twenty-nine of the 62 synthetic dyes were first used before 1909. Certification of synthetic dyes includes on the label of the dye bottle a statement as to dye content. See Fig. 10.3. No biological stain or dye is 100% pure. To be certified, all dyes must meet or exceed the minimum dye content standards. For the Papanicolaou counterstain dyes, these are: orange G, 80%; eosin Y, 90%; light green SF yellowish, 65%; and Bismarck brown Y, 45%. Unless these variations are taken into account when preparing dye solutions, the dye content will be less than expected.

C.I. No. means Colour Index Number. These are 5-digit numbers assigned by the Society of Dyers and Colourists in England to uniquely identify stains that are the same chemically but have different names.

FIG. 10.3. The Pap stain's 4 counterstain dyes are among those certified by the Biological Stan Commission. The total dye content is printed on the label provided by the BSC. Hematoxylin per se is not a dye and, for this reason, has no dye content that can be measured and reported.

C.I. Numbers must be specified when purchasing dyes or publishing articles in which the dyes are cited to ensure using the same dye, even if identified by different names.
Fifty-two certified dyes have Colour Index (C.I.) numbers, 12 do not. The Biological Stain Commission and the Society of Colourists and Dyers are unrelated to one another.

The Biological Stain Commission began testing stains in 1922. Its first book, Biological Stains, was published in 1925. Papanicolaou was apparently unaware of this information.

Dissolve a corrected amount of orange G in heated distilled water up to a liter or whatever volume is desired. To obtain the corrected weight, divide the desired amount by the percent dye content printed on the label of the bottle of orange G. See Fig. 10.3.

For example, 125 g orange G must be weighed out for a lot of dye that contains 80% dye: 100 g/0.80 = 125 g. Dissolving 125 g 80% orange G content dye in water up to a liter produces a solution that contains 100 g orange G and 25 g of impurities. The impurities include unknown salts and other compounds used in dye synthesis.

To prepare a liter of working Gill modified OG, combine the following ingredients at room temperature:

Orange G, 10% (TDC) aqueous stock solution	20 mL
95% ethyl alcohol	970 mL
Glacial acetic acid	10 mL

Gill Modified EA

This modified EA is dissimilar in significant ways to Papanicolaou's EAs and therefore, is not numbered.

To prepare an alcoholic solution of EA, begin by preparing 1 liter each of aqueous stock concentrates of 3% TDC light green SF yellowish (C.I. No. 42095, certified) and 20% TDC eosin Y (C.I. No. 45380, certified).

To prepare a liter of working Gill modified EA, combine the following ingredients at room temperature:

95% ethyl alcohol	700 mL
Absolute methyl alcohol	240 mL
Glacial acetic acid	20 mL
Light green, 3% TDC aqueous stock solution	10 mL
Eosin, 20% TDC aqueous stock solution	10 mL
Phosphotungstic acid, 20% (w/v) alcoholic stock solution	20 mL

Phosphotungstic acid is deliquescent (i.e., absorbs atmospheric water). If a lot of PTA contains an inordinate amount of water when you open the jar, which it often does, 2 g PTA/L EA will contain far less than 2 g actual PTA. Sometimes, the PTA can resemble the school paste of years gone by. In that case, PTA appears to be absent, as evidenced by little to no visible light green uptake in cells.

In consideration of the PTA's hygroscopicity, maintain a ready supply of stable water-free PTA that can be dispensed volumetrically *instead* of gravimetrically. Place a 100 g uncapped bottle of PTA in a 120° hot air oven overnight to evaporate the water. Prepare a 20% (w/v) alcoholic solution by dissolving 20 g dried PTA in 100 mL 95% ethanol. Alternatively, dissolve the entire weighed amount into a volume of alcohol that will result in a 20% solution. To dispense 4 g, add 20 mL of this 20% PTA stock solution to 990 mL EA. I found that doubling the PTA from 2 to 4 g produces an EA in which the light green stains satisfactorily longer.

Scott's Tap Water Substitute

Bluing can occur over a wide range of pH. Low pHs (e.g., 5–6, distilled water) blue slowly over several minutes. High pH (e.g.,

10–11, 1.5% NH_4OH in 70% alcohol) blues rapidly within seconds and causes cells to drop off the face of the slide. Moderate pH (e.g., pH 8, Scott's tap water substitute) blues single cells and thick tissue fragments and cellular clumps satisfactorily within 2 min. To prepare Scott's tap water substitute, combine:

Tap water	1 L
Magnesium sulfate	
$MgSO_4$, or	10 g
$MgSO_4 \cdot 7H_2O$	20 gm
Sodium carbonate	2 g

If you prepare this solution, be aware that dissolving magnesium sulfate is an exothermic reaction that can get unpleasantly warm. For safety, wear goggles and gloves. To minimize risks, add the magnesium sulfate slowly to the water so it dissolves rapidly and dissipates the heat produced. Alternatively, laboratorians who prefer to use Scott's TWS can purchase it readymade. Scott's TWS was described in an article published a century ago.[7]

The materials and methods of a modified Pap stain are shown in Fig. 10.4 and in Table 10.5 that also includes those for Enviro-Pap.[14]

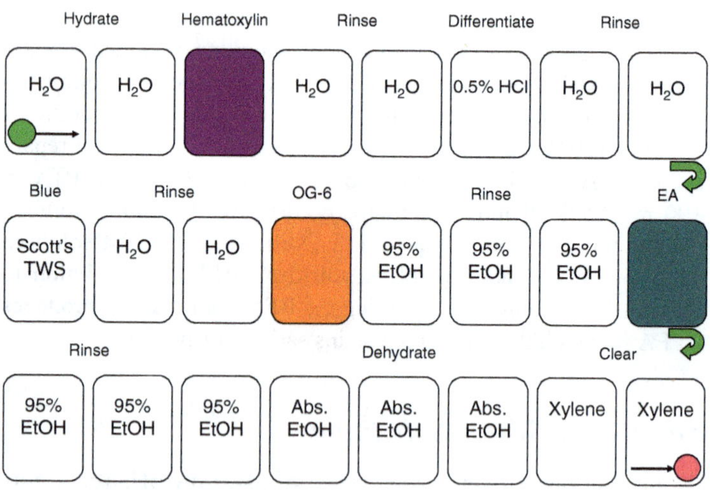

FIG. 10.4. This illustration reflects the steps in Table 10.5.

TABLE 10.5. Modified Papanicolaou stain and Enviro-Pap: side-by-side comparison of solutions.

No.	Modified Pap	Time	Enviro-Pap	Purpose
1.	Tap water	10 dips	Tap water	Rinses as needed
2.	Tap water	10 dips	Tap water	Rinses as needed
3.	Gill hematoxylin No. 2	2 min	Gill hematoxylin No. 2	Colors chromatin red
4.	Tap water	10 dips	Tap water	Rinses hematoxylin
5.	Tap water	10 dips	Tap water	Rinses hematoxylin
6.	Scott's tap water substitute	2 min	—	Blues red Al-hematein
7.	Tap water	10 dips	Tap water	Rinses as needed
8.	Tap water	10 dips	Tap water	Rinses as needed
9.	Gill modified OG*	10 s	Gill modified OG	Colors keratin yellow
10.	95% ethanol	10 dips	0.5% HOAc (acetic acid)	Rinse, keeps OG in cells
11.	95% ethanol	10 dips	0.5% HOAc	Rinse, keeps OG in cells
12.	95% ethanol	10 dips	0.5% HOAc	Rinse, keeps OG in cells
13.	Gill modified EA	8 min	Gill modified EA	Colors cyto red & green
14.	95% ethanol	10 dips	0.5% HOAc	Rinse, keeps dyes in cells
15.	95% ethanol	10 dips	0.5% HOAc	Rinse, keeps dyes in cells
16.	95% ethanol	10 dips	0.5% HOAc	Rinse, keeps dyes in cells
17.	Absolute ethanol	10 dips	Absolute ethanol	Dehydrates
18.	Absolute ethanol	10 dips	Absolute ethanol	Dehydrates
19.	Absolute ethanol	10 dips	Absolute ethanol	Dehydrates
20.	Xylene	10 dips	Xylene	Clears
21.	Xylene	10 dips	Xylene	Clears
22.	Xylene	10 dips	Xylene	*Clears*

Notes

1. Carbowax in spray-fixed smears is removed by water, as confirmed by microscopic observation using crossed polarizing filters.
2. No distinction is necessary between gynecologic and nongynecologic cytological specimens with regard to staining times, especially in an age of liquid-based preparations. If staining, rinsing, and clearing are performed properly, differences in average thickness of the preparations rarely create problems in visualizing microscopic detail.
3. There is no single staining time that is best for a given stain or cytological preparation under all conditions to all users.
4. Use one-step hydration and dehydration. Series of graded percentage alcohols are unnecessary.
5. Maintain a ratio of about 15 ml of stain or rinse per each slide of a staining dish's total slide capacity (e.g., a 30-slide-capacity dish should be filled with 450 ml solution). Lesser ratios reduce the working life of the solution.
6. Rinses following OG and EA are most effective when used in sets of three, maintained deep for maximum dilution, and rotated when the third rinse of each series becomes colored.
7. A "dip" is defined as gently raising the staining rack until it clears the solution, and without jarring it against the sides or bottom of the dish, lowering it until it is totally submerged again. A dip requires about one second. Ten dips effectively replace one solution by another, as judged by the glistening surface of the slides, and are simply a practical minimum number.
8. Drain the slides well between solutions but do not allow the preparations to dry. Stain uptake continues during the interval between the removal of slides from a stain and its immersion into the following rinse. To avoid excessive uptake of OG, especially formulations with acetic acid, keep the interval to a few seconds.

9. Absolute ethanol and reagent alcohol can be used interchangeably. The absolute ethanol and xylene rinses should remain color-free. The presence of color indicates carryover of materials such as glacial acetic acid and phosphotungstic acid that may contribute to stain fading in the mounted preparations.
10. To avoid carryover of water into xylene, the level of xylene should exceed that of the preceding absolute alcohols. The absolute alcohols should have a level greater than the preceding 95% alcohol, and it in turn should have a depth greater than the water rinses. The xylenes can be kept water-free by filtering through laboratory grade filter paper. Water and xylene, which is an oil, do not mix. The water droplets are suspended in the xylene and are blotted out of suspension during filtration. Periodic replacement is necessary, however, perhaps at weekly intervals. Xylene appears to give more consistent, trouble-free results than do xylene substitutes.
11. When checking stain quality among the routine slides, evaluate several slides from a group of specimens known to be usually well preserved. Evaluating only one slide carries with it the risk of encountering a poorly preserved specimen that has stained unsatisfactorily and produces an erroneous impression of the stain's performance.
12. As practiced conventionally, the Pap stain is expensive. Of the 22 solutions in Table 10.5, only 3 are dye formulations. Of the remaining 19 solutions: 6 are water, 1 is a chemically defined bluing reagent, 6 are 95% ethyl alcohol, and 3 each are absolute alcohol and xylene. In addition, the alcohol and xylene solutions require space-consuming flammables safety storage and costly disposal. Disposing of xylene sometimes costs more than buying it. All of these non-dye solutions constitute a significant portion of the cost of specimen processing. Reengineering the Pap stain process offers the potential to significantly reduce operating costs without sacrificing quality. Enviro-Pap is an environmentally friendly, cost-effective modification—developed in 1995—of the "standard"

Papanicolaou stain. It is a set of products and a process that yields high-quality, reproducible staining results while saving money by eliminating chemically defined bluing reagents (e.g., Scott's tap water substitute), consuming less alcohol, reusing xylene indefinitely, and reducing hazardous waste disposal costs. Enviro-Pap uses the same stains as those used successfully by a laboratory. It should not be used to replace proprietary Pap stain materials and methods required by manufacturers of liquid-based Pap test imaging devices.

13. *Bluing occurs satisfactorily in tap water alone.* The timing must be sufficiently long to blue Al-hematein in thick groups, as well as in isolated single cells. The local timing must be confirmed microscopically initially, but is usually not more than 2 min. A hematoxylin-stained buccal smear can be blued conventionally and used as a 1-time control for visual comparison. Chemically defined bluing agents such as Scott's tap water substitute are unnecessary.
14. Pre-OG alcohol baths have been replaced by tap water. Subsequent water dilution of OG is inconsequential.
15. OG and EA staining times are interdependent. Too long in OG (e.g., 1 min) and too short in EA (e.g., 3 min) results in cells stained orange or green but not red. Depending on its concentration, OG may require dilution with equal parts of 95% ethanol to lengthen staining times that may otherwise be too brief (e.g., 5 s) to control reproducibly. This is the only potential pitfall I'm aware of.
16. One-half percent acetic acid rinses replace the post-OG and post-EA alcohols in Enviro-Pap. Acid pH keeps dyes in cells to produce the same effect as that by 95% ethanol. The pH of this very dilute acetic acid solution is too weak to remove hematoxylin as HCl does when used to differentiate regressive hematoxylin stained cells. To prepare 0.5% acetic acid, mix 1 part glacial acid with 200 parts tap water (e.g., 5 mL HOAc per 995 mL water; 100 mL HOAC per 19,900 mL water [nominal 5 gallons total]). Plain water alone extracts counterstain dyes rapidly and should not be used.

17. Acetic acid rinses do not require special disposal. Restaurants routinely dispose of vinegar, which is at least 10 times as concentrated, in public sewer systems.
18. Water-scavenging beads (i.e., molecular sieves) in xylene adsorb water in real time. In conjunction with daily filtration to clarify xylene baths, these beads extend the use of xylene indefinitely. The immediately preceding absolute alcohol must be maintained color-free to avoid contaminating xylene with dyes.
19. Staining is reversible. Whatever dyes are deposited can also be removed. Staining mistakes can be remedied.
20. Precipitates form in stains as a result of (a) the concentration of an ingredient exceeds its solubility limit in the solvent (e.g., aluminum–hematein precipitates on the surface of Harris hematoxylin or Gill hematoxylin No. 3 or orange G crystals when the solution is cooler than room temperature) or (b) an ingredient or combination of ingredients is not soluble at all in the solvent (e.g., phosphotungstic acid and Bismarck brown Y).
21. Rinses that become dye-laden are dilute dye solutions that are less effective for their intended purpose. Rinses should be changed and rotated when the last dish in a set of rinses becomes colored. This is one of the more common failings in any staining method and the one that will most lower overall contrast, robbing cells of a crisp appearance.
22. Use buccal smears with each stain independently ad hoc as a quality assessment method. This approach will show you what each stain looks like under the best conditions. If the results are unsatisfactory, determine whether the cause is the material or method. For example, if the hematoxylin is blue and optically dense in much of the cytoplasm, the staining time must be decreased, the stain diluted, or differentiation increased if applied. But if the color is brown, the hematoxylin is overoxidized and cannot be salvaged.
23. Stain in OG for approximately 15 s; EA, 6–10 min.
24. Good starting materials used with bad methods will produce poor results. Examples of bad methods include staining times too long or short, exhausted stains, and dirty rinses.

Results

When performing properly, the Pap stain is capable of producing a full spectrum of colors (i.e., red, orange, yellow, green, blue, and violet hues).

Chromatin should be blue; keratin, orange; superficial squamous cells, erythrocytes, nucleoli, and cilia, red or pink; and cytoplasm of all metabolic cells, green. See Fig. 10.5.

The site of primary interest is the nucleus, which should be blue. The optical density of hematoxylin should be light enough to allow appreciation of chromatin particles in the lobes of well-flattened PMNs and dark enough to make visible chromatin particles in intermediate squamous cell nuclei. See Fig. 10.6.

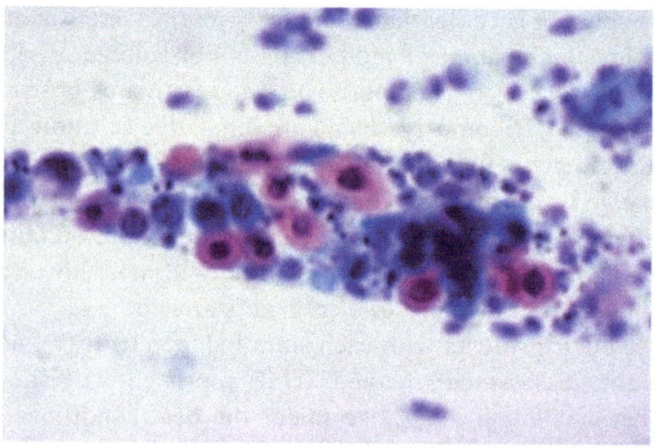

Fig. 10.5. Conventional Pap smear stained by Gill modified Pap stain.

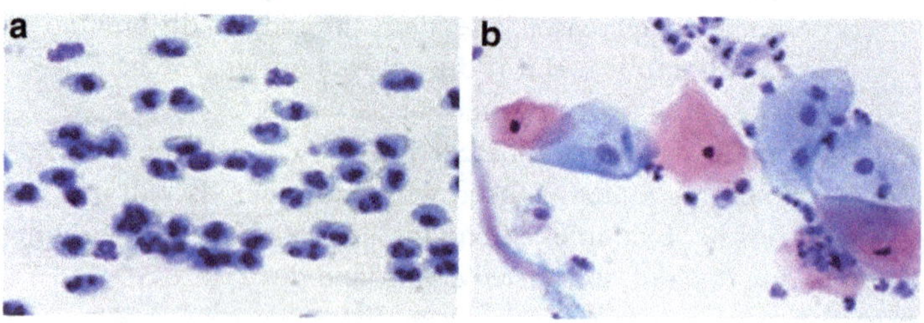

Fig. 10.6. The optical density of hematoxylin should be (**a**) dark enough to distinguish chromatin detail in well-flattened PMNs and (**b**) light enough to make visible fine chromatin particles in intermediate squamous cell nuclei.

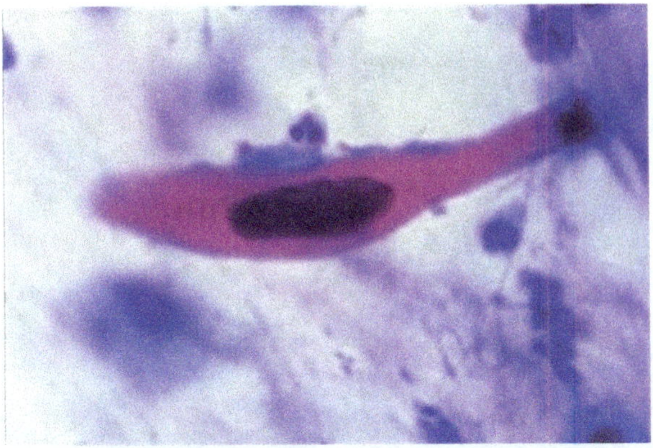

Fig. 10.7. Orange G in keratinizing squamous cell carcinoma in sputum.

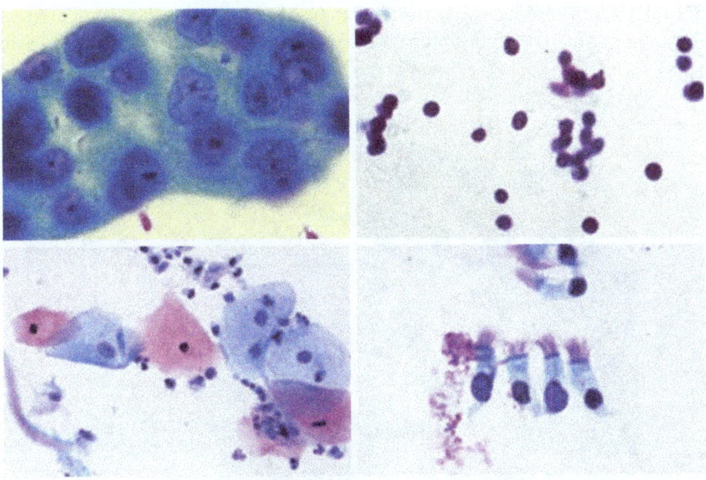

Fig. 10.8. Eosin colors superficial squamous cells, cilia in columnar cells, nucleoli, and erythrocytes.

Orange G colors keratin orange and was introduced initially to make visible the small cells of keratinizing squamous carcinoma in sputum specimens. See Fig. 10.7.

Eosin should color superficial squamous cells, nucleoli, RBCs, and cilia. See Fig. 10.8.

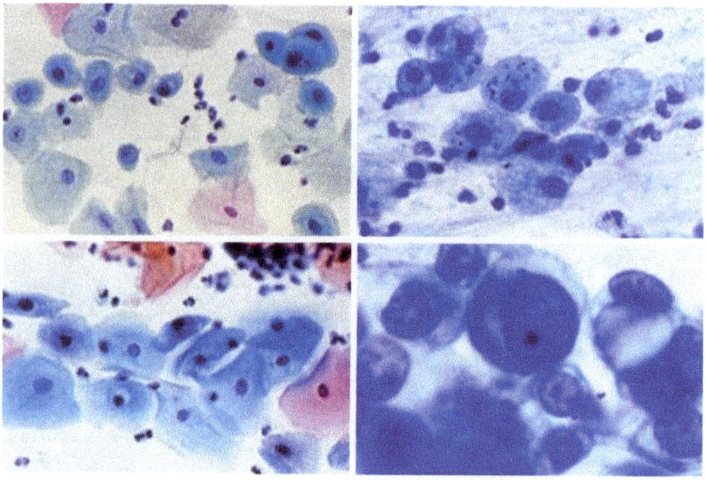

FIG. 10.9. Examples of the wide range of cells with cytoplasm stained by *light green*. The photomicrography did not capture the true color in all cases.

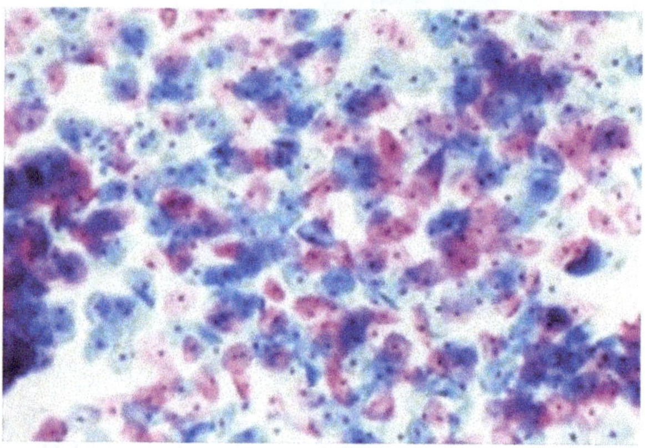

FIG. 10.10. Well-stained field of squamous cells in conventional Pap smear.

Light green should color the cytoplasm of all cells other than superficial squamous cells. See Fig. 10.9.

Bismarck brown Y colors nothing of interest in Pap tests. The yellow color of glycogen is difficult to explain but may be due to unmordanted hematein. Thick groups should be stained with the appropriate colors uniformly throughout. All cells should be transparent. See Fig. 10.10.

Discussion

A good Pap stain promotes the detection and interpretation of abnormal cells. It is a key part of the one-time investment in the quality of the preparation that repays multiple dividends with every microscopic examination. Poor staining quality, on the other hand, is often cited as a feature of litigated false-negative Pap smears. [16]

Hematoxylin

As a matter of convention, we say that hematoxylin stains chromatin. In fact, hematoxylin per se cannot stain any biological substrate. Aluminum–hematein, "hemalum", is the stain agent. Further, hematoxylin and the counterstains of the Pap stain are not stains at all but dyes. Stains color surfaces (i.e., 2-dimensional); dyes penetrate substrates (i.e., 3-dimensional).

Hematoxylins can be applied progressively (e.g., Gill) or regressively (e.g., Harris). See Table 10.6. Progressive hematoxylin

TABLE 10.6. Contrasting aspects of progressive and regressive hematoxylins.

	Hematoxylin	
Aspect	Progressive	Regressive
Hemalum concentration	Less (i.e., 1 to 4 g/L)	More (i.e., 5 g/L or more)
Acetic acid	Present	Absent
Rate of uptake	Slow	Rapid
Easily controlled?	Yes	No
Overstaining?	No	Yes
Differentiation required?	No	Yes

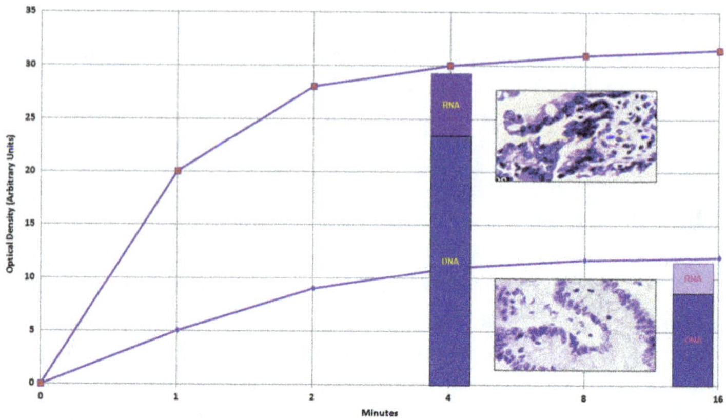

FIG. 10.11. Illustration of the difference between the uptake of progressive and regressive hematoxylins. Both sections have been blued. Differentiation for the regressive hematoxylin was deliberately omitted to demonstrate the need to differentially extract excess hematoxylin from the chromatin and the cytoplasm.

stains color primarily chromatin and to a much less extent cytoplasm to the desired optical density, regardless of the length of staining time. Regressive hematoxylin staining overstains chromatin and cytoplasm and requires subsequent immersion in dilute acid to pull out the excess color from the chromatin and cytoplasm. See Table 10.6 and Fig. 10.11. If differentiation is omitted or incomplete, residual hematoxylin visually obscures fine chromatin detail and can prevent the uptake of counterstain dyes entirely.

Differentiating Hematoxylin

Regressive hematoxylins require differentiation in dilute acid, progressive hematoxylins do not. Both types of hematoxylins require bluing. The differences between differentiation and bluing are listed in Table 10.7.

TABLE 10.7. Differentiation and bluing: comparisons and contrasts.

Property	Differentiation	Bluing
Purpose	Differentially extract excess hematoxylin from chromatin and cytoplasm; quantitative	Convert soluble red color to insoluble blue color; qualitative
Function	Attacks tissue/mordant bond	Oxidizes Al-hematein
Used with	Regressive hematoxylins	Progressive and regressive hematoxylins
Working pH	About 2.5	5–11
Common example	0.5% HCl in 70% ethanol	Scott's tap water substitute
Timing	Dips	Minutes
Timing accuracy	Critical	Forgiving
Risk if too brief	Hyperchromasia	Purple color
Risk if too long	Hypochromasia	Decolorization if pH ≥ 11
Possible negative impact	Low contrast = less detail	Cell loss if pH ≥ 11

Bluing Hematoxylin

Hematein, the oxidation product of hematoxylin, exists in solution in 3 forms: (1) free yellow hematein, (2) partially linked red hematein linked to 1 aluminum ion per molecule, and (3) fully linked blue hematein with each molecule attached to 2 aluminum ions. Below pH 5, hydrogen ions compete with aluminum ions. As the pH rises, competition by hydrogen ions decreases and the blue aluminum–hematein replaces the red. In my experience, tap water will blue hematoxylin in about 2 min. See Fig. 10.12.

Whether a hematoxylin is applied progressively or regressively, satisfactory results are obtained with chemically competent, properly applied hematoxylins

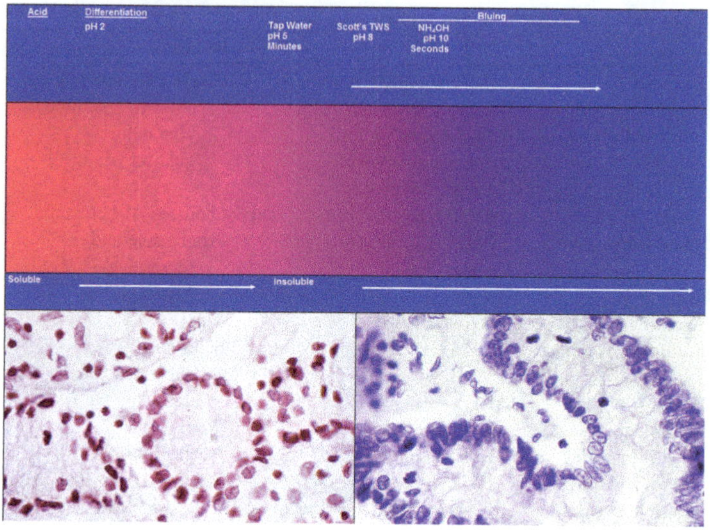

Fig. 10.12. Bluing is the process of converting the initially red soluble hemalum to a final blue insoluble form.

Orange G

Papanicolaou's OG formulations called for amounts of orange G that exceeded its solubility limit in alcohol more than twofold. Gill modified OG uses 2 g/L alcohol, which is within orange G's solubility limit. See Fig. 10.13.

OG and EA staining times are interdependent. The 1–1.5 min staining time recommended by Papanicolaou is too long relative to the 1.5–3 min recommended for EA. Orange G initially stains all cells, not just those that are keratinized as seen in the final staining results. Microscopically examining random slides upon immersion in the post-OG alcoholic rinses is the basis for this observation. During the initial time in EA, eosin must displace orange G from those cells for which eosin is intended. If the EA staining time is brief, the displacement is incomplete and results in a blend of orange G and eosin Y colors. For this reason, slides should be immersed in OG for about 15 s. This recommendation

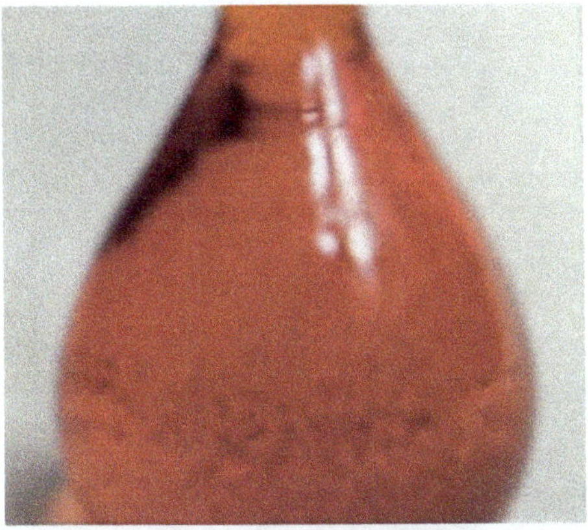

FIG. 10.13. Orange G precipitates as crystals when the amount exceeds its solubility limit in alcohol.

is the same for Gill modified OG with acetic acid and Papanicolaou's OG-6 without acetic acid.

EA

Papanicolaou's EA formulations called for eosin Y, light green SF yellowish, Bismarck brown Y, phosphotungstic acid, and lithium carbonate in alcohol. The formulations were not reproducible. Dye content variations were not addressed to enable quantitative reproducibility. Bismarck brown Y should never have been included, and the contribution of lithium carbonate is questionable at best. The recommended staining times were far too brief.

The major reason for the erratic uptake of light green in EA stains is the presence of PTA and Bismarck brown. Unlike eosin and light green, Bismarck brown is a basic dye, meaning that it carries positive charges. (Basic dyes are also known as cationic dyes. Eosin and light green are acid dyes and are also known as anionic dyes. Acid is somewhat misleading, as it suggests the dyes somehow make a solution acidic, which they do not. Acid dyes are

Fig. 10.14. Phosphotungstic acid added to an alcoholic solution of Bismarck brown Y precipitates both.

taken up much more vigorously in a solution to which an acid has been added [e.g., acetic acid].) It combines with, and is precipitated to some degree from solution, when it combines with the negatively charged PTA molecule. The ability of PTA to form insoluble compounds with basic dyes had been described 11 years before Papanicolaou's 1942 paper.[8] See Fig. 10.14.

The absence of Bismarck brown is not missed, as apart from staining mucus—if at all—it stains nothing else. The diminution of the concentration of PTA, however, is an entirely different story. PTA serves as a dye excluder and enables light green and eosin to stain differentially. In the total absence of PTA, there is no differential staining. See Fig. 10.15.

There is another aspect of PTA that merits attention. PTA is deliquescent (i.e., absorbs atmospheric water). If a lot of PTA contains an inordinate amount of water, 2 g PTA/L EA will contain far less than 2 g actual PTA. In severe cases, using 2 g PTA/L EA will appear as though no PTA was added to the stain at all! Instead of seeing clear-cut green and red cells, one sees dull green and dusty rose colors; eosin is not seen at all. In light of the hygroscopicity of PTA and its being precipitated from solution by Bismarck brown, it is not surprising that so many EA solutions

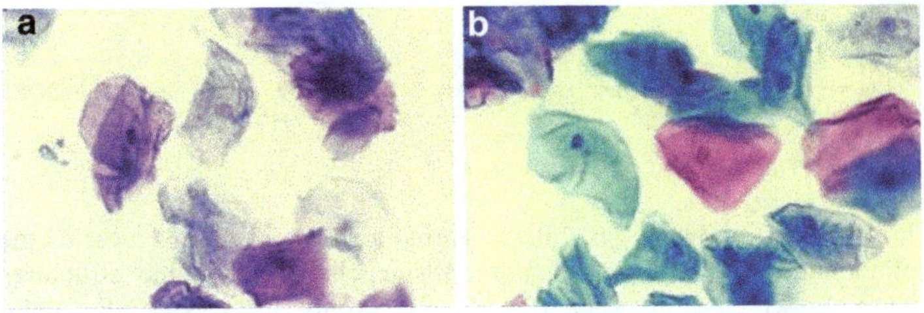

FIG. 10.15. Squamous cells in 2 buccal smears, each stained for 8 min in the identical EA solution: (**a**) without PTA and (**b**) with PTA.

perform poorly. Some perform unsatisfactorily at the outset, while some others may perform well for a short while initially and then suddenly perform unsatisfactorily.

PTA plays different roles in OG and EA. PTA acidifies OG; it does not act as a dye excluder. PTA unavoidably acidifies EA as well, but its primary purpose is as a dye excluder, as mentioned above.

Mallory introduced the use of PTA to staining in 1900 without providing any insight into his novel choice of this useful chemical.[17]

Glacial acetic acid added to EA acidifies more effectively than PTA. See Fig. 10.16.

The 1–1/2 to 3 min staining times recommended by Papanicolaou do not allow enough time for eosin Y to displace orange G and to promote uniform staining in single cells and thick groups. Nonuniform penetration of dyes is referred to as poor leveling. See Fig. 10.17.

In a solution that contains two dyes, the dye with the greater concentration penetrates first as predicted by the law of mass action. Given enough time, the second dye will eventually displace the first dye in bonding sites to which it is attracted. In the case of EA, eosin is far more concentrated than is light green. Thus, eosin penetrates cells first. To offset this phenomenon, stain for 6–10 min. See Fig. 10.18.

Every silver lining has a cloud, and so it is with staining longer. Staining longer promotes uniform staining in thick areas, but it

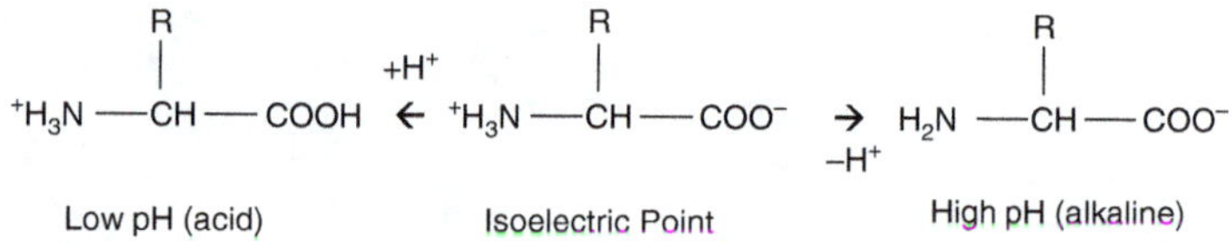

FIG. 10.16. Shows a generalized amino acid. In a protein, most of the amino and carboxyl groups are on side chains (R in the structures shown)—which outnumber the alpha-NH_2 (only at the N-terminal) and the C-terminal—COOH. Amino acids united by peptide linkages make up proteins, which are all that remain in cytoplasm following fixation in alcohol. If an amino acid in solution is placed in an electric field, as in electrophoresis, the molecules will migrate to one pole or the other in accordance with the pH of the solution. At a certain pH, which is unique to the particular protein, the amino acid does not migrate to anode or cathode. This pH is the isoelectric point. Adding glacial acetic acid (i.e., low pH), neutralizes the COO—groups and leaves relatively more positively charged H_3N groups. As a result, eosin Y and light green SF yellowish dye molecules, which are negatively charged, are attracted to the positively charged groups and thus are taken up faster and in greater total amounts per given amount of time.[18]

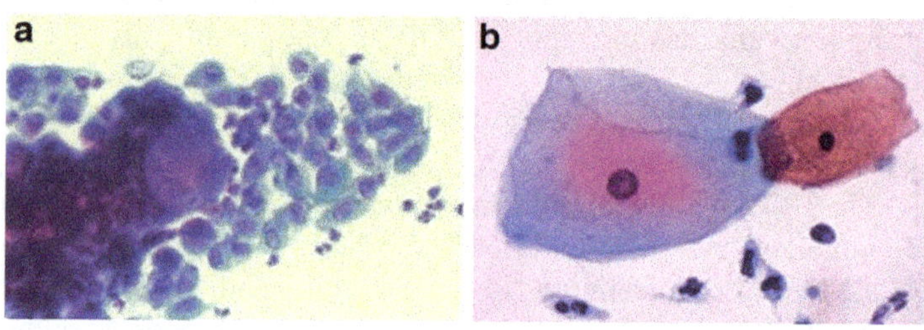

FIG. 10.17. Poor leveling is the nonuniform penetration of dyes in a solution with two or more dyes of unequal concentration. (**a**) The cells in the left-hand tissue fragment are the same types of cells as single cells on the right in the same field, and should be stained the same. They are stained differently, however, because more time is needed to allow the light green dye molecules to displace the eosin dye molecules in the center of the fragment. (**b**) Poor leveling in a single cell. Poor leveling can be avoided by staining longer. Additional rinsing—more dips in more clean baths—may be required to extract the excess dye molecules.

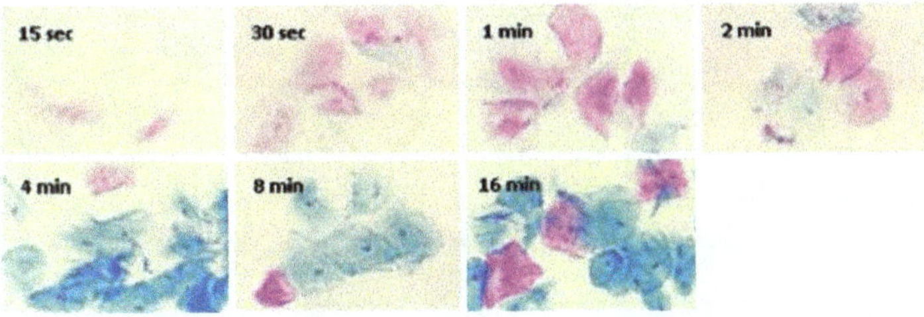

FIG. 10.18. This 7-slide series of buccal smears was stained for the times indicated in Gill modified EA to illustrate the limitations of staining briefly and the value of staining longer. At 2 min or less, light green has not had the time to displace eosin from its bonding sites. Longer times allow light green to displace eosin and to minimize poor leveling. Note the times have been doubled successively, much like bracketing exposure times in photography.

can also increase the optical density so much that discerning chromatin details becomes impossible. As light passes through stacked cells, the transmission falls off exponentially as seen in Fig. 10.19.

Successful Pap staining is a consequence of "preemptive occupation." When chemically competent stains are applied in the prescribed sequence for the right times and rinsed appropriately, hematoxylin colors chromatin; orange G colors keratin; eosin colors superficial squamous cells, cilia, and erythrocytes; and light green colors everything else. If hematoxylin overstains greatly, it blocks the uptake of subsequent dyes. If orange G stains excessively, it cannot be displaced by eosin. If the staining time in EA is too brief, eosin will not be displaced by light green.

Among the benefits to be derived from technically satisfactory Pap staining results are the following:

- Differences among gray-level values and hues heighten contrast between nucleus and cytoplasm of individual cells and among normal and abnormal cells, thereby increasing the detectability of abnormal cells (which is why stop signs aren't green, no contrast to separate the tree from the forest) during screening.
- Chromatin patterns are more amenable to interpretation.
- Interpretation of cells from borderline lesions becomes less difficult.

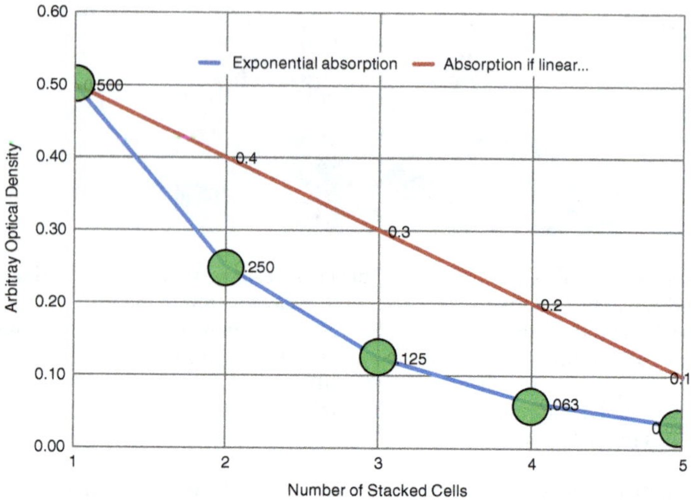

FIG. 10.19. Physics makes impossible maintaining visual clarity in thick stained cells. As seen in this hypothetical example, the amount of light escaping through the topmost cell in a stack of 5 identically stained cells is 1/16th as much as the light escaping through the bottommost cell with an optical density of 0.5. Each of the green circles represents a cell. The leftmost cell absorbs half the light that enters it from below and transmits the remaining half to the observer's retinas. Moving to the right, each of the cells transmits half the light as the one preceding it, until Cell No. 5 transmits almost none. In other words, the fall-off in light transmission is exponential. If the fall-off were linear, as depicted by the red line, one might expect that Cell No. 5 would transmit 1/5th as much light as Cell No.1 (i.e., $0.2 \times 0.5 = 0.10$) and therefore, not appear as dark as it does. However, that is not the case. Physics is always right.

- Screening is more satisfying.
- Teaching others to interpret cytomorphology becomes easier.
- Descriptions of cytomorphology by different individuals will agree more closely and consistently with one another.
- Well stained cells are visually pleasing.
- Photomicrographs are more instructive.

Staining and rinsing results are more easily controlled when the rate of dye deposition or removal is slow (i.e., the staining method should be "forgiving"). Dye uptake or removal that changes visibly within seconds is too difficult to control for good reproducibility of results.

The differential action of stain is a function of the chemical affinity between dye and substrate, the density of the cellular protein, and the permeability of the cells. The amount of dye that remains in a cell after staining equals the *difference* between the amount deposited by the stain solution and the amount removed by subsequent rinses or displaced by competing dyes.

The type of fixation that cells experience will influence the penetration rate of dyes. Shrinkage that results in more closely woven protein with reduced intermolecular spaces can be caused by isopropyl alcohol and air-drying. Longer staining times may be required to produce optical densities comparable to those that result in ethyl alcohol wet-fixed cells.

Rinses

Rinses are the neglected stepchild of staining. Defined here somewhat arbitrarily as all the non-coloring solutions, rinses constitute 80% or more of all the solutions in most stains and outnumber the stains more than 10 to 1 if the initial xylene and alcohol rinses are included, yet their purposes vary and their contributions often go unrecognized. Consequently, rinses not uncommonly are overused to extremes—even becoming dilute staining solutions themselves, usually in an effort to save money. Not appreciated, unfortunately, is the hidden cost of extended, tedious microscopic examinations.

Among the many purposes of rinses are the following:

- Effect transition from organic solvents to aqueous solutions and vice versa (i.e., dehydration and hydration)
- Stop action of previous solution (post-hematoxylin water rinses)
- Differentially extract excess hematoxylin (i.e., differentiation)
- Convert hematoxylin from red to blue color (i.e., bluing)

- Promote redistribution of dyes within tissue (i.e., uniformity)
- Allow expression of differential staining
- Dehydrate (i.e., absolute alcohol)
- Clear (i.e., xylene)

Of the post-stain rinses, therefore, it may generally be said that *the amount of stain that remains within cells represents the difference between what the staining solutions put in and the rinses take out.* The post-eosin rinses perform most effectively when *clean. Clean* simply means that there is the maximum difference in concentration gradient between the dyes in the cells and in the rinse. When stained tissue is immersed in clean alcohol, the dyes diffuse effectively into the surrounding rinse. As the rinses become dye-laden, the concentration gradient is reduced and diffusion slows. When the concentration of dye in the tissue equals that in the rinse, there is zero concentration gradient, and the benefits of rinsing are lost.

To promote effective rinsing: (1) keep the rinses deep for maximum dilution (not just simply covering the tops of the slides when the slides are oriented horizontally, as opposed to vertically), (2) use in sets of three, (3) dip racks at least ten times in each, and (4) change *as needed.*

As needed occurs when the third rinse becomes colored. Under such circumstances, discard the contents of the first dish, move rinses 2 and 3 back one step to become rinses 1 and 2, and replace the third rinse. The third dish in each series of three post-eosin rinses should remain color-free. Maintaining this level of quality allows the absolute alcohols and xylene rinses to remain color-free.

Substituting 0.5% acetic acid for 95% ethanol baths, using tap water instead of chemically defined bluing agents to blue hematoxylin, and using water-scavenging aluminosilicate beads in the clearing xylenes can be applied successfully to H&E staining as well.

Absolute alcohol can be recycled by using commercially available recycling systems.[19,20] The same vendors also sell xylene recycling systems.

STAT-Pap: A 12-Step, 2-Min, 4-Color Papanicolaou Stain for Rapid Evaluation

The same aqueous stock solutions of the counterstain dyes are used to make the combined OG and EA stain.

Materials

- Gill hematoxylin No. 2
- Scott's tap water substitute
- STAT-OG/EA
- 95% ethanol
- Absolute alcohol
- Xylene
- 12 Coplin jars
- Paper towels
- Slide forceps

Prepare a liter of STAT-OG/EA by combining these ingredients:

10% orange G (TDC), aqueous	6 mL
3% (TDC) light green SF yellowish, aqueous	20 mL
20% (TDC) eosin Y, aqueous	20 mL
95% ethanol	680 mL
Absolute methanol	250 mL
Glacial acetic acid	20 mL
20% (w/v) phosphotungstic acid in 95% ethanol	20 mL

Methods

Stain fixed slides as follows:

1.	Tap water	5–10 dips until surface is smooth
2.	Gill hematoxylin No. 2	30 s to 1 min
3.	Tap water	5 dips
4.	Scott's tap water substitute	15 s
5.	Tap water	5 dips
6.	STAT-OG/EA	30 s with 10 initial dips

(continued)

7.	Tap water	5 dips
8.	Tap water	5 dips
9.	Absolute ethanol	10 dips
10.	Absolute ethanol	10 dips
11.	Xylene	5 dips
12.	Xylene	5 dips and coverslip

Elapsed time about 3 min.

Notes

(a) Solutions should be deep enough to cover slides.
(b) Dip rapidly, about 2 per second, completely in-and-out with occasional agitation between dips.
(c) As a slide is moved from one solution to the next, complete displacement of the first by the second is evidence when the entire slide surface glistens smoothly after several dips.
(d) Drain slides after baths 1, 5, 6, 8, 9, and 10 by touching the slide end to paper towels.
(e) To avoid overstaining, precisely control the stain times and move the slides quickly from each stain into its rinse.
(f) Slides usually appear reddish following steps No. 7 and 8 and must be dipped sufficiently in clean alcohol rinses to remove the excess eosin, thereby revealing those cells stained by light green.

The results are comparable to those obtained with Gill modified Pap stain: blue chromatin and mucus; orange erythrocytes, keratin, and granules of eosinophils occasionally; red superficial squamous cells, erythrocytes, cilia, nucleoli; and green cytoplasm of all other cells. Thin cell spreads stain best.

Quality Assessment Using Buccal Smears

Alcohol wet-fixed buccal smears are invaluable probes to determine the performance of each new lot of stain, to select suitable staining times, to find out how many slides can be stained satisfactorily by a given volume of each stain, to learn when rinses should be changed, and to troubleshoot whether a given stain already in use is the cause of an observed staining problem. Once experience imparts confidence to selected staining times, stain, and rinse change schedules, the use of buccal smears is not necessary. However, they should be used when new containers of the same stain with different lot numbers are opened to confirm that the stain does indeed perform as expected. Manufacturers occasionally make bad batches of stain.

When attempting a complete staining procedure, *frequent microscopic inspection throughout the staining process* allows one to follow the progress of stain uptake, to detect trouble areas, to troubleshoot staining problems, and to understand how the final staining picture evolved. This recommendation educates users quickly about a stain's performance. It is not intended for routine daily use.

It cannot be assumed that manufactured hematoxylin, OG-6, and EA stains will perform satisfactorily when received in the laboratory. These stains vary too much in their age and composition (i.e., quality, quantity, and specific dyes and other chemicals) from manufacturer to manufacturer. The use of alcohol wet-fixed buccal smears as a simple quality assessment method is recommended to confirm that each batch of a given stain performs as expected before it is put into use. Staining each buccal smear in a separate stain allows one to see the true colors produced by each staining solution, without possible interference from other competing stains.

The recommendations that follow constitute true quality assessment. Simply looking at one of the first slides stained daily and initialing a stain quality log sheet are valueless if a laboratory has not defined its standards. It is not uncommon to see such sheets dutifully maintained and also to see unsatisfactory staining results.

Quality Assessment. A series of buccal smears should be used to answer three questions for each of the three staining solutions of the Pap stain: (1) what color is produced, (2) what staining time is required to produce the desired optical density, and (3) what is the distribution vis-à-vis the chromatin and cytoplasm:

- The *color* of a dye or stain is a function not only of the dye itself but also the presence of any colored impurities, possible oxidation effects, and coexisting uptake of a subsequently applied stain.

- The *optical density* of a stain is influenced by its concentration, pH control, solvent, duration of staining, and potential decolorization due to subsequent rinses.
- The *distribution* of a stain is due to the factors that influence its optical density, as well as the order of application in a sequence of dyes and its concentration relative to the concentration of a competing dye.

To confirm that each new lot of stain is performing as expected, stain different sets of alcoholic wet-fixed buccal smears in hematoxylin for 30 s, 1, 2, and 4 min; OG, 15 and 30 s and 1 min; EA, 1, 2, 4, 6, and 8 min:

- Rinse the hematoxylin stained slides in two changes of tap water (10 dips each), blue per lab convention, rinse in two changes of tap water (10 dips each), dehydrate, clear, and mount.
- Rinse the OG and EA stained sets of buccal smears in 95% alcohol, dehydrate in absolute alcohol, clear, and mount. Note that the successive staining times differ from one another by a factor of 2. Seconds-long differences don't change optical density enough to be appreciated visually.

Hematoxylin should be blue, with slight cytoplasmic coloration. The optical density should be light enough to show chromatin detail within the lobes of well-flattened PMNs and dark enough to make visible chromatin particles of intermediate squamous cell nuclei. See Fig. 10.6. If the stain is too light, stain longer or use a stronger hematoxylin. If the stain is too dark, stain less time or use a weaker hematoxylin. Gray to brown colors indicate overoxidation; discard the stain, don't use it.

Orange G looks yellow in thin areas and orange in thicker ones. Light green and eosin are distinctly green and red in properly formulated stains. If the PTA concentration is too low, there may be little or no differential staining—the green and red colors will be muddy and dull.

Preparing buccal smears in this way lets you see the true color of each of the four major dyes without any possible interference by inadvertent misuse of the others. The colors observed in this reference set should also be seen after Pap smears are stained by a complete Pap stain. If not, then it has problems that should be identified and corrected.

Staining Times. Proceed as above, but stain multiple buccal smears in each stain: hematoxylin for 1, 2, and 4 min; OG for 15 and 30 s and 1 and 2 min; and EA for 15 and 30 s and 1, 2, 4, 6, and 8 min. Choose the staining time that appears best (see the immediately preceding descriptions). Usually, the times are close to those recommended. The EA buccal smear series is most instructive, as it shows the shift in stain uptake from the red to green with longer staining times.

Stain Duty Cycles. Duty cycle is the number of slides that can be stained per unit volume of stain before a decline in quality is detectable. Buccal smears can be used to estimate semiquantitatively how many slides can be stained satisfactorily per given volume of hematoxylin or any other stain for that matter. Since each of the three stains contains different concentrations of dyes, you may find that more slides can be stained in one stain than another, so that all three stains are not changed at the same time. This remains to be determined empirically in your own laboratory.

Simply label a series of alcoholic wet-fixed buccal smears 0, 1, 2, 3, etc. and date the label ends. Put 1 buccal smear in every fifth rack of slides as a stain is changed. Slide 0 is the first rack, slide 1 is the fifth rack, and so on. Remove the test slides as the rack passes through the counterstains. Examining the entire set of slides will reveal how many slides can be stained satisfactorily. This process need only be done occasionally, as needed, to confirm the continued validity of the practice.

To determine objectively how many slides can be stained satisfactorily per mL, prepare 3 sets of alcohol wet-fixed buccal smears for every 5 racks of slides. Identify each slide with the stain name, date, and number, beginning with 1 and continuing by fives (i.e., 1, 5, 10, 15…). Put hematoxylin-1 in the first rack of slides to pass through fresh hematoxylin, put OG-1 in fresh OG, and EA-1 in fresh EA. Remove the hematoxylin slide before it enters OG, dehydrate in absolute alcohol, clear, and mount. Rinse the OG slide in the usual three changes of 95% alcohol, skip the EA and its rinses, complete dehydration in absolute alcohol, clear, and mount. Rinse the EA slides in its three changes of 95% alcohol, dehydrate completely in absolute alcohol, clear, and mount.

You will end up with three sets of buccal smears stained only in hematoxylin, OG, or EA at 5-rack intervals. Compare the slides in the first and last racks of each set to see whether there is a visible difference. If there is, find the slide at which you can begin to see an unfavorable difference. That numbered slide represents how many racks you can put through the stain. If you do not see a difference, more slides can be stained before the stain should be changed. Use buccal smears to determine how many more. Usually 1–2 slides can be stained per mL stain, depending on the concentration of the stain and the cellularity of the preparations. For example, 250 mL of most EAs will stain 250 Pap smears satisfactorily but not 375. Part of the savings realized by using less alcohol can be applied to the cost of changing the counterstains more frequently.

Rinse Duty Cycles. Pre-hematoxylin water rinses should be changed when cellular debris is visible. First two post-hematoxylin water rinses should be changed after each rack, as the hematoxylin rapidly colors the waters. Next two water rinses can be rotated back two positions and replaced by clean water. The two sets of post-OG/EA alcohols should be changed when the third bath in each set becomes lightly colored: discard the contents of the first dish in each series, move dishes 2 and 3 to positions 1 and 2 in the 3-bath series, and add fresh alcohol to dish 3.

Alcohol rinses work best when clean, deep, and are used for sufficient time to effect removal of excess dye from the Pap smears. Dirty rinses become more like a stain than a rinse and prevent effective rinsing. Buccal smears are also compelling indicators of overused alcohol rinses following OG and EA stains. Simply place a wet-fixed unstained buccal smear in the staining rack as it leaves OG. Remove the buccal smear after it passes through the three alcohol rinses, dehydrate, and clear. Do the same for EA. Dirty rinses are weak stains and can stain the unstained buccal smears dramatically. The cells will look as though they have been immersed in a primary staining solution, rather than in a rinse. They are terrific reality checks.

Destaining

Destaining and restaining allows one to see what he's missing. See Fig. 10.20.

Destaining works by immersing the slide in solutions that have a pH opposite that of the pH that promoted stain uptake (i.e., acid removes hematoxylin; alkaline removes counterstain dyes)

Remove the cover glass. Several methods are available; each one works:

1. Submerse the tilted slide in a Coplin jar of xylene, covered side down. Soak the slide in xylene until the cover glass falls away from the slide. This may take hours to days, depending on the age of the slide and the associated hardness of the mounting medium.
2. Briefly "cook" the slide on a foil-covered 120 °C hot plate to soften the mounting medium almost instantaneously. Remove

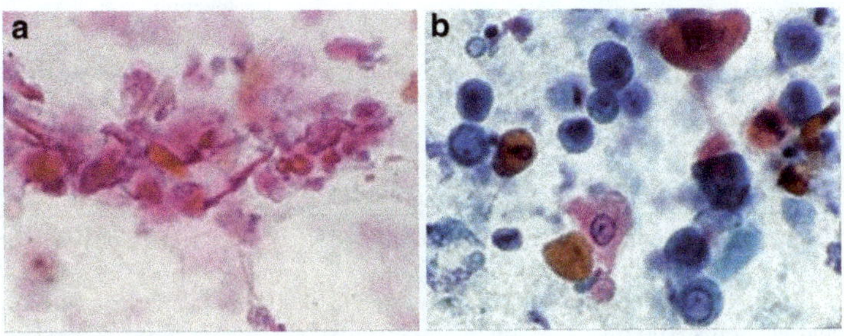

Fig. 10.20. Cells in sputum on a Millipore filter: (**a**) before destaining and (**b**)after destaining and restaining.

the cover glass with forceps and immerse the cooled slide in xylene.

3. Put the slide on a metal slide tray in a freezer for a short while. The cold temperature will cause the cover glass to "pop off" the slide.

To destain, see Table 10.8.

Table 10.8. After removing the cover glass, destain as follows.

No.	Solution	Time	Note
1–2	Xylene	1 min/each	Removes traces of mountant. Dip repeatedly; inspect surface. A wavy rather than a smoothly glistening surface denotes incomplete rinsing and indicates further dipping
3–5	Absolute alcohol	1 min/each	Prepares slides for next step
6	1.0% HCl[a]	min–1 h	Removes hematoxylin. The exact time depends particularly on the type of cytological material, its thickness, and fixation history. Monitor the progress of colorization by periodic microscopic inspection
7	1.5% NH_4OH in 70% ethanol[b]	1 min	Time required may vary and should be adjusted as needed. Repeated dipping aids uniform decolorization
8–9	Tap water	1 min/each	To Papanicolaou stain

[a]0.23 mL HCl concentrated (i.e., 36.5–38% w/w, S.G. 1.1854–1.1923, or 5.5 mL N/2 HCl) in water q.s. 100 mL.

[b]5.7 mL NH_4OH concentrated, 29.2% w/w, S.G. 0.900, in 75 mL 95% ethanol, and water q.s. 100 mL.

Troubleshooting

Problems do not occur when quality stains are used for the appropriate staining times, changed at the proper intervals, and used in conjunction with suitable rinses that are also changed at the proper intervals. If a problem does arise, look to the stains for the cause(s). Obscure causes (e.g., the pH of the tap water often being the putative cause) that cannot replicate the untoward results experimentally are never the cause in my experience.

Fix poor staining results immediately. The effects of poor staining are pervasive; the costs, insidious. In a properly controlled operation, satisfactory staining results can be achieved routinely. "While slide staining (preparation) is seldom the direct target of litigation, it is in fact one of the things repeatedly mentioned when slides from cytology claims are reviewed by experts in the course of evaluating and defending these claims."

When microscopically examining a preparation, one must remember that one is looking not at the object itself but an image of the object that is projected onto the retina. Therefore, separate in your mind's eye the effects of the materials and methods that interact with the object per se (i.e., salt solutions, unintended air-drying, fixative, stain) and those that influence the image of the object (i.e., mounting medium thickness, cover glass thickness, microscope cleanliness, and optical alignment [i.e., Kohler illumination]).

A knowledgeable observer can assess whether the preparation is technically satisfactory and/or functionally satisfactory. If deficiencies are noted, one should be able to identify the likely cause and implement a solution. A technically satisfactory preparation exhibits no technical deficiencies. Such preparations are also likely functionally satisfactory but not always. A functionally satisfactory preparation may exhibit technical deficiencies but still be useful for its intended purpose. This means the preparation does not have to be redone, but a solution should be implemented to assure technically satisfactory preparations in the future. Examples of technical deficiencies include incomplete differentiation, no eosin, excessively thick mounting medium, and cover glass that cause image-degrading spherical aberration.

TABLE 10.9. Troubleshooting the Papanicolaou stain: *hematoxylin.*

Complaint	Cause	Correction
Color absent	Hematoxylin exhausted from overuse	Replace hematoxylin
	Differentiation in HCl included mistakenly for progressive stain	Omit HCl
	Chlorine in tap water	Use distilled water
Wrong color present: red–purple	Bluing in acid tap water	Use Scott's tap water substitute (TWS)
	Bluing in running tap water too briefly	Blue longer. To reduce cell loss, use a standing bluing bath such as Scott's TWS
	Bluing solution exhausted	Change bluing solution daily
	Bluing too briefly	Blue longer
	Inadequately rinsed after EA or eosin Y	Use clean rinses. Rinse longer
	No "daylight" filter in microscope	Use "daylight" filter
Gray	Overoxidized when prepared	Use less oxidizing agent
	Overoxidized after preparation	Replace hematoxylin
	Colored impurities	Use certified hematoxylin
Brown	Overoxidized hematoxylin during preparation	Use less oxidizing agent
	Overoxidized hematoxylin by long term exposure to air	Replace hematoxylin. Prevent recurrence by storing hematoxylin in container with no air space
Optical density: hyperchromatic	Hematoxylin too concentrated	Use less concentrated hematoxylin. Dilute with ethylene glycol. Stain for less time. Differentiate in 0.25% HCl
	Staining time too long	Stain for less time
Optical density: hyperchromatic	Inadequate differentiation in HCl	Differentiate for longer time. Use more concentrated HCl. Replace HCl at least daily

Optical density: hypochromatic	Air-drying before fixation	Wet-fixed cells
	Air-drying after fixation	Keep cells wet
	Fixed, air-dried, Carbowax-coated cells	Immerse in alcohol for 10 min before staining and double staining time
	Cellular degeneration	Obtain repeat fresh specimen
	Hematoxylin nearly exhausted	Replace hematoxylin
	Staining too briefly	Increase staining time
	Overdifferentiation in HCl	Differentiate for less time. Use less concentrated HCl
	Differentiation in HCl mistakenly used for progressive stain	Omit HCl
	Paraffin sections very thin	Cut thicker sections
Distribution in wrong site: cytoplasm	Hematoxylin too concentrated	Stain for less time or dilute
	RNA-rich cytoplasm	Stain for less time or dilute
Nucleoli	Hematoxylin too concentrated	Dilute or stain for less time
	Staining time too long	Stain for less time
	Ineffective EA or eosin Y	Use effective EA or eosin Y

Whether hematoxylin, OG, EA, or any stain, wayward results can be categorized as (1) too much stain, (2) too little stain, (3) wrong color, or (4) wrong site. See Tables 10.9, 10.10, and 10.11.

Table 10.10. Troubleshooting the Papanicolaou stain: *OG*.

Complaint	Cause	Correction
Color absent	No keratinized cells	None
	OG-6 exhausted	Replace OG-6
	OG-6 bypassed	Stain in OG-6
	Staining time too brief	Stain longer
	Excessive rinsing in alcohol	Use recommended number of dips
	Slides dipped mistakenly in water or bluing bath	Restain in OG-6
	Cytoplasmic bonding sites blocked by hematoxylin applied regressively and not differentiated	Differentiate completely or use progressive hematoxylin
Wrong color present: grayish or purplish orange	Cytoplasm has retained hematoxylin applied regressively and differentiated incompletely	Differentiate completely or use progressive hematoxylin
Optical density: hyperchromatic	Staining time too long	Stain for 15 s–1 min
	Staining time too long in OG-6 acidified with glacial acetic acid	Limit staining time in acetified OG-6 to 15 s
Hypochromatic	OG-6 exhausted	Replace OG-6
	Staining time too brief	Stain longer
	No glacial acetic acid	Use 10 mL/L OG-6
Distribution in wrong site: cytoplasm of nonkeratinized cells	Cells stained too long in OG-6 acidified with acetic acid	Stain for 15–30 s in OG-6
	EA staining too brief	Stain in EA for 6–10 min

Table 10.11. Troubleshooting the Papanicolaou stain: *EA*.

Complaint	Cause	Correction
Color absent	EA exhausted	Replace with fresh EA
	Cytoplasmic bonding sites blocked by hematoxylin applied regressively and not differentiated	Differentiate completely or use progressive hematoxylin
Wrong color: grayish or purplish red	Cytoplasm has retained hematoxylin applied regressively and differentiated incompletely	Differentiate completely or use progressive hematoxylin
Orangish	Adulterated by residual OG	Assess true colors of OG and EA by staining one buccal smear in each solution exclusively. Stain less time in OG and/or more time in EA
	EA dye lot impurities	Same as above
	Low voltage illumination	Increase voltage
	No daylight filter	Use daylight filter
	Insufficient subsequent alcohol rinses	Increase rinse time, dip more
	Stain-laden rinses	Use clean alcohol rinses
Optical density: hyperchromatic	Optical density exceeds user expectations	Accustom use to more vivid hues or dilute EA with 95% ethanol
	Insufficient subsequent alcohol rinses	Increase rinse time, dip more
	Stain-laden rinses	Use clean alcohol rinses
Optical density: hypochromatic	EA nearly exhausted	Replace EA
	EA staining time brief	Stain for 10 min in EA
Wrong color and distribution in wrong site: single pink cells and pink throughout cell sheets that should be stained green	Poor leveling, a nonuniform penetration of cells by eosin and light green. Light green does not penetrate cells as rapidly	Stain in EA for at least 10 min. Dip slides during EA staining. Dip slides more in alcohol rinses following EA

If the problem is too much stain, put less in by using a less concentrated stain for the same staining time or staining for less time with the same concentration. It's vice versa when the problem is too little stain.

References

1. Böhmer F. Zur pathologischen Anatomie der Meningitis cerebromedularis epidemica. Aerztl Intelligenzb (Munich). 1865;12:539–50.
2. Fischer E. Eosin als Tinctionsmittel für mikroskopische Präparate. Arch Mikrosk Anat. 1875;12:349–52.
3. Wissowzky A. Ueber das Eosin als reagenz auf Hämoglobin und die Bildung von Blutgefässen und Blutkörperchen bei Säugetier und Hühnerembryonen. Arch Mikrosk Anat. 1876;13:479–96.
4. Bracegirdle B. A history of microtechnique. Ithaca, NY: Cornell University Press; 1978.
5. Mann G. Physiological histology. Oxford, England: Clarendon; 1902.
6. Harris HF. On the rapid conversion of haematoxylin into haematein in staining reactions. J Appl Microsc. 1900;3(3):777–80.
7. Scott SG. On successive double staining for histological purposes—(preliminary note). J Path Bact. 1911–1912;16:390–8.
8. Trotman SR, Frearson TB. New methods for the quantitative analysis of dyestuffs. J Soc Dyers Colourists. 1931;47(12):344–7.
9. Papanicolaou GN. A new procedure for staining vaginal smears. Science. 1942;95:438–9.
10. Papanicolaou GN. Atlas of exfoliative cytology. Cambridge, MA: Harvard University Press; 1954.
11. Papanicolaou GN. Memorandum on staining. In: Atlas of exfoliative cytology. Cambridge, MA: Harvard University Press, Supplement II; 1960. p. 12.
12. Gill GW, Frost JK, Miller KA. A new formula for a half-oxidized hematoxylin solution that cannot overstain and does not require differentiation. Acta Cytol. 1974;18(4):300–11.
13. Gill GW. Bismark [*sic*] brown and Papanicolaou EA stains. Scanner. 1975;14(3):2.
14. Gill GW. Enviro-Pap: an environmentally friendly, economical, and effective Pap stain. LabMedicine. 2006;37(2):105–8.

15. Gill GW. Gill hematoxylins: first person account. Biotech Histochem. 2010;85(1):7–18.
16. Scott MD. Liability issues with the Papanicolaou smear: an insurance industry perspective. Arch Pathol Lab Med. 1997;121(3):239–40.
17. Mallory FB. A contribution to staining methods. J Exp Med. 1900;5:15–20.
18. Singer M. Factors which control the staining of tissue sections with acid and basic dyes. Int Rev Cytol. 1952;1:211–56.
19. Creative Waste Solutions, Inc. Alcohol Recycling—BTARS (Bench Top Alcohol Recycling System). Available at http://www.cwsincorp.com/products/ars.htm. Accessed 22 Apr 2012.
20. Suncycle Technologies, Inc. GRAC (Gravity Recycling Alcohol Cartridge) System. Available at http://www.suncycletech.com/Products.htm. Accessed 22 Apr 2012.

Chapter 11
Cross-Contamination Control

I have known a great many troubles, but most of them never happened.
Mark Twain

PRINCIPLE

Specimens represent patients: get the right answer on the right slide to the right patient.

PRACTICE

Be vigilant to the possibility of cross-contamination.

Historic Milestones

- 1954—Papanicolaou describes contamination of smears and criteria to recognize "floaters" and recommends ways to reduce the chance of contamination.[1]
- 1963—Graham: *"…likeliest place for contamination to occur is in the mounting of the slides."*[2]
- 1967—CLIA '67 mentions cross-contamination.[3]
- 1968—Koss identifies three common causes of contamination.[4]
- 1970—Barr investigates mechanism of contamination.[5]

G.W. Gill, *Cytopreparation: Principles & Practice*,
Essentials in Cytopathology 12, DOI 10.1007/978-1-4614-4933-1_11,

© Springer Science+Business Media New York 2013

- 1975—Gill describes a cross-contamination control and stain storage system.[6]
- 1987—Wall Street Journal publishes three Bogdanich articles that spark CLIA '88.[7–9]
- 1988—CLIA '88 continues to include to cross-contamination.[10]

Cross-contamination in cytology means cells being transferred from their original slide preparation to another "foreign" slide. It is particularly likely when Pap test slides are stained in the same set of stains and rinse solutions as non-gyn specimens, particularly malignant effusions. Cross-contamination introduces the unlikely possibility of a true negative Pap test becoming a false positive.

In my experience, the possibility of cross-contamination is real. However, the likelihood of cellular cross-contamination contributing to a false positive is vanishingly small. In addition, I am unaware of literature reports of cross-contamination-based false-positive result being reported for any specimen. Nonetheless, that cross-contamination control practices are memorialized in CLIA '88 and in the College of American Pathologists cytopathology checklist compels me to address the issue.

Papanicolaou on Cross-Contamination[1]

In his 1954 Atlas of Exfoliative Cytology, page 12, Papanicolaou wrote "*Contamination of smears.*" In the process of staining, cells, cell clusters or even small parts of the film of a smear ("floaters") may become detached from a slide and adhere to other slides stained in the same container. This is a rare occurrence yet it may lead to a false-positive evaluation if positive "floaters" adhere to a negative smear. Detachment of cells or of cell clusters occurs more often when smears are unusually rich, as is often the case with smears from exudates. One should always be aware of the possibility of contamination when malignant cells are found in a relatively small area of an otherwise negative smear and in only one of the slides prepared from a specimen.

The following are some of the criteria by which one can recognize a "floater":

1. It is usually on a plane of focus higher than that of the remainder of the smear, and often on top of other cells.
2. Its staining often differs from that of the rest of the smear.
3. It may carry with it a substratum which contrasts with that of the host smear.

The following recommendations may help to reduce the chance of contamination.

1. Any movement of slides in and out of solutions should be done slowly and gently and with as little agitation as possible.
2. All solutions used in the staining procedure should be replaced frequently and the jars washed thoroughly.
3. Crowding of the slides within the staining jars should be avoided.
4. Stains should be filtered before re-use.
5. Large and deep staining jars allowing free space beneath the slides are recommended.
6. Since detachment of floaters occurs more frequently in exudates, it is advisable to stain exudate smears in separate jars and renew the solutions more frequently."

On page 20 of the same 1954 Atlas, Papanicolaou described three instances of false positives due to contamination that was unrelated to technical processing: (1) a ureteral specimen that had been contaminated with cells from a small papillary carcinoma of the bladder situated near a ureteral orifice, (2) a nasopharyngeal washing was contaminated by bronchial discharge from a reticulum cell sarcoma of the lung, and (3) a gastric specimen was contaminated by cancer cells of the lung. "Dust cells intermixed with cancer cells was the only clue to the bronchial origin of the malignant cells."

CLIA '88 §493.1274 Standard: Cytology[10]

As a practical matter, cross-contamination has always been non-problematic and certainly did not merit regulatory attention. Nonetheless, CLIA '88 addresses cross-contamination as follows:

(2) Effective measures to prevent cross-contamination between gynecologic and nongynecologic specimens during the staining process must be used.

(3) Nongynecologic specimens that have a high potential for cross-contamination must be stained separately from other nongynecologic specimens, and the stains must be filtered or changed following staining.

How Did CLIA '88 Become "Cross-Contaminated"?

Given that Papanicolaou had described cross-contamination as a potential problem in 1954, and that no published reports have ever identified it as a real problem, why was it included in CLIA '67 and retained in CLIA '88? My best guess is that when the 1987 publication of the 3 Wall Street Journal articles highlighted previously unpublicized, serious lab deficiencies,[7–9] CMS officials decided to leave well enough alone. Why rock the protect-the-public boat?

Those articles shined unflattering light on some cytology laboratories that "pay screeners on a piecework basis that encourages them to rush the analysis."[8] Some cytotechnologists screened hundreds of conventional Pap tests in a short time to earn a living wage. Mistakes were made; women died from undetected cervical cancer (i.e., false negatives). None died from false positives caused by cross-contamination, which wasn't even mentioned in the articles. CLIA '88 ignored piecework pay but continued to include cross-contamination control in the regulations. President Ronald Reagan signed CLIA '88 into law on Halloween, Wednesday, October 31, 1988.

Parenthetically, the letter A means Act in CLIA '67 and Amendments in CLIA '88.

A brief review of pertinent literature justifies removing cross-contamination control from the next revision of CLIA '88.

Graham[2] wrote, "In our laboratory, contamination has been a problem, since **50%** [emphasis added] of the slides that go through the staining procedure are positive.... We believe the likeliest place for contamination to occur is in the mounting of the slides. If the technician touches the balsam rod to a smear, cells that are not firmly fixed to the slide may stick to the rod, to be placed back in the balsam bottle, where they remain perfectly preserved. For this reason, it is most important that the balsam be placed on the underside of the cover slip and *not* on the slide."

Koss[4] wrote 5 years later: "Contamination of one specimen with cells from another, which may infrequently occur in a cytology laboratory, should be investigated at once. It may mean faulty or careless techniques, insufficient cleansing of glassware or other reasons. Some of the common causes are:

1. Failure to filter solutions daily or as often as required.
2. Poor adherence of specimens to slides due to lack of albumin coating or to spreading the specimen too thickly.
3. Contamination of mounting media (the dropper or glass rod used to dispense the mounting media may have been allowed to touch the surface of the smears)."

Barr, Powell, and Raffan[5]—2 years after Koss and 3 years after CLIA '67—investigated possible mechanisms of contamination. "Slides smeared with glycerine/egg albumen were used as receptors. Routine cervical smears or peritoneal fluid known to contain numerous adenocarcinoma cells were used as sources." He found contaminated slides in batch manual staining and automated staining. "However, the incident of contamination described above, and the disturbing frequency of cell transfer in the automatic staining machine baths prompted us to abandon its use, that is, until manual batch staining methods were subjected to the same scrutiny, which showed that the incidence of transfer was appreciably higher. It is clearly impossible to process each slide individually, so that an automatic machine method is justifiable, if not essential, provided precautions are taken."

Gill[6] wrote, "In the course of mid-1960s seeding experiments, we found we had a problem with cross-contamination by floaters. We devised an experiment that without question showed our filtration of staining solutions through qualitative filter paper (Eaton-Dikeman, Grade 617) only partially filtered out cells. Many cells were not removed but remained suspended in the "filtered" stains. Reacting to these findings, we devised a filtration system to completely removed cells and debris from stains and other solutions. The system has other advantages."

The qualitative grade filter paper we had been using consists of cellulosic fibers pressed together, leaving gaping pores through which cells and occasional tissue fragments can pass. See Fig. 11.1. I filtered an alcoholic suspension of tissue-cultured cancer cells (HeLa) through laboratory grade filter paper and refiltered the filtrate through a 5-μm pore size Millipore filter. The tip of the filter paper was cut out and together with the whole Millipore filter, stained, mounted, and examined microscopically. The filter paper contained some cells, but the Millipore filter contained more. See Fig. 11.1.

The cross-contamination control system described below uses Millipore filters to effectively remove all floaters from suspension. It also allows evaporation-free storage of the filtered stains and subsequent convenient dispensing. See Fig. 11.2.

Materials

Vacuum Source

- In-house vacuum line, portable electric pump, or aspirator filter pump
- 1-L filter flask as trap bottle
- Connecting plastic tubing
- 2 Y-connectors

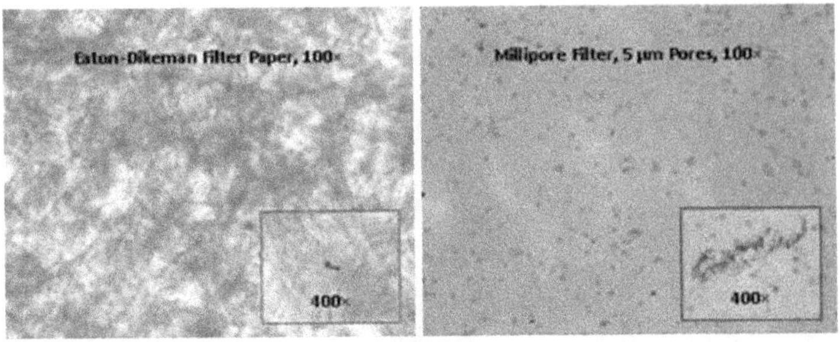

Fig. 11.1. The pores in filter paper vary in size and are larger than the largest cells. A suspension of 600,000 HeLa cells was added to the filter paper and allowed to filter at gravity. The filtrate was refiltered through a 5-µm pore diameter Millipore filter, which retained all the cells that had passed through the coarsely porous filter paper. The original filters were prepared on April 18, 1968, and photomicrographed on February 10, 2012.

Stain Storage

- 1-L separatory flask, globe or pear-shaped, glass or Teflon, with Teflon plug
- 2-hole stopper
  - First hole: 3-way stopcock, side arm connected to trap bottle; free end for vent
  - Second hole: right-angle tubing, connected to Millipore Swinnex 47-mm filter holder
- Support (5 × 7-in. base with 20-in. rod, ring support, small utility clamp—not shown)

Filtration

- Millipore Swinnex 47-mm filter holder (Cat. No. SX 047 00) with:
  - Millipore type AP 47-mm prefilters (Cat. No. AP25 047 00)

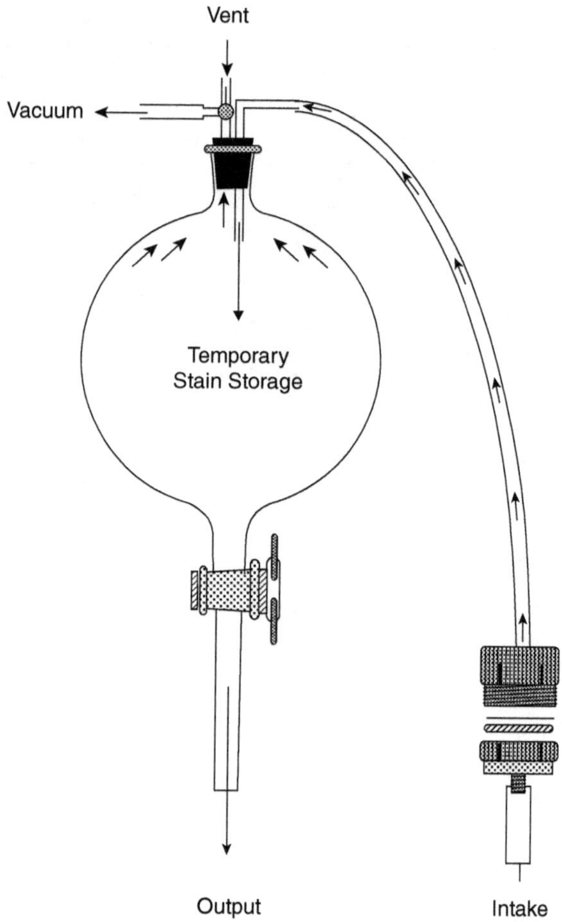

FIG. 11.2. This cross-contamination control system relies on Millipore filters to remove all cells and particles.

- Millipore hydrophilic Durapore 47-mm filters, 5-μm pores (Cat. No. SVLP 047 00)
- Millipore parts kit, Swinnex 47 (Cat. No. SX 00 047 RK)

• 3-in. long plastic tubing to dip into stain for intake into system

The parts were purchased separately from several vendors. Enough parts were ordered to assemble three stain storage subsystems, one each for hematoxylin, OG, and EA.

Methods

The apparatus conveniently: (1) absolutely removes cells and debris from any solution, (2) stores the filtered solution air-tight until ready for use, and (3) dispenses the particle-free stain into a clean staining dish.

To filter stains, turn the stopcock of the separatory funnel to its horizontal shutoff position, and rotate the lever of the 3-way stopcock to the 3 o'clock position to access vacuum. Immerse the intake tubing end in the stain to filter the stain into the separatory funnel. Stop filtration by turning the lever of the 3-way stopcock to the 12 o'clock position to vent the system and/or by turning off the vacuum source.

Once filtered into the separatory funnel, a stain may remain there until needed again. Air-tight storage is assured by turning the lever of the 3-way stopcock off to stop vacuum. If exposed to vacuum overnight, half the volume will evaporate.

To empty the separatory funnel, turn the lever of the 3-way stopcock to the vent position, and open the lower stopcock as shown in Fig. 11.2. The staining dishes are kept under the separatory funnels, not in line with the other staining dishes.

Filtration may be slow to start when filters are still wet from previous use. Wet pores require substantial vacuum to break loose the liquid held in the pores. Once the pores are open again, filtration proceeds smoothly. In other words, slow-starting filtration may mimic an overloaded filter that needs replacement. Replace the filters only when filtration becomes labored after several uses. Prefilters extend the useful life of the more expensive Durapore filters.

The price of the component parts has increased substantially since this system was first described. In 1975, $1.00 bought as much as $4.18 buys in 2011. Nonetheless, this cross-contamination control system fulfills the spirit and the intent of the related CLIA regulation.

Discussion

Conventional Pap test preparations were the standard cervical cytology preparation until 1996, when liquid-based preparations were first FDA-approved. Conventional Pap tests were often thick and served as magnets or docking stations for floaters. Cross-contamination was considered something new and interesting and attracted attention. Keep in mind, however, that cross-contamination is a potential problem *only* when a Pap test is contaminated with adenocarcinoma tissue fragments or malignant cellular clusters from an effusion. Under those circumstances, the morphology of the malignant floaters is so unlike any adenocarcinoma that arises in the female genital tract that the floaters are instantly recognizable for what they are: floaters.

As is often the case historically, understanding was incomplete, and overreaching solutions were normal. Graham wrote that "50% of the slides that go through the staining procedure are positive." She believed that touching the tip of a balsam rod to a preparation contaminated the balsam bottle when the rod was put back into the bottle and used for another preparation. She suggested two possible prevention strategies: (1) coating slides with celloidin and (2) placing the balsam on the underside of the coverslip and *not* on the slide. Given our current understanding, her explanation is off the mark and unconfirmed by microscopic observation. Microscopically examining a few unstained drops applied to a slide would have allowed her to see whether cells were present.

Koss supported Graham's notion in his third point about common causes of cross-contamination and also introduced filtering solutions daily. Barr introduced glycerine/albumin-coated slides as receptors to monitor cross-contamination.

Gill described a cross-contamination control system that was developed in an experimental environment that was attempting to quantify cell recovery. The system was transferred into the routine cytopreparatory laboratory, where it's still used 40 years later.

Husain incorporated Barr's receptor slides into his study of cross-contamination of cytological smears, with automated staining machines and bulk manual staining procedures. Like Gill, he used Millipore filters to demonstrate "that the earlier wash fluids and solutions contained large number of all types of cells, but after the haematoxylin wash the fluids contained much smaller numbers of cells." Also "The findings could be interpreted as suggesting that the cells that will come off will do so in the earlier pots, but it has been found that rapid dehydration tends to pull off cells more than if the dehydration is gradual."[11]

Floaters Happen

Floaters are associated virtually exclusively with malignant effusions. Why? See Fig. 11.3.

CLIA '67 devoted 61 words and 112 characters to preventing cross-contamination; CLIA '88, 46 words and 117 characters. CLIA '88 interpretive guidelines attempt to clarify what labs should do to prevent cross-contamination.[12]

§493.1274 Standard: Cytology.

(b)(2) Effective measures to prevent cross-contamination between gynecologic and nongynecologic specimens during the staining process must be used.

Interpretive Guidelines §493.1274(b)(2)

The laboratory must develop its own policies and procedures for the prevention of cross-contamination between gynecologic and nongynecologic specimens. The majority of gynecologic specimens are fixed prior to transport to the laboratory. Staining times may differ between gynecologic and nongynecologic specimens. Commonly used methods include separate staining dishes for various specimens (i.e., gynecologic specimens, CSF, sputa, other body fluids) or separate staining

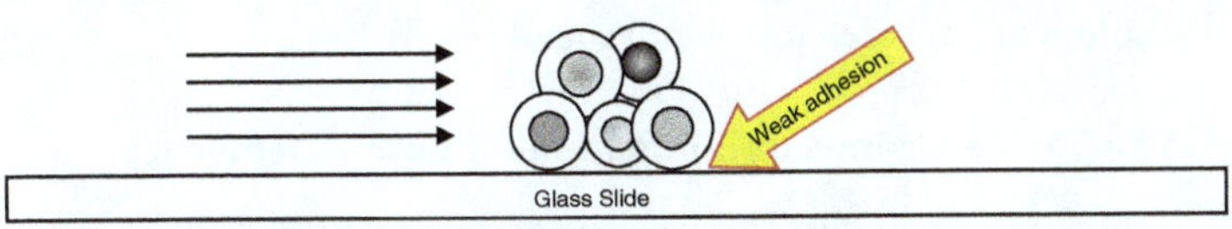

Fig. 11.3. Floaters are thick, not thin. Their physical thickness resists flattening onto a glass surface, which translates into a weak bond. Coupled with their high profile, cell clusters are easily sheared off the face of the slide in the process of staining and rinsing and begin their floater existence.

times (i.e., gynecologic specimens in the morning and nongynecologic specimens in the afternoon), with the staining dishes washed and stains filtered between staining times.

Probes §493.1274(b)(2)

What does the laboratory do to ensure that cross-contamination between gynecologic and nongynecologic specimens does not occur?

§493.1274 Standard: Cytology.

(b)(3) Nongynecologic specimens that have a high potential for cross-contamination must be stained separately from other nongynecologic specimens, and the stains must be filtered or changed following staining.

Interpretive Guidelines §493.1274(b)(3)

A monochromatic stain such as toluidine blue may be used to determine the cellularity of nongynecologic specimens. Once a specimen has been concentrated, usually by centrifugation, a small drop of specimen is placed on a slide. A drop of stain is placed next to the specimen, allowed to mix, and coverslipped. Cellularity is evaluated microscopically. Highly cellular specimens have a high potential for cross-contamination. One option would be for the laboratory to stain these specimens after routine staining has been completed.

Laboratories which use automated staining methodologies must follow the manufacturer's instructions. Use D5411.

Probes §493.1274(b)(3)

How is the cellularity of nongynecologic specimens checked prior to cytopreparation (staining)?

What procedure does the laboratory use to determine which specimens must be stained separately?

In its cytopathology checklist, the College of American Pathologists includes this requirement:[13]

CYP.07680—Cross-Contamination—Phase I Deficiency

There are procedures to prevent cross-contamination of specimens during processing and staining.

NOTE: Procedures must prevent cross-contamination between gynecologic and non-gynecologic specimens.

Also, procedures must prevent contamination among non-gynecologic cases when highly cellular specimens are processed. Methods to minimize this potential problem may include cytocentrifuge, filter, and monolayer preparations. Direct smears made from the sediment of highly cellular cases should be stained after the other cases, and the staining fluids must be changed or filtered between each of the highly cellular cases. One procedure to detect highly cellular specimens is to use a toluidine blue, or other rapid stain, on a wet preparation. One procedure to detect possible contamination is to insert a clean blank slide in each staining run and examine it for contamination.

In 2009, CYP.07680 deficiencies were the third most often cited in CAP laboratory inspections. "Is there a documented policy for ensuring that non-gynecologic specimens with a high potential for cross-contamination are processed and stained separately from other specimens?" was asked 1591 times. There were 41 citations (2.7%).[14]

Particularly note that logical follow-up questions aren't asked, such as: do you keep records of how many times cross-contamination was observed? Are you aware of any false-negative results being reported? Were any patients harmed? Details?

Cross-contamination impacts patient care if, and only if, a clinician treats his patient based on false-positive cervical cytology results. A highly unlikely string of even more unlikely events would have to happen: (1) a cluster of cancer cells, usually adenocarcinoma in an exudate, must fall off a slide during processing; (2) attach to a NILM cervical cytology preparation, which is probably liquid-based; (3) survive processing in a recognizable form; (4) not be recognized as an uninvited guest; and (5) be reported falsely by a pathologist. Detecting possible contamination by inserting a clean blank slide in each staining run and examining it for contamination add work without value. A positive receptor slide doesn't guarantee a false-positive Pap test.

What is the likelihood of that chain of events happening? Take Graham's scenario, for example. How many cancer cells would have to be present only in the area that the balsam rod touched to enable that chain of events? Nonetheless, we were taught to avoid touching the tip of a glass balsam bottle rod to any preparation when we were mounting slides.

The cited publications describe time and circumstances different than today.[2–5] All described cellular cross-contamination as a phenomenon, but none explained how it happens. Continued inclusion of preventing cross-contamination in CLIA '88 is unwarranted and unjustified. It attempts to micromanage a nonproblem. It is highlighting a perceived problem that does not impact patient care. By so doing, it undermines the credibility of CLIA '88 and its anonymous authors. If they're wrong about this, what else are they wrong about?

Conclusion and Recommendations

The best way to prevent cross-contamination is to constantly recognize its possibility. Cytotechnologists alerting one another in the same laboratory when contamination is suspected works. All the regulations in the world pale in effectiveness compared to a well-trained staff. My recommendations include:

Don't Do Unnecessary Work

- If only gyn preparations are processed, cross-contamination is not a problem.
- Any slides being evaluated immediately for adequacy aren't a threat.
- Acellular non-gyn specimens don't need special treatment.
- Non-gyn specimens that don't have abnormal cells also don't need special care.
- Adding dye of any kind[15] is unnecessary to see whether a non-gyn specimen has high potential for cross-contamination.

Do Necessary Work

- Microscopically examine a drop of coverslipped, unstained non-gyn cell suspension. Close the microscope's substage condenser's aperture diaphragm slightly to increase diffraction that makes the cells visible. Using a 10× objective with 10× eyepieces, one can readily evaluate the risk of cross-contamination: are abnormal cell clusters or tissue fragments present? If not, don't worry, be happy.

- If cross-contamination is a risk, use an absolute filtration system such as that seen in Fig. 11.2. Don't use coarse filter paper. It's ineffective.
- Avoid undue forceful agitation (e.g., running water, rapid dipping).
- Replace non-stain solutions frequently.
- Stain gyn specimens before non-gyn preparations.
- Stain high-risk shedder specimens (e.g., body cavity fluids) last.
- Stain known shedders in separate miniature Pap stain setups—using nearly exhausted solutions that can be discarded afterward where appropriate.
- Filter xylene baths through laboratory grade filter paper to remove particulates.

Cross-contamination-related language should be expunged from the next revision of CLIA '88.

References

1. Papanicolaou GN. Atlas of exfoliative cytology. Cambridge, MA: Published for The Commonwealth Fund by Harvard University Press; 1954. p. 12.
2. Graham RM. The cytologic diagnosis of cancer. 2nd ed. Philadelphia, PA: W. B. Saunders Company; 1963. p. 347.
3. §493.1257 Standard; Cytology. (a)(2) and (3). Federal Reg. 1990; 55(50):9599.
4. Koss LG. Diagnostic cytology and its histopathologic bases. 2nd ed. Philadelphia, PA: J. B. Lippincott Company; 1968. p. 604–5.
5. Barr WT, Powell DE, Raffan JB. Cellular contamination during automatic and manual staining of cytological smears. J Clin Pathol. 1970;23(7):604–7.
6. Gill GW. A cross contamination control and stain storage system. Cytotechnol Bull. 1975;12(3):12–3.

7. Bogdanich W. Physicians' careless with Pap tests is cited in procedure's high failure rate. Wall St J. 29 Dec 1987.
8. Bogdanich W. Lax laboratories. The Pap test misses much cervical cancer through labs' errors. Cut-rate 'Pap mills' process slides with incentives to rush. Misplaced sense of security? Wall St J. 2 Nov 1987.
9. Bogdanich W. False negative. Medical labs, trusted as largely error-free, are far from infallible. Haste, misuse of equipment, specimen mix-ups afflict even best labs at times. Regulation; weak and spotty. Wall St J. 2 Feb 1987.
10. Laboratory Requirements. Code of Federal Register. Title 42, Ch. IV, Part 493.1274: Cytology, 10-1-2010 ed.
11. Husain OA, Grainger JM, Sims J. Cross contamination of cytological smears, with automated staining machines and bulk manual staining procedures. J Clin Pathol. 1978;31(1):63–8.
12. Centers for Medicare and Medicaid Services. Interpretive guidelines for laboratories and laboratory services, Subpart K, Part 2. http://www.cms.gov/CLIA/downloads/apcsubk2.pdf. Accessed 8 Feb 2012.
13. College of American Pathologists. Cytopathology checklist. Northfield, IL: CAP Accreditation Program; 07.11.2011 Revision.
14. Mody DR, Henry M. The rules, tools and jewels of the cytopathology lab inspection. LAP Audioconference. 06.17.2009. http://www.cap.org/apps/docs/education/lapaudio/pdf/061709_presentation.pdf. Accessed 10 Feb 2012.
15. Fischer AH. Complying with cross-contamination question. CAP Today. 2010;24(5).

Chapter 12
H&E Stain

PRINCIPLE NO. 5

Stain to facilitate cell visibility, detection, and interpretation.

PRACTICE

Use chemically competent dyes and rinses for times that promote desired outcomes and replace as needed.

Historic Milestones

- 1865—Böhmer combined hematoxylin with mordant[1]
- 1875—Fischer introduced eosin Y[2]
- 1876—Wissowzky introduced H&E[3]

Böhmer and Fischer independently introduced the stains hematoxylin and eosin in 1865 and 1875, respectively.[1,2] In 1867, Schwarz introduced the first double-staining technique using successive solutions of picric acid and carmine.[3] With the idea of a double-staining technique already published, it wasn't difficult for Wissowzky to describe the combination of the hematoxylin and eosin (H&E) dyes in 1876.[4] All four authors published their articles in German journals, two in the same journal, which may account for the relative rapidity of communication and development in those pre-Internet times more than a century ago.

G.W. Gill, *Cytopreparation: Principles & Practice*,
Essentials in Cytopathology 12, DOI 10.1007/978-1-4614-4933-1_12,
© Springer Science+Business Media New York 2013

"Simple is best," and so it is that H&E has stood the test of time. Even today, 136 years after being introduced, H&E is still the most frequently used staining method in anatomical pathology worldwide.[5] Simple though it may be, however, H&E staining doesn't always produce satisfactory outcomes. Successful results are based on successful processes, which include the stains, rinses, and staining times.

Hematoxylin

See Chap. 10.

Eosin

Eosin Y is the basis for eosin stains, see Fig. 12.1. Although classically named eosin (*eos*, meaning dawn; Y, yellowish), it is also known as eosin G, Bromo acid, J, TS, XL, or XX, Bronze bromo ES, and tetrabromofluorescein.[6] Its solubility in water at room temperature far exceeds the amount used in any eosin stain solution. Similarly, its solubility in alcohol also exceeds the amount ever used in alcohol but is a fraction of its solubility in water (i.e., 2.18% vs. 44.2%). This difference can be used to advantage

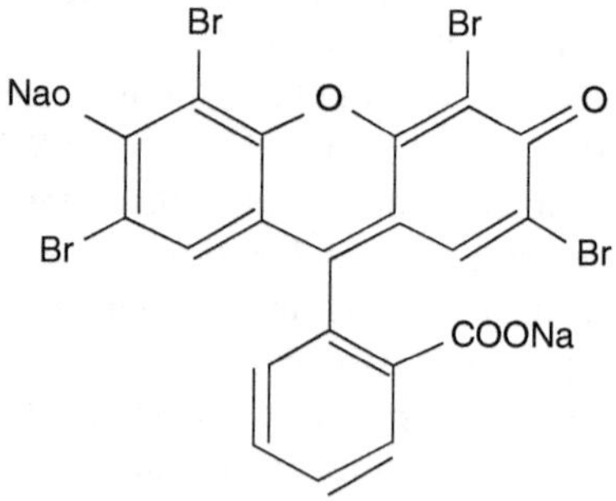

FIG. 12.1. Eosin Y (tetrabromofluorescein), C.I. 45380, molecular weight 691.863.

when preparing stain solutions by making concentrated aqueous stock solutions of eosin. American companies have access to eosin dyes that have been certified by the Biological Stain Commission, which requires—among other things for eosin Y—minimum dye content of 90%.

To ensure quantitative consistency, the amount of dye used for any particular eosin formulation must be based on total dye content (TDC) and be adjusted as needed. For example, if 100 l of 0.5% TDC eosin Y is being prepared commercially using 90% dye content eosin, 555 g—not 500 g—must be added (i.e., 500/0.9 = 555 g). The extra 55 g is a variable mix of unidentified salts and impurities that are found in most, if not all, biological stains and dyes. Otherwise, using 500 g of 90% dye content will result in 450 g eosin Y being used, which equals a 0.45% (w/v) eosin Y solution. I don't know whether vendors correct for dye content.

To those rare individuals who prepare their eosin solutions from scratch, it is recommended they make no less than 500 mL of an aqueous 20% (w/v) TDC eosin Y stock solution. Using aqueous stock solutions saves time and facilitates dissolving the dye in alcohol. Table 12.1 shows a variety of eosin formulations.

TABLE 12.1. Eosin Y stain solution variants in order of increasing staining strength.

Reference	Eosin Y (g)	Water (mL)	95% ethanol (mL)	Acetic acid (mL)	Staining time
McManus[7]	0.5	30	70	0	2–3 min
Lillie[8]	0.5	100	0	0	1 min
Disbrey/Rack[9]	1.0	100	0	0	5–10 dips
AFIP[10]	0.25	20	80	0.5	15 s
JHMI[11] [a]	0.5	30	70	0.5	1 dip
Carson/Hladik[12]	2.5	33	67	0.5	10–20 dips

[a]JHMI = The Johns Hopkins Medical Institutions, Baltimore, Maryland. References are in superscript.

Composition differences that promote eosin uptake in cells:

- Increasing eosin concentration (from 0.5 to 2.5 g [note a 5-fold range])
- Water rather than alcohol
- Acetic acid
- Staining times (1 dip to 3 min). Acetic acid acts as an accentuator that dramatically shortens staining times. If eosin overstains, it can be removed by differentiation in alcohol until the desired color density is reached

Note that glacial acetic acid is included in 3 of the 6 eosin stain formulations in Table 12.1. The impact of acetic acid on the uptake of acid (negatively charged) dyes such as eosin is immense. Fig. 12.2 illustrates the mechanism.

Phloxin B (C.I. No. 45410)[1] is sometimes added to eosin formulations (e.g., 0.5 g/L eosin stain) to increase the range of red

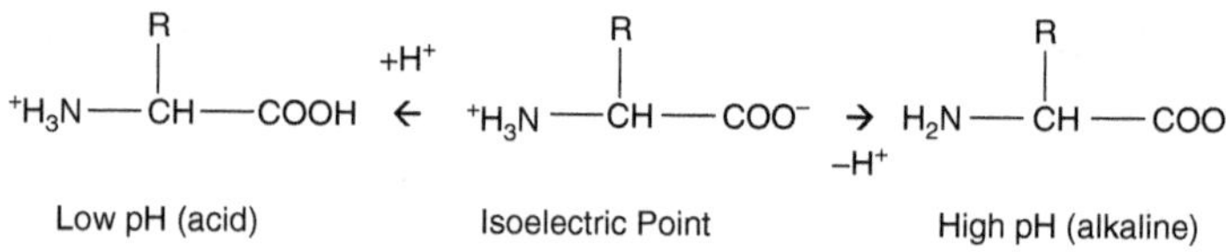

FIG. 12.2. Amino acids united by peptide linkages make up proteins, which are all that remain in cytoplasm following fixation in formalin. If an amino acid in solution is placed in an electric field, as in electrophoresis, the molecules will migrate to one pole or the other in accordance with the pH of the solution. At a certain pH, which is unique to the particular protein, the amino acid does not migrate to anode or cathode. This pH is the isoelectric point. Adding glacial acetic acid (i.e., low pH) neutralizes the COO– groups and leaves relatively more positively charged H_3N groups. As a result, eosin Y molecules, which are negatively charged, are attracted to the positively charged groups and thus are taken up faster and in greater total amounts per given amount of time.[13]

[1] C.I. numbers are 5-digit numbers assigned by The Society of Dyers and Colourists (http://www.sdc.org.uk/) to uniquely identify stains with the same chemical composition but different names. These 5-digit numbers must be specified when publishing or purchasing dyes to ensure using the same dye, even if identified by different names.

colors. However, phloxin B is exceedingly "bright" and can be visually overpowering if too much is used. Therefore, one needs to be cautious when using phloxin B. One successful formula for preparing eosin is as follows:[11]

Biebrich scarlet (ws [water soluble])	C.I. No. 26905	0.4 g
Eosin Y	C.I. No. 45380	5.0 g
Phloxin B	C.I. No. 45410	2.1 g
95% ethanol		200 mL
Distilled water		800 mL

To see what H&E stain colors should look like when applied together to the same tissue section at the same time, stain a section in hematoxylin only and another in eosin only using the same solutions and times as in the routine method. This approach allows one to see what each stain looks like without any interference from the other. Sections stained in H&E that don't display the pure colors seen in singly stained sections should trigger troubleshooting of the method. When H&E outcomes go awry, it is usually because too much of one stain or the other has been taken up, or removed, or a combination of the two.

Hematoxylin may be applied progressively or regressively, depending on the concentration of the hematoxylin formulation. Apart from the particular hematoxylin formulation and associated differences in staining times, as well as the addition of an acid bath and related rinses, the 2 H&E staining methods are almost identical (Table 12.2). Table 12.2 begins with paraffin-embedded fixed tissue sections that have already been deparaffinized (also referred to as dewaxed or decerated) and are ready to be stained:

Notes

- Rinses may also be referred to as baths or washes.
- Rinses remove traces of previous solutions; they prepare the sections for the next solutions that are different.
- Dipping sections in each rinse promotes the exchange of solutions. Standing rinses are discouraged. A dip is fully submersing sections in, and removing them from, each rinse.

TABLE 12.2. Progressive and regressive H&E staining methods.

Step	Progressive times	Solution	Regressive times	Purpose
1.	10 dips	Tap water	10 dips	Hydrate
2.	10 dips	Tap water	10 dips	
3.	10 dips	Tap water	10 dips	
4.	Gill-2 × 2 min	Hematoxylin	Harris × 6 min[a]	Color nuclei
5.	NA	Tap water	10 dips	Rinse
6.	NA	Tap water	10 dips	
7.	NA	0.5% HCl in 70% EtOH[a]	10 dips	Differentiate
8.	10 dips	Tap water	10 dips	Rinse/blue/rinse
9.	10 dips	Tap water	10 dips	
10.	10 dips	Tap water	10 dips	
11.	10 dips	Tap water	10 dips	
12.	1–2 dips	0.5% (w/v) eosin Y[a]	1–2 dips	Color tissue and nucleoli
13.	10 dips	Tap water	10 dips	Rinse
14.	10 dips	Tap water	10 dips	
15.	10 dips	Tap water	10 dips	
16.	10 dips	Absolute ethanol	10 dips	Dehydrate
17.	10 dips	Absolute ethanol	10 dips	
18.	10 dips	Absolute ethanol	10 dips	
19.	10 dips	Xylene	10 dips	Clear
20.	10 dips	Xylene	10 dips	
21.	10 dips	Xylene	10 dips	

[a]See Notes for details.

- For maximum effectiveness, rinses should be in sets of three, kept deep, and clean.
- If used repeatedly without being changed, rinses become less effective. For example, rinses unchanged following eosin become dye solutions themselves. When the concentration of dye in the rinse equals that in the tissue, the eosin cannot escape the tissue, which results in "muddy" staining results.

- The amount of stain that remains in a tissue represents the difference between the amount deposited by the stain solution and the amount removed by the rinse.
- One-step hydration and dehydration work satisfactorily. Graded alcohols are unnecessary.
- Gill hematoxylin No. 2 is recommended. It is a progressive stain with high staining capacity.
- Harris hematoxylin[2] is available in four different formulations of decreasing strength in the following order: (1) full-strength without acetic acid, (2) full-strength with acetic acid, (3) half-strength without acetic acid, and (4) half-strength with acetic acid. The stronger formulations (1 and 2) stain regressively; the weaker formulations (3 and 4), progressively.
- Regardless of the hematoxylin, whether it is Gill, Harris, Mayer, Ehrlich, Delafield, etc., the finished results should be virtually identical in terms of color, optical density (i.e., light, dark), and distribution (i.e., nucleus vs. cytoplasm).
- Gill and Harris hematoxylins are used as examples because I'm more familiar with them.
- Thin sections will stain less optically dense than thick sections when both are stained for the same length of time.
- Differentiation is a portmanteau for differential extraction.
- 0.5% HCl in 70% ethanol is prepared by adding 5 mL concentrated HCl to 995 mL 70% ethanol. Using a higher concentration of HCl (e.g., 1%) can extract excess hematoxylin rapidly and result in understaining, especially if the acid is mixed with water only. Seventy percent ethanol slows the rate of decolorization.
- Overdifferentiation (removing too much dye) in HCl is a potential limitation when using regressive hematoxylin formulations.
- Most tap water sources will "blue" hematoxylin. A chemically defined bluing agent (e.g., Scott's tap water substitute) isn't necessary.

---

[2] Harris and Gill are the only currently marketed hematoxylin formulations in America.…Further, their names are the only ones cited in the Clinical Laboratory Improvement Amendments (CLIA '88) interpretive guidelines: "Stains used (i.e., Harris, Gill or other type of hematoxylin…"[14]

- There are many eosin stain formulations (Table 12.1). The one described here is comprised of 5 g (total dye content) eosin Y (C.I. No. 45380), 5 mL glacial acetic acid, and 995 mL 70% ethanol.
- No appreciable fading occurs in preparations stained and rinsed well. Slides stored in the dark do not fade even after more than 35 years. Fading is defined as any change in color and not merely a weakening of the shade.[15]

Results

Ideally, hematoxylin should color chromatin blue. Depending on the mordant, mucin may also be colored blue. The depth of color (i.e., optical density) should be deep enough to make small particles visible and shallow enough to not obscure fine details. Cytoplasm should be colored scarcely at all.

Eosin should color nucleoli red and stain cytoplasmic structures with varying shades of red to pink. When present, erythrocytes and cilia should also be colored with varying shades of red to pink (Fig. 12.3).

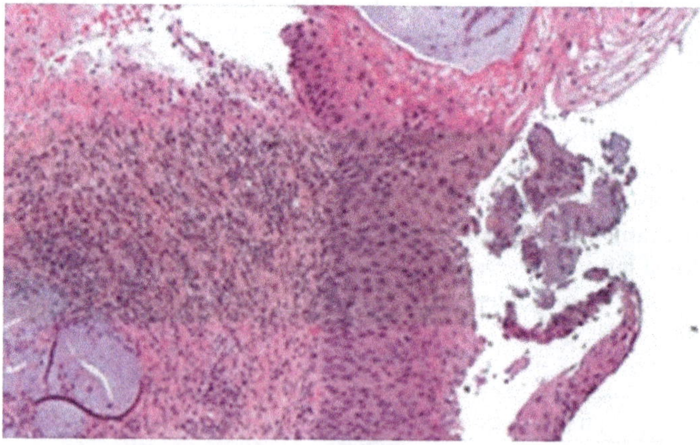

Fig. 12.3. H&E stained sectioned biopsy of uterine cervix with marked dysplasia (precancerous changes). ×100 (original magnification).

It is interesting that few laboratories, if any nowadays, prepare hematoxylin and eosin staining solutions from "scratch." Stains are bought readymade. They are prepared by vendors with varying degrees of staining knowledge and quality assessment programs. Not all stain solutions with the same name prepared by different vendors perform the same. It is also to be noted that those who perform H&E staining are not the same individuals who interpret the microscopic morphology. Given these two practical realities, the opportunities for things going wrong are plentiful.

While the exact appearance of an H&E-stained section will vary from lab to lab, results that meet the individual user's expectations are considered satisfactory. This means, of course, that others with different expectations may conclude otherwise. Quality has many definitions and is context dependent, but a practical working definition is "the result useful for its intended purpose." If the user can see what s/he needs to see to interpret the tissue, the H&E results are functionally satisfactory. This is not necessarily the same as technically satisfactory, in which an experienced observer can see no technical deficiencies in the results.

Conclusion

The widespread use of commercially prepared stain solutions such as hematoxylin and eosin has increased user reliance on the manufacturers and decreased user reliance on basic knowledge. An unintended consequence has been a reduced recognition of satisfactory results, an increased tolerance for marginal satisfactory or unsatisfactory results, and an inability to troubleshoot problems. Hence, it is essential that users immerse themselves in basic knowledge about staining materials and methods, so they can control the quality of results.

References

1. Böhmer F. Zur pathologischen Anatomie der Meningitis cerebromedularis epidemica. Aerztl Intelligenzb (Munich). 1865;12:539–50.
2. Fischer E. Eosin als Tinctionsmittel für mikroskopische Präparate. Archiv für mikroskopische Anatomie. 1875;12:349–52.

3. Schwarz E. Uber eine Methode doppelter Färbung mikroskopischer Objecte, und ihre Anwendung zur Untersuchung der Musculatur des Milz, Lymphdrusen und anderer Organe. Sitz Akad W math naturw Cl. 1867;55:671–91.
4. Wissowzky A. Ueber das Eosin als reagenz auf Hämoglobin und die Bildung von Blutgefässen und Blutkörperchen bei Säugetier und Hühnerembryonen. Archiv für mikroskopische Anatomie. 1876;13: 479–96.
5. Cook HC. Origins of tinctorial methods in histology. J Clin Pathol. 1997;50(9):716–20.
6. Lillie RD, editor. H.J. Conn's biological stains. 9th ed. Baltimore, MD: The Williams & Wilkins Company; 1977.
7. McManus JFA, Mowry RW. Staining methods: histologic and histochemical. New York: Paul B. Hoeber Inc.; 1960.
8. Lillie RD. Histopathologic technic and practical histochemistry. 3rd ed. New York: The Blakiston Company; 1965.
9. Disbrey BD, Rack JH. Histological laboratory methods. Edinburgh and London: E. and L. Livingstone; 1970.
10. Luna LG, editor. Manual of histologic staining methods of the Armed Forces Institute of Pathology. 3rd ed. New York: The Blakiston Division; 1969.
11. Gill GW. Troubleshooting eosin staining, and a brighter eosin stain. Micros Today. 1999;8:27.
12. Carson FL, Hladik C. Histotechnology: a self-instructional text. 3rd ed. Chicago, IL: ASCP Press; 2009.
13. Singer M. Factors which control the staining of tissue sections with acid and basic dyes. Int Rev Cytol. 1952;1:211–56.
14. Interpretive Guidelines for Laboratories – Subpart K, Part 2, p 35. Centers for Medicare and Medicaid Services (CMS) Web site http://www.cms.hhs.gov/CLIA/downloads/apcsubk2.pdf. Accessed 5 Feb 2012.
15. Venkataraman K. The chemistry of synthetic dyes, vol. 1. New York, NY: Academic; 1952.

Chapter 13
Romanowsky Stains

The color purple: from royalty to laboratory with apologies to Malachowski.

Krafts, Hempelmann, and Oleksyn

Principle No. 5

Stain preparations to facilitate cell visibility, detection, and interpretation.

Practice

Prepare cell spreads as thinly as the specimen will allow, air-dry, fix in methanol, and stain as recommended for the particular stain.

Historic Milestones

- 1880—Ehrlich prepares the first neutral stain mixture that allows differentiation of blood cells.[1]
- 1891—Malachowski and Romanowsky independently developed stains composed of eosin and "ripened" methylene blue that not only differentiated blood cells but also demonstrated the nuclei of malarial parasites.[2, 3]
- 1902—Wright publishes a "rapid method for the differential staining of blood films and malarial parasites."[4]
- 1904—Giemsa introduces his eponymous stain.[5]

G.W. Gill, *Cytopreparation: Principles & Practice*,
Essentials in Cytopathology 12, DOI 10.1007/978-1-4614-4933-1_13,

© Springer Science+Business Media New York 2013

- 1927—Babeş uses Giemsa stain to diagnose cancer cells in cervical smears.[6]
- 1930—European hematologists use Romanowsky-type stains for fine needle aspirates.[7]
- 1944—Lillie uses Romanovsky as alternative to Romanowsky.[8]
- 1970—Witlin modifies Wright–Giemsa stain that forms basis for Diff-Quik, a commercial Romanowsky stain variant.[9]
- 1972—American Hospital Supply Corporation trademarks Diff-Quik.[10]
- 1975—Marshall uses purified dyes to standardize Romanowsky stain.[11]
- 1978—Lillie shows that Malachowski's 1891 stain preceded Romanowsky's 1891 stain.[12]
- 1993—Baxter Diagnostics renews Diff-Quik trademark. Trademark is currently "dead."[10]

This chapter could have been titled any one of at least 15 names associated with Romanowsky-type stains.[13] The name Romanowsky may properly be Romanovsky. If the historical record were used as the basis for the generic name of these stains, Malachowski would replace Romanowsky. The list of historic milestones is severely abbreviated and is intended to demonstrate how far back in history the development of blood stains can be traced.

Romanowsky stains are similar to the Papanicolaou stain in several ways:

- Both trace their roots to other stains.
- Both were introduced when little was known about biological dyes, their properties, and impurities.
- Both are applied to important medical problems.

Far more specimens of all kinds are Romanowsky-stained than are Pap-stained for obvious reasons. The need is greater, and it's less expensive. Romanowsky stain is included as a chapter in this volume because Diff-Quik, its commercial equivalent, is used so widely in clinical cytopathology for nongynecological cytology specimens and fine needle aspirations.

Advantages of Romanowsky-Type Stains in Routine Cytological Practice

"Common practice dictates the use of both a Romanowsky-type and the Papanicolaou (or hematoxylin and eosin) stain to evaluate most cytological specimens. Because the information provided by each stain is unique and complementary, both types are essential for accurate cytological diagnosis. Although Papanicolaou and hematoxylin and eosin stains show better nuclear detail and generally perform better on thick or extensively necrotic smears, Romanowsky-type stains allow better estimation of relative cell and nuclear sizes and superior visualization of cytoplasmic details, smear background elements, and intercellular matrix components."[7]

Unlike the Pap stain, Romanowsky stains can be standardized. They can be prepared from a mixture of pure dyes in constant proportions.[14] Of all the various dyes used to prepare Romanowsky-type stains, only three are necessary: (1) methylene blue (C.I. 52015), (2) azure B (C.I. 52010), and (3) eosin Y (C.I. 45380).[11]

Whether Diff-Quik is standardized quantitatively is uncertain, as the composition is proprietary. The generic compositions of Diff-Quik's 3 solutions are seen in Table 13.1:

The specific dyes, whether they're pure, and the amounts are usually not provided by manufacturers. Regardless, users of all staining methods—not just Romanowsky-type stains—need answers to three questions:

TABLE 13.1. There are at least 40 xanthene dyes, which include eosin Y, and 10 thiazine dyes, which include azure B and methylene blue.

Solution	Solvent	Solute no. 1	Solute no. 2
Fixative	99.99% methanol	<0.01% triarylmethane dye	–
Solution I (eosinophilic)	pH buffer	Xanthene dye	Sodium azide
Solution II (basophilic)	pH buffer	Thiazine dye	–

The triarylmethane dye colors the fixative, but not cells. Sodium azide is not a dye.

1. What colors are possible?
2. What staining times will produce the desired optical densities?
3. What parts of cells will be colored by the component dyes?

The questions can be answered by the same approach as taken with the Pap stain. Use a simple, readily available "standard" preparation to familiarize yourself with the stain's capabilities. For the Pap stain, buccal smears are satisfactory. For Romanowsky/Diff-Quik stains, buccal smears may be satisfactory, but given that blood cells are the intended target, I recommend using blood films. Blood films are thinner than fine needle aspirations and contain cells that will demonstrate what to expect under the best circumstances. If they don't look good here, they can't look good elsewhere.

For quality assessment, I recommend:

- Begin with clean slides. Label each explicitly (e.g., date, stain, time).
- Make fresh air-dried blood films.
- Fix in methanol.
- Stain individual smears in separate dye solutions to see what colors to expect under the simplest circumstances. Rinse as required.
- Also stain a set in the complete stain to permit metachromasia.
- Stain individual smears over a range of times (e.g., 15 s, 30 s, 1 min, 2 min, etc.).
- Rinse thoroughly.
- Clear, mount, and examine microscopically.

When done in advance of putting the stain solutions into routine use, the recommended protocol constitutes quality assessment of the stains. Using stains with confirmed performance characteristics constitutes quality control. Without knowing in advance what stains should look like under the best circumstances, how can one evaluate the quality of the staining results? Interpreting cell morphology and its meaning in terms of cellular health and disease status should not be compromised by unsatisfactory staining results.

Romanowsky/Diff-Quik stains work best on well-behaved samples that resemble whole blood in terms of composition and ability to be spread thinly. Such behavior is the exception in the case of fine needle aspirations that are often thick. Thick cell preparations thwart satisfactory stain uptake.

To the extent possible, those who prepare FNA specimens should be knowledgeable about the prerequisites for successful preparations:

- Use clean slides. Clean slides are wettable. Wettable slides retain water on the surfaces when immersed in water. Dirty slides are not wettable. An invisible film of dirt neutralizes the positive charges on glass that are essential to the adhesiveness of the surface for the negatively charged cell surfaces. Dip slides in alcohol and wipe dry to clean them as needed. The contribution of clean slides to retaining and flattening cells cannot be overemphasized. Commercially available positively charged slides may be equally satisfactory. I've not used them and cannot comment.
- Prepare FNA slides as described by Dr. Glant in Chap. 5.
- Let the slides air-dry completely to promote cell adhesion and flattening. Air-drying per se does not fix cells.
- Fix in methanol.
- Stain as usual.
- Rinse thoroughly in clean solutions. See Fig. 13.1.

Dye solutions add color; rinses remove the excess. Dirty rinses resemble dilute dye solutions and leave unwanted dye that reduces the usefulness of the preparation. Table 13.2 systematically addresses troubleshooting problems encountered in Romanowsky staining.

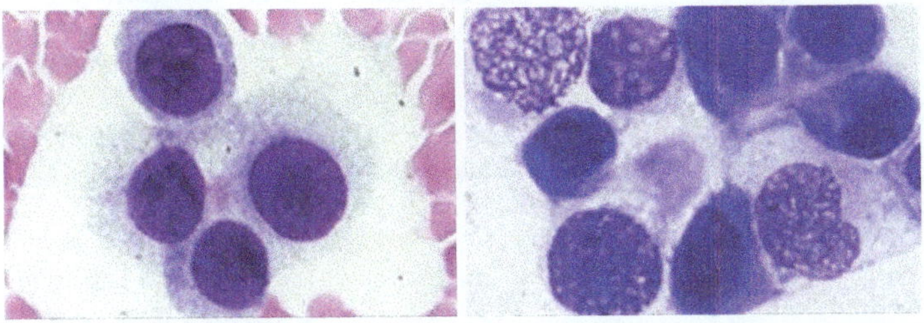

Fig. 13.1. Romanowsky-type staining of well-flattened cells. The composition of most FNA specimens prevents this degree of cell flattening and dye expression. Original magnification ×1,000.

TABLE 13.2. Troubleshooting Romanowsky staining of blood smears and related specimens.[17]

Group	Complaint	Cause	Correction
A	Stain precipitates on slides or within pipework of autostainer	Buffer concentration too high	Reduce
		Methanol (glycerol, dimethyl sulfoxide) content too low	Increase
		Staining solution too old	Replace
		Temperature too high	Keep cooler
B	Overall staining too pale, but color balance satisfactory	If "pure" azure B, dye content too low; due to impure lot of dye or stock solution	Replace
		If due to error in dilution or weighing	Check and correct
C	Cell nuclei blue, not purple	If "pure" azure B, dye content much too low, see B	
		If "polychrome methylene blue," azure B content of lot is low…	Replace
		If either dye type: Methanol (etc.) content too high…	Reduce
		pH of staining solution too low…	Check and correct
		Staining time too short…	Lengthen
		Staining temperature too low…	Check
		Eosin concentration in stock solution too low…	Make fresh
		Specimen too thick…	Lengthen staining time
		If histological section, formalin fixation too extended…	Lengthen staining time or increase dye concentration
		If histological section, acetic acid differentiation or alcoholic dehydration too long…	Reduce or change solvents
		Occlusion of specimen…	Check under the microscope for presence of any overlying material.

(continued)

TABLE 13.2. (continued)

Group	Complaint	Cause	Correction
D	Neutrophils appear agranular in blood smears and collagen fibers pink, not purple, in histologic sections	Any of possible causes for C	Any of possible corrections for C
E	Neutrophils appear "toxic" (i.e., grossly enlarged granules); general purple tint to basiphilic cytoplasm	pH too high… Staining time too long… Azure B too concentrated	Check and correct if necessary Reduce staining time Check and reduce
F	Erythrocytes and eosinophil granules too blue	pH too high… Wrong buffer used… Staining time too long…	Check Check Reduce
G	Erythrocytes and eosinophil granules too brownish orange, not pink	Some standardized stains give this appearance… pH too high…	Rinse briefly in distilled water Check and adjust
H	Basophil granules in blood smears fail to stain	Azure dye concentration to low or absent	Fix with azure dye dissolved in methanol

A 2011 issue of *Biotechnic & Histochemistry*, formerly *Stain Technology*, was devoted to Romanowsky stains and staining.[7, 15–18]

References

1. Ehrlich P. Methodologische Beiträge zur Physiologie und Pathologie der verschiedenen Formen der Leukocyten. Z klin Med. 1880;1: 553–60.
2. Malachowski E. Zur Morphologie des Plasmodium malariae. Centbl f klin Med. 1891;31:601–3.
3. Romanowsky D. Zur Frage der Parasitologie und Therapie der Malaria. St Petersburg Med Wochenschr. 1891;16:297–302. 307–315.
4. Wright JH. A rapid method for the differential staining of blood films and malarial parasites. J Med Res. 1902;7:138–44.
5. Giemsa G. Eine Vereinfachung und Vervollkommnung meiner Methylenazur-Methylenblau-Eosin- Färbemethode zur Erzielung der Romanowsky-Nochtschen Chromatinfärbung. Centralbl f Bakt etc. 1904;37:308–11.
6. Babeş A. Diagnostic du cancer du col utérin par les frottis. Presse Méd. 1928;29:451–4.
7. Krafts KP, Pambuccian SE. Romanowsky staining in cytopathology – history, advantages and limitations. Biotechn Histochem. 2011;86(2):82–93.
8. Lillie RD. Factors influencing the Romanovsky staining of blood films and the role of methylene violet. J Lab Clin Med. 1944;27: 1181–97.
9. Diff-Quik. Available at http://en.wikipedia.org/wiki/Diff-Quik. Accessed 28 Apr 2012.
10. USPTO. Diff-Quik trademark Registration Number 0939394. Available at http://tess2.uspto.gov/bin/showfield?f=doc&state=4009:jgr60.2.1. Accessed 28 Apr 2012.
11. Marshall PN, Bentley SA, Lewis SM. A standardized Romanowsky stain prepared from purified dyes. J Clin Pathol. 1975;28(11):920–3.
12. Lillie RD. Romanowsky-Malachowski stains. The so-called Romanowsky stain: Malachowski's 1891 use of alkali polychromed methylene blue for malaria plasmodia. Stain Technol. 1978;53(1): 23–8.
13. Woronzoff-Dashkoff KP. The Ehrlich-Chenzinsky-Plehn-Malachowski-Romanowsky-Nocht-Jenner-May-Grünwald-Leishman-Reuter-

Wright-Giemsa-Lillie-Roe-Wilcox stain. The mystery unfolds. Clin Lab Med. 1993;13(4):759–71.
14. Marshall PN, Bentley SA, Lewis SM. Standardization of Romanowsky stains. The relationship between stain composition and performance. Scand J Haematol. 1978;20(3):206–12.
15. Krafts KP. The Malachowski-Wright-Giemsa stain: a many-splendored thing. Biotechn Histochem. 2011;86(1):2–6.
16. Krafts KP, Hempelmann E, Oleksyn BJ. The color purple: from royalty to laboratory with apologies to Malachowski. Biotechn Histochem. 2011;86(1):7–35.
17. Horobin RW. How Romanowsky stains work and why they remain valuable—including a proposed universal Romanowsky staining mechanism and a rational troubleshooting scheme. Biotechn Histochem. 2011;86(1):36–51.
18. Dunning K, Safo AO. The ultimate Wright-Giemsa stain: 60 years in the making. Biotechn Histochem. 2011;86(2):69–75.

Chapter 14
Special Stains

Special stains are "special" because they are not routine. They are applied to tissue sections in addition to hematoxylin- and eosin (H&E)-stained sections to answer questions that arise above and beyond those that can be answered by interpreting H&E-colored tissue morphology. Special stains that are applied as a matter of routine are not special and run the risk of compliance violations. The term "special stains" is of uncertain provenance but must have come into use after 1876, when H&E was introduced.[1]

Special stains can answer these questions:

- Is a certain class of molecules present or absent?
- Where are the molecules located in the preparation?
- How much of the molecules are present?

Answering the last question requires sophisticated instrumentation and computation methods. To my knowledge, this aspect of special stains is neither well documented nor understood.

In this chapter, I describe some commonly used non-immunohistochemical stains. For readers who have not used special stains, I first compare key aspects between H&E and special stains (Table 14.1) and then discuss classification of "special stains" by the Food and Drug Administrations (FDA) and certification of "special stains" by the Biological Stain Commission.

G.W. Gill, *Cytopreparation: Principles & Practice*,
Essentials in Cytopathology 12, DOI 10.1007/978-1-4614-4933-1_14,
© Springer Science+Business Media New York 2013

TABLE 14.1. A comparison of general aspects of H&E with special stains.

Aspect	H&E	Special stains
Basis of design	Empirical	Empirical to scientific
Questions that can be answered	Many	One to many
Primary interest	Nucleus <u>and</u> cytoplasm	Nucleus <u>or</u> cytoplasm
	Medical diagnosis (e.g., growth activity)	Mostly in the diagnosis of infectious diseases and cancer based on chemical composition
Basis of interpretation	Morphology	Morphology and color
Frequency of use	Routine	As needed
Quantitative	No	No
Controls needed	No	Yes
Substrate specific	No	Yes

Classification of Biological Stains by FDA and Certification of "Special Stains" by the Biological Stain Commission

According to the US Food and Drug Administration (FDA), dye and chemical solution stains are identified as "mixtures of synthetic or natural dyes or non-dye chemicals in solutions used in staining cells and tissues for diagnostic histopathology, cytopathology, or hematology."[2] These substances (special stains) are not immunohistochemistry (IHC) reagents and kits, are classified as class I devices, and are exempt from the premarket notification procedures.

There are 61 biological stains that are on a certification basis with the Biological Stain Commission.[3] Among its objectives, the Biological Stain Commission strives to insure the quality of dyes through independent testing according to appropriate rigorous chemical and performance criteria. There are no comparable efforts elsewhere in the world to ensure the quality of dyes used as special stains and other applications. Sixty-one dyes are on a certification basis with the Biological Stain Commission. See Table 14.2. All but two, hematoxylin and orcein, are synthetic dyes. Twenty-three of the 61 synthetic dyes were first used before 1909. Those 23 dyes are highlighted by their blue color.

TABLE 14.2. Biological stains certified by The Biological Stain Commission.

Only 5 (**bold**) of 61 certified dyes are used as special stains

1. Acid fuchsin, C.I. 42685	22. Erythrosin, C.I. 45430	42. Nile blue A, C.I. 51180
2. **Alcian blue 8 GX, C.I. 74240**	23. Ethyl eosin, C.I. 45386	**43. Oil Red O, C.I. 26125**
3. Alizarin red S, C.I. 58005	24. Ethyl green, C.I. 42590	44. Orange G, C.I. 16230
4. Aniline blue WS, C.I. 42755	25. Fast green F C F, C.I. 42053	45. Orange II, C.I. 15510
5. Auramine O, C.I. 41000	26. Fluorescein isothiocyanate	46. Orcein
6. Azocarmine G, C.I. 50085	27. Giemsa stain	47. Pararosaniline, C.I. 42500
7. Azure A, C.I. 52005	28. Hematoxylin, C.I. 75290	48. Phloxin B, C.I. 45410
8. Azure B, C.I. 52010	29. Indigo carmine, C.I. 73015	49. Protargol S
9. Azure C, C.I. 52002	30. Janus green B, C.I. 11050	50. Pyronine B, C.I. 45010
10. Basic fuchsine, C.I. 42510	31. Jenner stain	51. Pyronine Y, C.I. 45005
11. Bismarck brown Y, C.I. 21000	32. Light green SF, C.I. 42095	52. Resazurin
12. Brilliant cresyl blue, C.I. 51010	33. Malachite green, C.I. 42000	53. Rose Bengal, C.I. 45435
13. Brilliant green, C.I. 42040	34. Martius yellow, C.I. 10315	54. Safranine O, C.I. 50240
14. Carmine, C.I. 75470	35. Methylene blue thiocyanate	**55. Sudan black B, C.I. 26150**
15. Chlorazol black E, C.I. 30235	36. Methyl orange, C.I. 13025	56. Sudan III, C.I. 26100
16. Congo red, C.I. 22120	37. Methyl violet 2B, C.I. 42535	57. Sudan IV, C.I. 26105
17. Crystal violet acetate	38. Methylene blue, C.I. 52015	58. Tetrachrome stain (MacNeal)
18. Crystal violet, C.I. 42555	39. Methylene violet (Bernthsen) C.I. 52041	59. Thionine, C.I. 52000
19. Darrow red	40. Neutral red, C.I. 50040	**60. Toluidine blue O, C.I. 52040**
20. Eosin B, C.I. 45400	41. Nigrosin, C.I. 50420	61. Wright's stain
21. Eosin Y, C.I. 45380		

Of the 61 dyes, 52 have color index (C.I.) numbers; 9 do not. C.I. numbers are 5-digit numbers assigned by The Society of Dyers and Colourists[4] to uniquely identify stains with the same chemical composition but different names. These 5-digit numbers must be specified when publishing or purchasing dyes to ensure using the same dye, even if identified by different names.

The Biological Stain Commission uses the words stains and dyes as though they're synonymous. Strictly speaking, however, they are not. "It is a strange fact that the housewife is more careful in her terminology of colouring agents than many microscopists are. She distinguishes clearly between *staining and painting* the floor, while they often used the word 'staining' without regard for the diversity of the processes grouped by them under this single name. The 'staining' of specimens by electron-microscopists does not involve dyeing. The word was formerly used as a synonym for 'dyeing', but has come to be treated so loosely in microtechnique that it is avoided in this book."[5] And "the process of dyeing is distinguished from all others by the fact that the tissues are exposed to the action of a *solution* of a *dye*. A dye may be defined, for the purposes of microtechnique, as an aromatic-salt-like compound having these characters:

1. It ionizes in the presence of water.
2. Either the cations or the anions are colored (sometimes both).
3. The colored ions are able to make chemical linkages with the proteins (and generally also with other constituents) of the fixed tissues of organisms (and in some cases with constituents of living cells as well).
4. When the colored ions make their linkages with the tissues, they do not lose color, and generally, they do not change it."[6]

The US-based Biological Stain Commission was an indirect consequence of World War I. During the Great War there was a blockade of German products, including dyes. By 1920, the supply of prewar dyes was almost exhausted, foreign supplies were erratic, and the domestic dyes were still often unsatisfactory. As a consequence, several concerned groups and individuals came together, which resulted in two key conferences in 1921 on the standardization of stains. From this activity came the Commission on the Standardization of Biological Stains. By 1923, the Commission already had a constitution that is recognizably the forerunner of the aims of the present commission. In parallel with this, cofounder Dr. Harold J. Conn, while Chairman of the Commission, published in 1925 the first edition of *Biological Stains*. This book has become a standard source of reference in technical and research histopathological and biological laboratories using dyes. The book has been kept up to date by regular revisions, with a 10th edition (2002) being the most recent version.[7] In 1944, the Commission on the Standardization of Biological Stains became the Biological Stain Commission.

The objectives of the Biological Stain Commission are the following (1) to insure uninterrupted supply of dyes used in biological and medical applications, (2) to promote cooperation and dialogue among manufacturers, vendors and users of dyes for histochemical applications, (3) to insure the quality of dyes through independent testing according to appropriately rigorous chemical and performance criteria, (4) to educate users of biological stains about sources of reliable dyes and how they might best be used, and (5) to publish information concerning new or improved uses for biological dyes and related histochemical techniques.

These objectives are met by the following: (1) analysis in the Commission's laboratory of dye content and composition of samples supplied voluntarily by dye manufacturers or vendors; (2) testing performance of dye samples in rigorous standardized procedures known to be discerning tests of the staining quality of the dye; (3) issuing certification labels to be attached to the containers used by companies marketing accepted dyes to assure consumers that these dyes have met the performance criteria of the Biological Stain Commission; (4) conducting and supporting research on biological dyes and histochemical techniques requiring them; (5) publishing books concerning biological dyes and histochemical techniques, and Biotechnic & Histochemistry, a bimonthly journal of microtechnic and histochemistry; and (6) maintaining through correspondence and annual meetings, active dialogue among scientists, manufacturers, and vendors concerned with biological stains.

Special Stains

Special stain is a term is used mostly in a laboratory setting. Special stains have two broad areas of application: research and diagnostic. In research, special stains are used as probes to identify certain chemical constituents in normal and abnormal cells. The information so obtained is used as a basis for further study and also as a baseline against which the results of special staining can be compared in diagnostic applications. On the basis of such a comparison, the significance of the findings can be interpreted.

Special stains can be applied to cell biology and histology. Among the mutually useful applications are the following (1) the determination of DNA and RNA content, (2) the mode of actions of drugs, hormones, or of potentially toxic food additives, (3) metabolic biochemistry, (4) the biochemistry of disease processes, (5) the primary sites of many metastatic tumors, (6) identifying nonpigmented metastatic melanomas, (7) detecting early invading tumors, (8) defining the margins of surgically resected tumors, (9) identifying Barr bodies, (10) staining cells in ways that can be used as a basis for cell separation by appropriate instrumentation (e.g., fluorescence), and (11) identifying microorganisms (e.g., *Cryptococcus neoformans*, *Helicobacter pylori*[8]).

Special stains described in this chapter do not include immunochemical stains and those used in molecular cytopathology.[9, 10]

Table 14.3 lists 47 special stains with their applications and staining specificity. For simplicity, the stains are grouped according to clinical applications. Color photomicrographs are available elsewhere.[11]

The materials, methods, and interpretation of these special stains can be found in Churukian[12] and Carson and Hladik.[13] When working with special stains, keep in mind these considerations:

- Special staining often requires the use of unusual stains and reagents that are available from only a few sources. Knowledge of such sources (vendors) is essential to overcome technical bottlenecks.

Table 14.3. Common special stains.

Special stain	Clinical application	Staining specificity
1/7. Detecting microorganisms and *Helicobacter pylori*		
Acid fast (Ziehl-Neelsen stain)	Detect nocardioform actinomycete group of bacteria, including *Mycobacterium* spp. (acid fast) *Rhodococcus equi and nocardia* spp. (weakly acid fast)	Acid-fast bacilli retain a cationic dye that is extracted from all other types of bacteria and animal cells by acidified alcohol. The waxy wall (with mycolic acid) of mycobacteria retains the dye
Alcian yellow/toluidine blue (Leung) stain	*Used for the detection of H. pylori*	The yellow dye stains oxidized and sulfonated gastric mucus, providing contrast for suspended Helicobacter organisms that are stained with toluidine blue
Dieterle's stain	Identify *Borrelia burgdorferi, Legionella pneumophila, Treponema pallidum*	Stains whole organisms
Diff-Quik stain (Diff-Quik is the formerly trademarked name for a proprietary rapid Romanowsky-like stain)	*Detect H pylori and some fungi* (e.g., *Pneumocystis jiroveci)*	*H. pylori and Pneumocystis jiroveci*
Genta stain	*Detect H. pylori*	*H. pylori*
Giemsa	*Used in the staining of H. pylori, Plasmodium vivax, Rickettsia prowazekii, Rickettsia rickettsii, Rickettsia tsutsugamushi, Trypanosoma cruzi*	Stains polyanions blue and polycations pink. Bacteria show up blue on account of their nucleic acids. Acidic capsules (e.g., anthrax bacilli, Cryptococcus) would be expected to be blue or purple

(continued)

Table 14.3. (continued)

Special stain	Clinical application	Staining specificity
Gram stain (named after its inventor Danish scientist Hans Christian Gram who developed the technique in 1884 to discriminate between 2 types of bacteria with similar clinical symptoms)	Used in the detection of Gram-positive (*Clostridium botulinum, Clostridium tetani, Staphlococcus aureus and Corynebacterium diphtheriae)* or Gram-negative bacteria (Salmonella, *Shigella dysenteriae*, *Escherichia coli,* and *Pseudomonas aeruginosa)*. Also used in the detection of *Actinomyces israelii, Legionella pneumophila, Neisseria gonorrhoeae, Neisseria meningitidis, Nocardia asteroides*	Stains whole organisms: "positive" = black/blue "negative" = red
Grocott's methenamine silver (GMS)	Useful in the identification of a variety of pathogenic fungi, including *Aspergillus fumigatus, Blastomyces dermatitidis, Candida albicans, Coccidioides immitis, Cryptococcus neoformans, Histoplasma capsulatum, Nocardia asteroides, Pneumocystis carnii, Pneumocystis Jiroveci* (human), and *Sporothrix schenckii*	Polysaccharide components of the fungal cell wall
H. pylori silver stain (HpSS)	*Detect H. pylori in gastric biopsies*	*H. pylori*

Mayer's mucicarmine stain	Detect encapsulated yeast-like fungus—*Cryptococcus neoformans*	Polysaccharides on the capsule
Periodic acid-schiff (PAS)	Used in the identification of *Aspergillus fumigatus, Blastomyces dermatitidis, Candida albicans, Coccidioides immitis, Cryptococcus neoformans, Sporothrix schenckii*	Polysaccharide components of the fungal cell wall
Sayeed's stain	*Detect H. pylori*	*H. pylori*
Steiner & Steiner staining method	*Detect spirochetes and legionella and pneumophila bacteria, examples: Borrelia burgdorferi, H. pylori, Legionella pneumophila, Treponema pallidum*	Stains whole organisms
Warthin-Starry	*Identify Alipia feles, Bartonella henselae, Borrelia burgdorferi, H. pylori, Legionella pneumophila, Treponema pallidum*	Stains whole organisms
2/7. Demonstrate connective tissue, muscle, collagen, lipid, and fibrin		
Gomori's 1-step trichrome stain	Used for distinguishing collagen and smooth muscle fibers	Collagen and smooth muscle fibers
Jones' basement membrane stain	Used for the identification of basement membranes (of the glomerulus in the kidney or) in tissue samples	Basement membranes

(continued)

TABLE 14.3. (continued)

Special stain	Clinical application	Staining specificity
Masson's Trichrome stain (TRI)	Used for distinguishing cells from surrounding connective tissue, which has several variants and is probably the trichrome most used in histopathology	Muscle, collagen fibers, fibrin, and erythrocytes Black nuclei, red cytoplasm (including muscle), blue or green collagen (including fine fibers), cartilage, and mucus
Modified Russel-Movat Pentachrome	Used for simultaneous demonstration of muscle, elastic fibers, collagen/reticular fibers, ground substance, and fibrinoid in tissues	Muscle, elastic fibers, collagen/reticular fibers
Oil red O and Sudan black B stains	Used for staining neutral triglycerides and lipids in frozen sections and some lipoproteins on paraffin sections	Neutral triglycerides and lipids
Orcein stain	Used for staining elastic fibers	Elastic fibers
Picro-Mallory	Fibrin	Fibrin
PTAH or phosphotungstic acid-hematoxylin stain	Used in demonstrating striated muscle fibers. Also used to stain abnormal neuroglia (reactive astrocytosis)	Muscle fibers, collagen
Reticulin-nuclear fast red stain	Used for the identification of reticulin fibers in tissue samples	Reticular fibers

Verhoeff-Van Gieson (VVG) stain	Used for identification of collagen and elastic fibers in tissues	The Van Gieson stain is specific for collagen. The Verhoeff component is specific for elastic fibers
3/7. Detect nucleic acids		
Ethyl green-pyronine Y stain	Used in the differential demonstration of DNA and RNA	A buffered mixture of the 2 dyes gives blue-green DNA and red RNA (rRNA in cytoplasm, nucleoli)
Feulgen stain	Used in the identification of chromosomal material or deoxyribonucleic acid (DNA in paraffin-embedded tissue or cell specimens)	Deoxyribonucleic acid (DNA)
4/7. Neuropathology		
Bielschowsky silver stain	Used in the diagnosis of Alzheimer's disease to show neuritic components of plaques and tangles	Amyloid and neuritic components of the senile plaques and neurofibrillary tangles
Congo red	Used in the detection of amyloid plaques in brain	Extracellular amyloid deposits
Cresyl violet stain	Useful in identifying cell bodies of neurons in tissue sections	Nissl substance in neurons. Cresyl violet shows cell bodies of neurons, by virtue of their abundant rough ER and ribosomes (rRNA)
PTAH or Phosophotungstic acid-hematoxylin stain	Used to stain abnormal neuroglia (reactive astrocytosis)	Abnormal neuroglia (reactive astrocytosis)

(continued)

Table 14.3. (continued)

Special stain	Clinical application	Staining specificity
5/7. Demonstrate myelin		
Luxol fast blue MBS	Used for demonstrating myelin	Myelin
Page's eriochrome cyanine R	Used for demonstrating myelin	Myelin
6/7. Dermatopathology, hematology, pigment detection, minerals and bone		
Alizarin red S stain	Calcium detection in tissues	Complexes with calcium
Chloroacetate esterase (Leder) stain	Useful as a marker for neutrophils	Neutrophils
Hall's stain	Detection of bile pigment	Bilirubin
Masson-Fontana stain	Melanin detection	Melanin
Perls' Prussian blue stain	Demonstrates hemosiderin in bone marrow macrophages and within erythroblasts	Hemosiderin (iron storage complex)
p-dimethylaminobenzylidene-rhodamine stain	Copper detection in tissues	Copper or copper-associated proteins
Villanueva osteochrome bone stain	Gives uniform and reproducible results for mineralized or undecalcified bone	Mineralized or undecalcified bone
7/7. Miscellaneous and multipurpose stains		
Alcian Blue pH 2.5	Used in the identification of sulfated mucins. At pH 2.5, stains sulfated and non-sulfated acidic carbohydrates. At pH 1.0, stains only sulfated carbohydrates	Mucins
Giemsa	Used in hematology, example: detection of erythroid colonies, binucleate normoblast, megaloblasts, mast cells, etc. Also used in chromosome staining	Specific for phosphate groups of DNA

Gomori's silver stain	Used in the detection of reticulin in the bone marrow	Reticulin
Jones' methenamine silver stain	Staining method for demonstrating glomerular basement membranes	Glomerular basement membranes
Mucicarmine stain	Detect mucins	Mucins
Periodic Acid-Schiff (PAS)	Used for staining structures containing a high proportion of carbohydrate macromolecules (glycogen and glycoprotein), basement membranes, collagen, primary cell types	Carbohydrate macromolecules
Periodic-acid silver methenamine (PEM) stain	Used for the delineation of basement membranes	Basement membranes

- One must be aware of special stains that contain colored and colorless impurities (e.g., salts), as these substances may interfere with the staining.
- Special staining requires a good knowledge of the tissue or cells that are the target of interest.
- When working with special stains, care should be taken so that the specimen is collected, fixed, and prepared in ways that maintain the molecule of interest within cells or tissues. For example, one must work with frozen sections when attempting to identify enzymes, or one must avoid fat solvents such as alcohol and xylene when attempting to identify lipids.
- With cell suspensions, it is essential to determine up-front by microscopy whether cells are present, and how many cells are to be used when making the slides. Using this quality control step can result in good cellular preparations and reliable results.
- Control preparations must be run in parallel with experimental preparations for one or more of the following reasons (1) to determine if the special stain is working, (2) to assess the degree of nonspecific staining, (3) to determine whether a reagent is still active, and (4) to serve as a standard in fractional reduction of staining procedures. If a positive reaction is noted when a control is not used, it can still be determined that the reaction is at least working (how well or how specifically is open to speculation). However, a negative reaction in the absence of a control can mean either that the sought constituent is not present or the reaction is not working and therefore unreliable.
- Designated positive and negative control slides should be the following: (1) sections of tissue/cells high in a particular molecule/constituent, (2) purified samples of a particular molecule in smears, (3) samples of the same specimen pretreated with solvents or enzymes to remove the sought constituent, (4) samples of the same specimen with essential reagents or steps in the staining procedure omitted, or (5) running a duplicate cell spread in the same manner as the experimental minus 1 essential step.
- The amount of special stain within a cell or tissue represents the difference between the amount taken up during staining and

the amount removed by the rinses following staining. To ensure the optimal amount, the user must employ those materials and methods that promote stain uptake during and after staining (e.g., dye concentration, suitable solvent, control of favorable pH, addition of salts if necessary, control of ionic concentration if necessary, time, and temperature).

- To maintain the right amount and hue of the special stain, one must mount the stained specimen in a medium that does not promote bleaching or leaching.
- To ensure optimal image quality of the stained specimen, one must use the right amount of mounting medium, cover glass with the right thickness (No. 1 cover glass), and a clean microscope in which the illumination is adjusted according to the method of Köhler.

Manual Versus Automation of Special Stains Protocol

Depending on the economic situation of the laboratory, specimen sample size, and the number of personnel available, special stain protocols are performed either manually or by automated systems. Manual staining of slides works well in a research setting, especially when the numbers of processed slides are few per day. However, with increasing numbers of slides to be stained, the manual method becomes prone to error resulting in decreased flexibility and productivity. With the medical community demanding faster turnaround time, increased flexibility, productivity, and greater standardization, automated instruments have replaced some manual methods of staining, thus becoming an integral part of the laboratory. With automation combined with specialized software applications and connectivity, many instruments are now capable of multiprogramming runs resulting in standardized protocols, manageable work schedules, enhanced workflow, cost effectiveness, and the ability to adapt to regulatory requirements.

Conclusion

Special stains belong to an assorted family of stains for microscopic visualization and general identification of cells, tissues, and microorganisms. Special stains remain an important tool for many pathologists and technologists, providing a powerful complement to immunohistochemistry, flow cytometry, in situ hybridization, and other diagnostic technologies that define a patient's medical profile. With the medical community demanding greater standardization and quality control, special stain protocols have become increasingly automated, resulting in higher levels of productivity and flexibility. Automation is no substitute for a solid understanding of the principles and practices of good staining. In a nutshell, this introduction was intended to provide guidance to help interested readers acquire proficiency in selecting and performing special stains faster than they might have otherwise done.

References

1. Wissowzky A. Ueber das Eosin als reagenz auf Hämoglobin und die Bildung von Blutgefässen und Blutkörperchen bei Säugetier und Hühnerembryonen. Archiv für mikroskopische Anatomie. 1876; 13:479–96.
2. 21 CFR Ch. I (4–1–09 Edition), p. 228–9. http://edocket.access.gpo.gov/cfr_2009/aprqtr/pdf/21cfr864.1850.pdf. Accessed 4 Mar 2012.
3. The Biological Stain Commission. http://www.biologicalstaincommission.org/ Accessed 4 Mar 2012.
4. The Society of Dyers and Colourists Home Page. Accessed 4 Mar 2012 at http://www.sdc.org.uk/.
5. Baker JR. Cytological technique. 5th ed. London: Methuen & Co. Ltd.; 1966.
6. Baker JR. Principles of biological microtechnique: a study of fixation and dyeing. Bungay, Suffolk: Methuen & Co., Ltd.; 1958.
7. Horobin RW, Kiernan JA, editors. Conn's biological stains: a handbook of dyes, stains and fluorochromes for use in biology and medicine. 10th ed. Oxford, UK: BIOS Scientific Publishers; 2002.
8. Rotimi O, Cairns A, Gray S, Moayyedi P, Dixon MF. Histological identification of *Helicobacter pylori*: comparison of staining methods. J Clin Pathol. 2000;53(10):756–9.

9. DeMay RM. Stains, chapter 26. In: DeMay R, editor. The art & science of cytopathology, vol. 3. 2nd ed. Chicago, IL: ASCP Press; 2011. p. 1503–26.
10. Lacroix L, Saada V, Valent A, et al. Molecular cytopathology, chapter 27. In: DeMay RM, editor. The art & science of cytopathology, vol. 3. 2nd ed. Chicago, IL: ASCP Press; 2011. p. 1527–38.
11. Kumar GL, Gill GW. Introduction to special stains. In: Special stains and H&E. Carpenteria, CA: Dako Connections. 2010;14:16–42. http://www.dako.com/us/28829_connection_14.pdf. Accessed 4 Mar 2012.
12. Churukian CJ, Method of the Histochemical Stains & Diagnostic Application, Department of Pathology and Laboratory Medicine, University of Rochester, Rochester NY, second web edition (2009). http://www.urmc.rochester.edu/path/zqu/StainsManual/index.html. Accessed 4 Mar 2012.
13. Carson FL, Hladik C. Histotechnology: a self-instructional text. 3rd ed. Chicago, IL: ASCP Press; 2009.

Suggested Reading

14. Garrett RH, Grisham CM. Biochemistry. 4th ed. Boston, MA: Cengage Learning; 2010.
15. Horobin RW, Bancroft JD. Troubleshooting histology stains. New York: Churchill Livingstone; 1998.
16. Horobin RW. Histochemistry: an explanatory outline of histochemistry and biophysical staining. London: Butterworths; 1982.
17. Kiernan JA. Staining, histochemistry, and histotechnology FAQ. Accessed 21 Aug 2009 at http://publish.uwo.ca/~jkiernan/faqlist.htm.

Part II
The Image

Chapter 15
Clearing

Xylene*The greatest care imaginable should be taken in preparing objects for examination; otherwise the best skill'd in magnifying glasses may be misled, if they give too sudden a judgement on what they see, without assuring themselves of the truth by repeated experiments.*

George Adams, 1746

Principle No. 6

Mount preparations to optimize microscope objective's performance.

Practice

Immerse fixed and stained preparations in an organic solvent that has a refractive index close to that of fixed protein.

Historic Milestones

- 1665—Hooke mounted specimens in olive oil for transparency: "you may clearly see…"[1]
- 1770—Hill used spirits of turpentine to clear tissues for microscopy.[2]
- 1877—Merkel introduces xylol as clearing agent.[3]
- 1979—Hemo-De introduced as first xylene substitute.

G.W. Gill, *Cytopreparation: Principles & Practice*, Essentials in Cytopathology 12, DOI 10.1007/978-1-4614-4933-1_15,
© Springer Science+Business Media New York 2013

To transition from staining to mounting, cytologic preparations must pass through an organic solvent that will displace any remaining alcohol between staining and mounting. In addition, the solvent should have a refractive index that is similar to that of fixed protein, which is 1.536,[4] and the resin of the medium in which it will be mounted. Such "matching" is an essential step in the chain of excellent imaging prerequisites, which include mounting medium, cover glass, clean microscope, and Köhler illumination.

Typically the solvent is xylene. It is infinitely miscible with absolute alcohol and itself, the usual solvent in mounting media. Xylene, and solutions used similarly such as toluene, is referred to as a clearant or clearing agent. The process is referred to as "clearing," which means not only making the preparation clear, but also clearing the remaining traces of water from the tissue.

The term "clearing" is somewhat misleading. Fixed protein per se is naturally transparent and colorless and does not require clearing. Unless the spaces in and around the proteins are occupied by something with a similar refractive index; however, the dry protein diffracts light and does not appear transparent.[5] Diffraction is light bending as it passes through a transparent medium and encounters an edge. In this case, the transparent medium is air, and the edge is fixed protein. Visually, the effect is seen as scattered light or unsharp images. Diffraction and refraction are similar in that both involve light bending, but dissimilar in that refraction occurs when light passes between two transparent media with different refractive indexes.

Xylene baths are changed at arbitrary intervals, depending on each laboratory's practice. The interval can be extended, but not indefinitely, by using all nonstaining rinses in sets of 3, filling the containers as much as is allowable to slow the rate of contamination and minimize carry-over into the next set of dishes. Disposal costs can exceed purchase price. Thus, labs look for less expensive xylene alternatives.

Xylene Alternatives

Tertiary Butanol. A seldom-used alternative to xylene as a clearing agent is tertiary butanol (C4H9OH).[6] Its use was described in a letter-to-the-editor of Acta Cytologica in 1983, just about the time xylene substitutes had begun to come onto the cytology market. However, as the letter authors pointed out, tertiary butanol had been reported by Celalier to dehydrate botanical chromosome smears 27 years earlier.[7] It is an interesting example of how much useful information is available in the "old" literature if one knows where to look. I have not used tertiary butanol and so cannot comment on its merits.

Xylene Substitutes. Some laboratories use xylene substitutes, which may or may not require special disposal. While popular, xylene substitutes are not used universally and each has limitations. The potential health hazards have not been characterized as well as those for xylene. Manufacturer's claims to the contrary, handle these substitutes as though they were xylene.[8] In general, these substitutes fall into four classes and are marketed under various trade names. The chemical components are one of the following:[9]

- Limonene reagents
- Aliphatic hydrocarbon mixtures
- Aromatic hydrocarbon mixtures
- Mineral oil mixtures

Both alternatives have finite life cycles, must be disposed of, and replaced, which is expensive.

Xylene

Xylene is miscible with absolute alcohol and the solvent of the mounting medium, for which the solvent is usually xylene but sometimes may be toluene. Both solvents are aromatic hydrocarbons, which mean they are based on benzene. See Fig. 15.1.

When immersed in a medium of similar refractive index, cells become transparent as light passes unimpeded optically. See Table 15.1.

One should buy histological grade of xylene, which is comprised of a mixture of the 3 isomers, rather than the more expensive purified xylenes of one isomer. Xylenes contain up to 15%

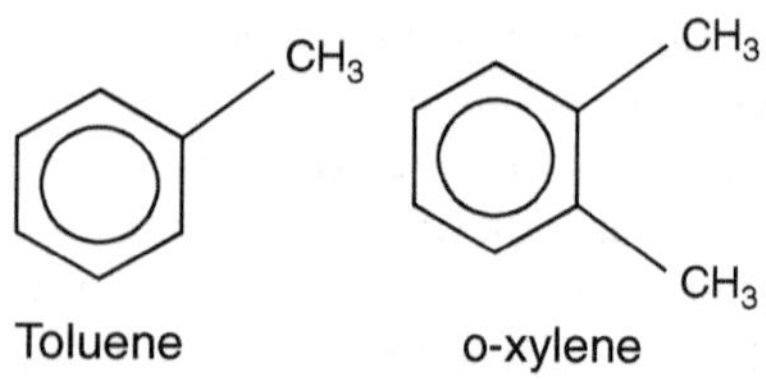

Fig. 15.1. Toluene has one methyl group; xylene, two. The two methyl groups in the illustration are located in the ortho-position, but they can also be separated from one another by one (meta-) or two carbons (para-) of the benzene ring. Collectively they are referred to as isomers.

Table 15.1. Cells become progressively transparent when immersed in increasingly similar refractive index media.

Environment	Refractive index	Cell appearance (R.I. 1.52–1.54)
• Air	1.00	Opaque white
• Water	1.33	Opalescent
• Alcohol	1.36	Less translucent
• Xylene	1.49	Transparent

Xylene, also known as xylol, is rooted in *xylon*, Greek for wood, and is similar in that respect to hematoxylin, bloodlike wood.

benzene, which is a bone marrow carcinogen. For health reasons, therefore, work with xylenes under a fume hood to avoid inhaling the vapors. Xylene is flammable.[10]

Xylene is chemically inert to biological dyes, so stained cytologic preparations can be safely immersed in xylene overnight if necessary. Xylene-immersed preparations experience conditions comparable to those encountered while coverslipped. The xylene carried over onto a coverslipped preparation remains there for at least a month or more. Prolonged immersion in xylene must be avoided, of course, if the slides are already identified with labels with adhesive backs that are xylene soluble.

Water is sometimes carried over from preceding water and alcohol baths into xylene, but it occurs primarily as a result of moisture adsorbed from the ambient air. The adsorption rate varies geographically (more in the south), seasonally (more often in the summer), and locally according to laboratory practices. Water in xylene is suspended as microscopic droplets that are not visible to

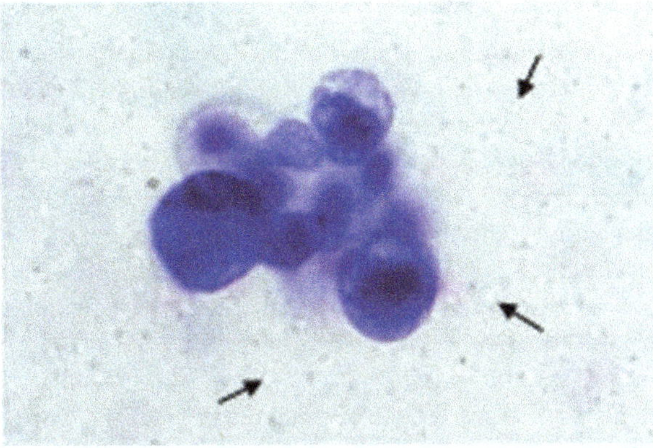

FIG. 15.2. When present in abundance in xylene, the water droplets can extract water-soluble dyes such as light green from cells, leaving cytoplasm colorless as shown. The droplets become colored. Microscopic water droplets in xylene are difficult to replicate deliberately by simply adding water to xylene.

the unaided eye. In coverslipped preparations, the water droplets are visible grossly as a whitish translucence. Microscopically the droplets are plainly visible as clear globules of various diameters. See Fig. 15.2.

Immortalizing Xylene[11]

Eliminating water from xylene introduces the possibility of using xylene indefinitely, thereby reducing costs and saving money. Many laboratories, if not most, continue to use xylene. For them the question is "what determines when xylene should be discarded?" The answer depends on one's understanding of what xylene does and second, understanding what might diminish its usefulness.

I have already described what xylene does. Its usefulness is diminished by anything that might cause xylene to fade stains, extract dyes from cytoplasm, or cross-contaminate other preparations. Xylene used repeatedly accumulates cellular debris and introduces the possibility of cross-contamination. Microscopic preparations that contain water droplets are neither stable nor permanent. The preparations must be de-coverslipped, taken back through xylene and alcohol, cleared again, and re-coverslipped.

Historically, water and particulates have been removed from xylene by filtering the xylene through laboratory grade filter paper. As water and xylene, like vinegar and oil, do not mix, the water droplets are blotted by the filter paper and water-free xylene is the filtrate.[12] Xylene baths should be filtered daily; often, however, they are not. *Consequently, xylene baths are not well maintained and are discarded prematurely.*

As a historical aside, "The original meaning of the word 'filter,' long obsolete, denoted a piece of felt, the principal device used during the Middle Ages for filtration. In its modern sense, the word 'filter' entered the English language before the birth of Shakespeare: it was variously spelled 'philtre,' 'philter,' 'filtre,' 'fylter,' or 'fylture.'"[12]

Water-Scavenging Beads. Zeolite adsorbents are synthetically produced molecular sieves that are microporous, crystalline, metal aluminosilicate bb-size beads that scavenge water in organic solvent systems (e.g., water in xylene). See Fig. 15.3. The uniform crystalline structure of molecular sieve adsorbents provides very predictable and reliable adsorptive properties. Metal cations contained in the crystalline structure of molecular sieve adsorbents balance the negative charge of the framework. These metal cations create an electrical field, hence their strong affinity for polar molecules such as H_2O. Depending on the type of crystalline structure and the occupying cation of the molecular sieve, adsorbent molecules may be readily adsorbed or completely excluded according to their relative molecular size. For example, a 4A molecular sieve is particularly useful for the dehydration of olefins. It will adsorb water but will exclude an olefin molecule.[13]

Avoiding costly rework and delays more than offsets the modest cost of these water-scavenging beads. Molecular sieves adsorb water admixed with xylene in real time and provide redundant assurance that water droplets will not accumulate in xylene.

Molecular sieves with indicator contain beads that have been chemically impregnated to display a blue color when fully acti-

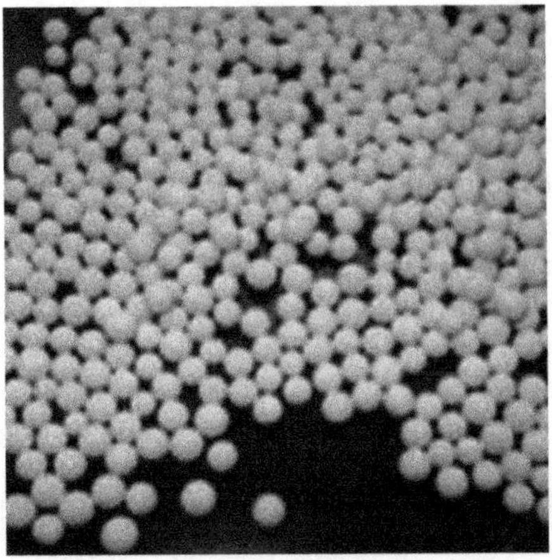

FIG. 15.3. The beads resemble 2-mm diameter BBs, though BBs are usually 4.5 mm in diameter.

vated by removing the water of hydration. When xylene is dried using this product, the color blue will turn to pink as water is adsorbed, which signals the sieve is saturated and should be discarded.

To use these beads, add a monolayer to the bottom of each xylene dish and fill with xylene. Pass racks of stained slides through them routinely. When the beads have adsorbed as much water as they can hold, the blue indicator beads become dark brown, though the manufacturer's literature describes the color as pink. At that point, replace the beads. Spread the used beads on a flat surface under a fume hood to exhaust the xylene as it evaporates. Discard as trash.

The recommended molecular sieve is type 4A with indicator beads, form 8 × 12 beads, manufactured by UOP (Universal Oil Products). Form 8 × 12 refers to standard mesh screen sizes though which the beads can pass. In this case, the beads are small enough to pass through the larger openings of the form 8 mesh screen (2.38 mm) but too large to pass through the smaller openings of the 12 mesh (1.4 mm). The nominal bead diameter is 2 mm.

Mesh is rooted in a word that refers to open space in a net. One well-known mesh series is the Tyler Equivalent created by the W.S. Tyler screening company. Tyler mesh size is the number of openings per (linear) inch of mesh. To calculate the size of the openings in a mesh, the thickness of the wires making up the mesh material must be taken into account. In practice, mesh openings are determined by referring to a chart.

For additional details such as quantities and pricing, contact Advanced Specialty Gas Equipment.[14] Do not contact UOP, as the company only sells its products in industrial strength quantities. A 5-lb can is affordably priced and will last a long, long time. Reusable xylene recycling pads are available.[15]

To filter xylene and recover the beads for re-use, these supplies are needed:

- Rectangular epoxyn stand, 5 x 8-in. base, 20-in. rod
- Ring support, 6-in. O.D.
- Castaloy adjustable-angle clamp, flat jaws
- Qualitative grade filter paper, Grade P8, 33-cm diameter
- Nalgene polypropylene funnel with long stem and external ridge

The following practices extend the working life of xylene solutions indefinitely:

- Following EA, use at least three changes *each* of 95% ethyl alcohol and absolute alcohol—10 dips each. To minimize rapid dilution, fill the dishes close to the brim. Immerse and withdraw the rack totally for each dip to promote thorough exchange of the excess dye with the rinse, approximately 1 s per dip—not too fast or too slow. When the third dish of 95% alcohol becomes slightly colored, discard the contents of the first rinse, move dishes 2 and 3 to dish 1 and 2 positions, and add fresh 95% alcohol to the third dish. Try to avoid coloring the absolute alcohol baths. Do not allow the xylene baths to become colored.
- Add a monolayer of molecular sieve beads to the bottom of each dish of xylene to adsorb water in real time. Change the beads as indicated by the change in color (e.g., once every second Friday); replace.

- Filter each dish of xylene as needed for Pap smears to remove cellular debris and adsorb any water the molecular sieves may have missed. For non-gyn cytologic specimens that shed floaters, use a cross-contamination control system. To filter xylene, place a wire mesh kitchen strainer over the filter funnel with the folded filter paper in place. Dump the entire contents of a xylene dish through the strainer, letting the strainer catch the beads and allowing the xylene to be filtered. Catch the filtrate in a clean water-free dish. Pour the filtered xylene over the beads in the strainer to wash them. Catch the refiltered xylene, re-add the rinsed beads to the dish; put the xylene back into service. Maintaining xylene in this way extends its useful life indefinitely. Absence of water carryover and dye fading over time confirms the usefulness of this practice. Distillation devices are unnecessary for removing water from xylene. However, such devices can separate xylene from paraffin in histological applications, which molecular sieves cannot.

  To accelerate the filtration rate:

- Use 58° funnels, rather than 60° funnels.
- Use long stem funnels.
- Use qualitative grade rapid flow filter paper.
- Make certain the tip of the funnel stem is not immersed in the filtrate.
- Sit the funnel loosely on the mouth of the vessel.

Dedicate a Coplin jar of xylene to removing cover glasses, thus avoiding the accumulation of mounting medium in the extended-use xylene that may shorten its useful life.

Check stained preparations for evidence of water and water-related deterioration of quality:

- Grossly, look for milky appearing translucence associated with massive water contamination.
- Microscopically, look for water droplets by closing the sub-stage condenser aperture diaphragm to exaggerate the appearance of water droplets.

- Microscopically, look for pale counterstains colors, which can result when water droplets extract dyes from cells and evidence of "floaters."
- Microscopically, look at a drop of xylene by itself by closing the substage condenser aperture diaphragm to see whether water droplets are present.

References

1. Hooke R. Micrographia: or some physiological descriptions of minutes bodies made by magnifying glasses. London: J. Martin & J. Allestry; 1665.
2. Hill J. The construction of timber, from its early growth; examined by the microscope, and proved from experiments, in a great variety of kinds. London: Printed for Hill; 1770.
3. Merkel FS. Purification of olefin and polymer process streams. Arch Mik Anat. 1877;14:622.
4. Crossmon GC. Mounting media for phase microscope specimens. Stain Technol. 1949;24(4):241–7.
5. Baker JR. Cytological technique. 6th ed. London, England: Chapman and Hall Ltd.; 1966.
6. Drijver JS, Boon ME. Tertiary butanol as a substitute for xylene as a clearing agent. Acta Cytol. 1983;27(2):210–1.
7. Celalier RP. Tertiary butanol alcohol dehydration of chromosome smears. Stain Histol. 1956;31($):155–7.
8. Gill GW. Xylene substitutes – CPT Forum. Cytotechnol Bull. 1984;21(2):32–3.
9. Kandyala R, Raghavendra SPC, Rajasekharan ST. Xylene. An overview of its health hazards and preventive measures. J Oral Maxillofac Pathol. 2010;14(1):1–5. http://www.ncbi.nlm.nih.gov/pmc/articles/PMC2996004/?tool=pubmed. Accessed 17 Feb 2012.
10. Sigma-Aldrich. Xylenes Material Safety Data Sheet. Available at http://www.sigmaaldrich.com/catalog/DisplayMSDSContent.do. Accessed 14 Mar 2012.
11. Gill GW. Using xylene forever: molecular sieves and daily filtration. Acta Cytol. 2000;44(5):925. Abstract 134.

12. Technical Staff of the Eaton-Dikeman Company. Handbook of filtration. Mt. Holly Springs, PA: The Eaton-Dikeman Company; 1960.
13. Purification of Olefin and Polymer Process Streams. http://www.uop.com/wp-content/uploads/2011/07/UOP-Adsorbents-for-purification-of-olefin-and-polymer-process-streams-brochure.pdf. Accessed 17 Feb 2012.
14. Advanced Specialty Gas Equipment. Molecular Sieves + Activated Alumina Adsorbents. Available at http://www.asge-online.com/pdf/MolecularSieves_AAAdsorbentsBrochure.pdf. Accessed 14 Mar 2012.
15. Creative Waste Solutions, Inc. (http://www.cwsincorp.com/).

Chapter 16
Mounting Media

Principle No. 6

Mount preparations to optimize microscope objective's performance.

Practice

Select a mounting medium with physical, optical, and chemical properties that will support short-term, mid-range, and long-term performance goals.

Historic Milestones

- 1830—Cooper first to use Canada balsam as a permanent mounting medium.[1]
- 1893—Friedrich Becke describes "Becke line" that accounts for optical membranes.
- 1936—Fisher Scientific introduces Permount as first synthetic mounting medium.
- 1950—Lillie et al. publish interim report on resinous mounting media.[2]
- 1953—Lillie et al. publish final report on resinous mounting media.[3]
- 1956—Seal introduces for cytological use Millipore filters that have a refractive index dissimilar to that of glass, and require

G.W. Gill, *Cytopreparation: Principles & Practice*,
Essentials in Cytopathology 12, DOI 10.1007/978-1-4614-4933-1_16,

© Springer Science+Business Media New York 2013

mounting medium with closely matching refractive index for maximum transparency.[4]

- 1957—Dakin patents 1-side fully frosted slides that require mounting medium with refractive index close to that of glass to avoid light scatter.[5]
- 1963—Hollander introduces an oil-soluble antioxidant in resinous media to inhibit fading of Romanowsky stains.[6]
- 1964—Seal introduces for cytological use Nuclepore filters that are birefringent (i.e., have two refractive indexes) and that make it impossible to make pore outlines "invisible."[7]
- 1970—Barr measures fading of hematoxylin and eosin dyes in sections mounted in 22 different mounting media exposed to sunlight for 10 h.[8]
- 1970—Hollander and Frost describe annual bands in synthetic-resin-mounted microscopic slides.[9]
- 1971—Hollander and Frost describe antioxidant inhibition of stain fading and mounting medium crazing.[10]

Baker wrote, "In fixation and dyeing the tissues are responsive; they react to what we do to them. In embedding and mounting they are more passive, allowing us to surround them with what we will."[11] Such a view may account for the relative dearth of published information about mounting media in cytological and histological techniques. Manufacturers of mounting media provide precious little useful information on their products' labels. MSDS list boiler-plate type information. Since 1970, only 182 articles are cited in the National Library of Medicine's holdings with "mounting media" (69) or "mounting medium" (113) in the title or body. None addresses the practical information presented in this chapter.

The 12 historic milestones span the years 1830 thru 1971. Lillie and Baker have published the most comprehensive information about mounting media that I've seen during that period.[2, 3, 11] I've cited additional references for the historical record.[12–29]

Working with Millipore filters while developing methods to collect circulating cancer cells in the peripheral blood of cancer patients introduced me to the "passive" properties of mounting media. Refractive index was the first property, which was followed by evaporative weight loss and so on. See Table 16.1.

TABLE 16.1. Property classes of mounting media and impact on coverslipping, microscopy, and filing/storage.

Property	Operations impacted					
Class	Coverslipping	Impacts	Microscopy	Impacts	Filing/storage	Impacts
Chemical	Glue-like properties	Glass adhesion			Chemical composition	Color fastness Annual band formation
Optical			Refractive index when dry	Cell and support medium transparency		
			Color	Background if yellow		
Physical	Viscosity	Flow	Thickness	Spherical aberration	Evaporative losses	Timely filing Retraction

Ideally, a mounting medium should appear water-white in color. It should flow easily when applied to a slide and not entrain or entrap air bubbles. (Entrain means the mountant doesn't flow as a uniform leading edge, thus entraining air bubbles. Entrap means any entrained air bubbles aren't easily released.). The solvent should evaporate relatively slowly to avoid air being aspirated under the cover glass. The proportion of solids to volume of solution should allow the solvent to evaporate without air being drawn under the cover glass (i.e., retraction). The refractive index when dry should closely match that of fixed protein to promote transparency and resolution. The chemical composition should not cause dye to fade or annual bands to form.

No mounting medium possesses all these properties and performance features, not for a single preparation type, and certainly not for all preparation types. All mounting media are a compromise. About the only time a mounting medium draws any attention is when a problem arises. Examples are abundant: (1) it's fouling the automated coverslipper, (2) it's taking forever to dry, causing filed slides to become one big glass brick, and (3) archived slides are faded.

The mounting media of interest in mounting cytologic and histologic preparations are classified as adhesive hydrophobes. See Table 16.2.

The first adhesive hydrophobe was Canada balsam, a natural resin. "Natural" implies that the composition varies from lot to lot. Like hematoxylin, some lots will perform better than others. Permount, the first synthetic resin mounting medium, was intro-

TABLE 16.2. Classification of mounting media according to Baker.[30]

Classification	Type	Example
Hydrophil	Nonadhesive (fluid mount)	Glycerol
	Adhesive	Farrant's medium
Hydrophobe	Nonadhesive (fluid mount)	Methyl salicylate
	Adhesive	Permount

duced in 1936 and is still marketed. Today, mounting media usually include a synthetic resin, a plasticizer to maintain pliability of the resin when the solvent has evaporated, and an oil-soluble antioxidant to minimize dye fading and prevent "annual bands" in certain resins.

Manufacturers of mounting media indicate the solvent, percent composition of solute, and refractive index. Such minimal information might leave users with the impression that mounting media are little more than expensive variations of ordinary glues. Such an impression would be wrong. Absent knowledge base for reference, how could users know otherwise? Lillie's 1953 paper recommended that manufacturers of mounting media provide the following information:

1. Source and nature of resin: chemical nature
2. Solubility in aromatic hydrocarbons and concentrations necessary to give viscosity similar to that of 60% (w/v) Canada balsam in xylene
3. Solutions: nature and amount of solvent and resin by weight
4. Drying rate of solution: rapid, medium, slow with time limits stated
5. Freedom from air aspiration. Satisfactory for 5–10-μm thick sections. Satisfactory for thick sections
6. Index of refraction of mounting solutions and of solid resin to 3 decimals
7. Acid number
8. Ester number, saponification number, or both
9. Iodine number
10. Conservation or bleaching of cobalt sulfide or FeS, or PbS, and of ferric ferrocyanide. Bleaching of rosanilin.

Items 1 thru 6 are included in Table 16.1. Items 7 thru 10 are not. Obtaining answers to those 4 implied questions isn't easy, and it's unclear whether they impact the dyes in the Pap stain at all.

Let's consider items 1 thru 6 in turn. In the early 1970s, many brands of resinous mounting media were available in America. See Fig. 16.1. I doubt this many are available today.

FIG. 16.1. With so many mounting media to choose from, how can one know which is best suited for one's applications?

Source and Nature of Resin: Chemical Nature

This information is often proprietary. Companies sometimes describe the chemicals vague in terms of classes. Today one can get some insight by examining the MSDS (i.e., Material Safety Data Sheet). MSDS was not generally available until 1986, which is when OSHA (i.e., Occupational Safety and Health Administration) began requiring that such information be provided for all hazardous materials.

Laboratorians primarily want to know whether the chemicals in a mounting medium will cause biological dyes to fade. Fading is any change in color, not simply a diminution in shade. Knowing the chemical nature of the resin is not predictive. In my experience, the best way to find out whether fading is likely is to run a use test similar to what the Biological Stain Commission (BSC) does for the dyes it certifies.

In addition to running qualitative and quantitative assays, BSC performs biological assays (i.e., use tests). A dye sample is given to an experienced individual who uses the dye in standardized materials and methods. If the results are as expected, the dye is certified. If the results are not as expected, the dye is not certified, even though it had passed "on paper."

Barr measured spectrophotometrically the degree to which hematoxylin-stained and eosin-stained formalin-fixed sections of adenocarcinoma of the colon faded when exposed to sunlight for 10 h intermittently.[8] Following his lead, I acquired a used spectrophotometer and modified the cuvette holder to accept a 3 × 1-in. slide. Histology Supervisor Donald Meyer prepared formalin-fixed liver sections that were large enough to occupy the window in the cuvette holder when positioned at one end of the slide. Richard-Allan prepared custom 3 × 1-in. No. 1 cover glasses that covered each entire slide, thus allowing the end not occupied by stained section to serve as a control. See Fig. 16.2.

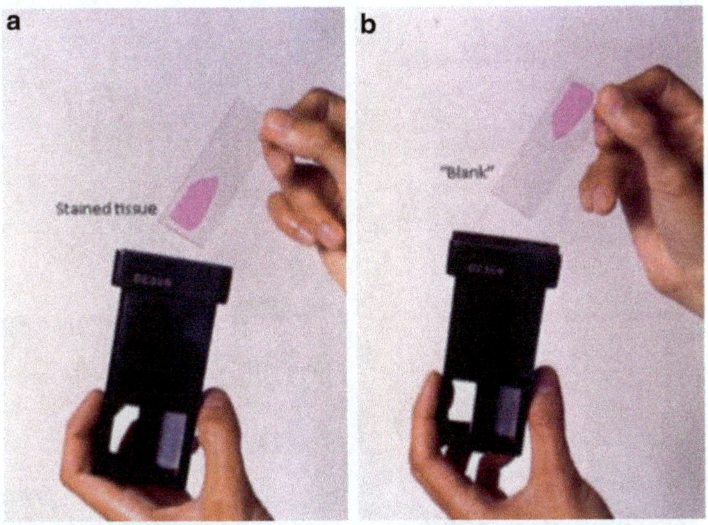

FIG. 16.2 Necessity is the mother of invention. Salvaging an old spectrophotometer and modifying the cuvette holder allowed me to measure changes in peak absorbance in formalin-fixed liver sections that had been stained in separate solutions of hematoxylin, orange G, eosin Y, or light green SF yellowish.

Figures 16.3 and 16.4 show the stained tissues after either having been stored in the dark for 60 days or exposed to sunlight for 5 days from noon to 2 o'clock on the roof of the Pathology Building.

Table 16.3 summarizes the data shown in Figs. 16.3 and 16.4.

Of the four dyes, hematoxylin and eosin faded least in storage for 2 months or 10 h sunlight exposure in 1 mounting medium. Overall, the take-home message is choose your mounting medium well. Absent supporting documentation, perform your own tests. Use tissue sections stained by one dye each to isolate the stain.

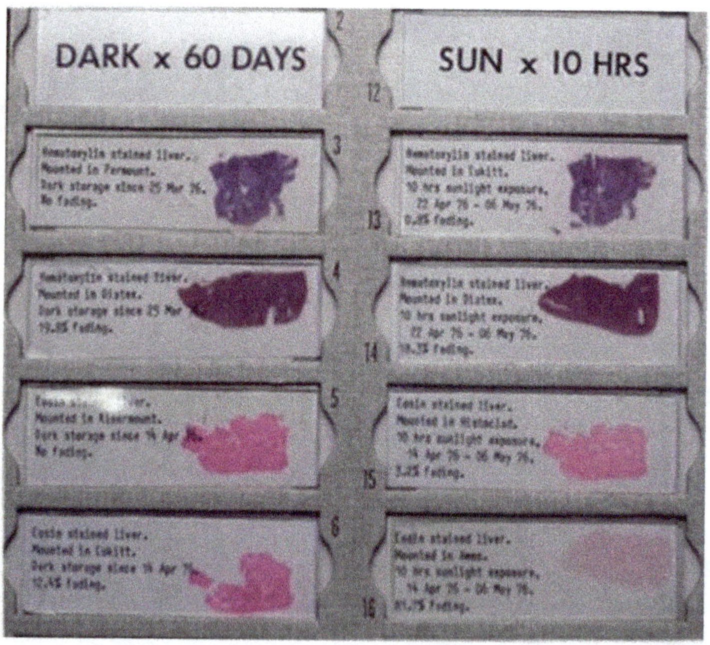

FIG. 16.3. None of the 25 tested mounting media totally prevented fading of hematoxylin- or eosin-stained sections under both the least and most extreme light exposures.

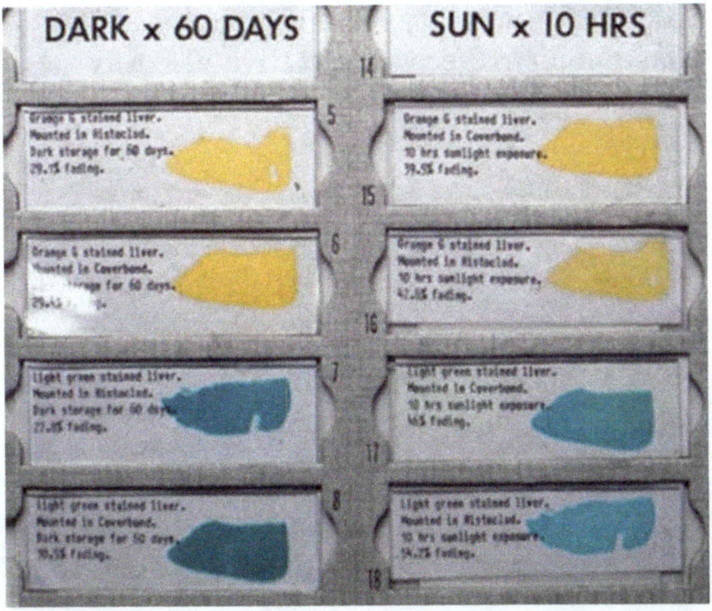

FIG. 16.4. None of the 25 tested mounting media prevented fading of orange G or light green SF yellowish stained sections under both the least and most extreme light exposures.

TABLE 16.3. Night and day changes in absorbance peaks.

	Percent decrease in peak absorbance			
	Storage in dark × 60 days		Sunlight exposure × 10 h	
Dye	Least	Most	Least	Most
Hematoxylin	0	19.8	0.8	18.3
Orange G	29.1	29.4	39.5	42.6
Eosin Y	0	12.4	3.2	81.7
Light green SF	27.8	30.5	46	54.7

Solubility in Aromatic Hydrocarbons and Concentrations Necessary to Give Viscosity Similar to that of 60% (w/v) Canada Balsam in Xylene

Absent a specification in terms of a standard measurement of viscosity such as centipoises (CPS) or milliPascal seconds, for example, this recommendation resists reduction to practice. Viscosity is resistance to shear and is a whole science unto itself.

Solutions: Nature and Amount of Solvent and Resin by Weight

Whether this information is provided is somewhat moot. What counts most is the total weight, and therefore volume, that is lost by evaporation. Table 16.4 shows how much weight is lost by each of 25 different mounting media. These losses were measured by first weighing blank slides, weighing each one again after a sample of mounting medium had been added and subtracting one weight of each pair from the other to equal how much weight of each medium was on the slides.

The slides were oven-dried overnight to drive the evaporative weight losses to the maximum. The slides were each weighed twice independently for accuracy. Fortunately, I had the benefit of having a Mettler analytical balance in my laboratory. Dividing the ending weights by the starting weights gave the results shown in Table 16.4. Whether these mounting media are available today is immaterial. In my view, this is among the basic information that all manufacturers of mounting media should provide with the products.

Laboratories are required to keep cytology and histology slides for 5, 10, or even 20 years. Archived slides that become unusable microscopically can be salvaged, but it takes know-how and work to do it right.

TABLE 16.4. Maximum evaporative weight loss among 25 tested mounting media.

No.	Product	Weight loss (%)	No.	Product	Weight loss (%)
1.	Damar	25.9	14.	Diaphane	51.8
2.	Bio-Mount	29.1	15.	Ames	52.0
3.	Lipshaw	32.1	16.	Eukitt	52.2
4.	Preservaslide	32.7	17.	Pro-Texx	55.4
5.	Cover Bond	33.4	18.	NAmount	57.3
6.	Technicon	33.6	19.	Permaslip	63.6
7.	Permount	33.8	20.	DPX	63.8
8.	HSR	34.3	21.	Flo-Texx	64.8
9.	Histoclad	36.1	22.	Trycolac	65.2
10.	Kleermount	36.7	23.	Gurr's MM	67.0
11.	Supermount	38.5	24.	Fixseal	71.1
12.	Wallabs' RMM	43.2	25.	Gelman	85.8
13.	Diatex	47.8			

Drying Rate of Solution: Rapid, Medium, Slow with Time Limits Stated

- Rapid: cover glasses set in 2–3 h. Slides will not stick to each other after 24 h drying time.
- Medium: cover glasses set in 1–2 days. Slides require more than 1 day but less than 2 weeks to dry to non-stickiness at 20–30° C.
- Slow: longer periods required.

This information is based on a laboratory's first-hand experience with a mounting medium. I'm unaware of any mounting medium that dries rapidly and allows slides to be filed within 24 h without first being oven-dried. Drying overnight in a 70° C hot air gravity convection oven works well. Since different mountants contain different solutes, be aware that some may darken in color if heated too long at that temperature.

Only cooking slides *à la méthode du* Graham as described in Chap. 18 allow immediate safe slide filing. Unfortunately, the method is laborious and impractical in today's world.

Freedom From Air Aspiration

Air aspiration means the solvent evaporates so rapidly that air is aspirated under the cover glass before an edge seal can form. An edge seal ordinarily forms around the perimeter of the cover glass at the interface of air and mounting medium. An edge seal forms a barrier that slows the evaporation of solvent, which is essential to a stable permanent mount. Xylene and toluene are the usual solvents in adhesive hydrophobes.

- Satisfactory for small sections of 5–10 μm
- Satisfactory for thick sections

Air aspiration is different than annual band formation, though both phenomena lessen the usefulness of the preparation for microscopic examination.

Annual bands are concentric granulations around the perimeter of the cover glass. They occur in some, but not all, synthetic resin-based mounting media at the rate of 1 band per year. Counting annual bands, like dendrochronology (counting tree rings), reveals the age of the preparation in years.[9] See Fig. 16.5. Adding 1 g of the oil-soluble food grade antioxidant 2, 6-di-tertiary butyl para-cresol to each 100 mL mounting medium inhibits this phenomenon.[10]

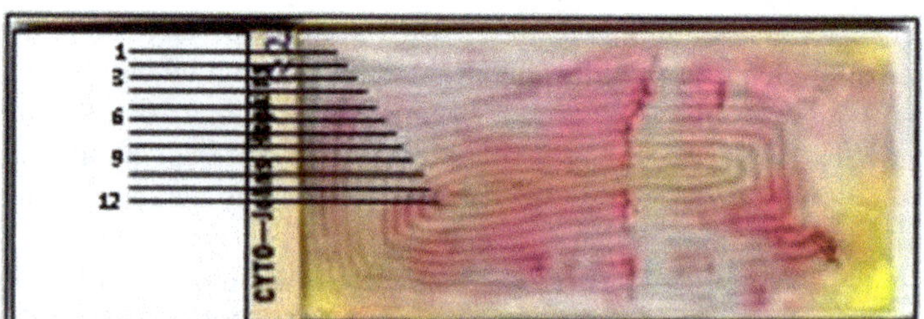

FIG. 16.5. This slide was prepared in 1960. Annual bands developed for 12 years until the resin was exhausted.

Index of Refraction of Mounting Solutions and of Solid Resin to 3 Decimals

Refraction is rooted in a word that means "to break." Refractive index is a measure of how much light bends as its passes from one transparent medium into another (e.g., air to glass, glass to mounting medium, mounting medium to fixed protein, mounting medium to membrane filter). See Fig. 16.6.

The refractive index of a mounting medium impacts the transparency and resolution of detail in microscopic objects. The ideal refractive index matches that of fixed protein, which is about 1.536. Minor differences are inconsequential. A simple way to appreciate the impact of refractive index is to examine microscopically any stained preparation that has been coverslipped with water as a temporary mounting medium.

All mounting media have two refractive indexes: one in the liquid state, which is a blend of the refractive indexes of the solvent and solutes, and the other in the dry state of the solutes that

Fig. 16.6. Refracted light "breaks" the straw at the interface of air (RI = 1.000) and water (RI = 1.333).

remain when the solvent has evaporated. Measuring the refractive index of the liquid is simple if one has a refractometer, which I did. See Fig. 16.7.

Measuring the refractive index of dried mounting medium is less straightforward. Two ways were available to me at the time: (1) mounting an object-of-interest (e.g., Millipore filters) in the mounting medium and (2) placing a drop of mounting medium on glass particles of known refractive index.

The first method was the simpler of the two. Remember, we were using Millipore filters exclusively in a research mode, trying to optimize the results in a study of circulating cancer cells in the blood. We were pioneers but didn't know it. Figure 16.8 shows blank Millipore filters mounted in 20 different mounting media and sitting on a piece of graph paper. Mounting media that let the lines be seen clearly were better matches than those that didn't.

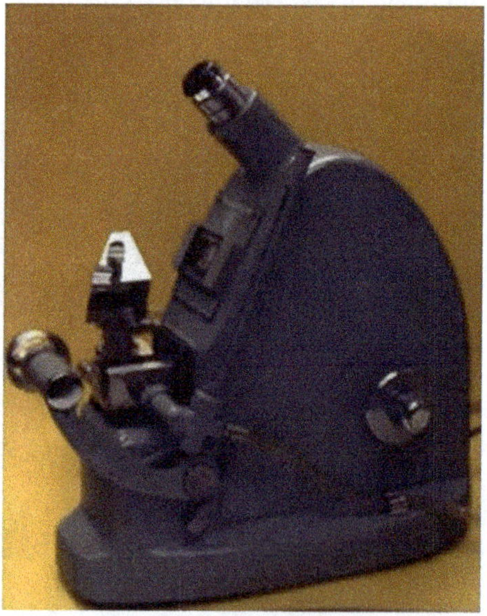

Fig. 16.7. The refractive index of a mounting medium for the D line of sodium was determined directly by using this Abbe refractometer.

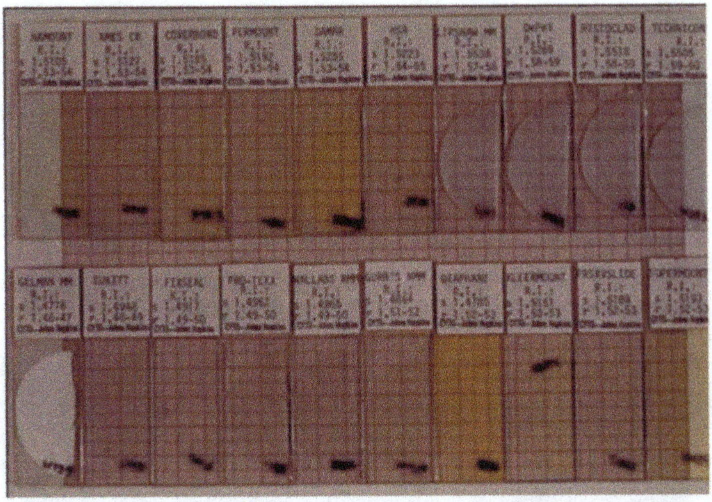

Fig. 16.8. This figure illustrates simply the impact of refractive indexes of different mounting media that closely match or don't match at all the refractive index of Millipore filters (1.495). We knew the refractive indexes because we had measured them. Knowing that information wasn't essential in this case. The mounted filter preparations had been dried in an oven, which explains why the filter in Gelman-mounting medium is white. The Gelman product loses so much volume to evaporation that there weren't enough solids remaining to occupy the filter's porous space. In other words, the Gelman filter became mounted in air.

Eukitt was the medium we used routinely. Its refractive index in solution (i.e., *s*) was 1.4948. As a dry resin (i.e., *r*), its refractive index was between 1.48 and 1.49. Clay-Adams Histoclad, on the other hand, couldn't make Millipore filters transparent. Its solution and resin refractive indexes were too high at 1.5518 and 1.59–1.60, respectively.

The second method of determining the refractive index of a dried mounting medium uses certified glass powders according to the method of Ludwig Kofler. There are 24 glass powders that range in refractive index from 1.3400 thru 1.6877 in intervals of approximately 0.01. Sprinkle a few crystals of each of several powders, 2 per slide, likely to bracket the expected refractive index of the solid mounting on a pre-labeled slide. See Fig. 16.9.

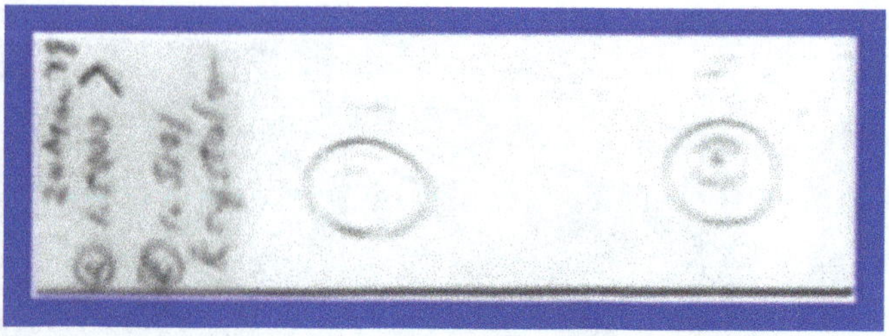

FIG. 16.9. This slide has glass powders with refractive indexes of 1.5000 on the left and 1.5101 on the right. A few drops of the same mounting medium have been added to each area. The slide was oven-dried. The glass particles were then examined microscopically using the Becke line test to determine the 0.01 refractive index window within which the dry mounting medium falls.

The Becke line is a relatively bright halo near the edge of a transparent particle immersed in a medium. It is named after Friedrich Johann Karl Becke who introduced it in the late nineteenth century for use in optical mineralogy. His method takes advantage of that fact that particles are usually thinner at the edges than at the center, thus behaving like a lens. Focusing above the plane of best focus always move the light into the material of higher refractive index. See Fig. 16.10.

The Becke line accounts for the refractile halo often associated with hyaline appearance of keratin as seen in keratinizing squamous cell carcinoma. The glassy boundary that appears to go in-and-out of the cytoplasmic border is due to the difference in refractive indexes between the protein and the mounting medium.

The refractive index of a mounting medium varies over time as its solvent evaporates, but it is a single value at any given moment. It can never have two different refractive indexes at the same time. Thus, the outline of pores in birefringent Nuclepore filters could not be made less visible by mounting them in a medium with matching refractive index. See Fig. 16.11.

Birefringence means having two refractive indexes as a consequence of a material's molecular structure. Figure 16.12 shows what birefringence looks like.

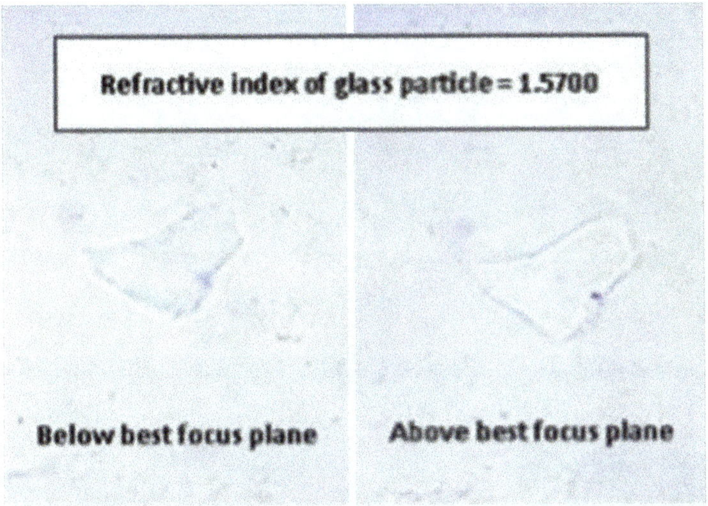

FIG. 16.10. The Becke line at work: The same mounting medium has been applied to a glass particle with a refractive index of 1.5700. Since the halo is within the particle when focused above, the refractive index of the glass is higher than that of the mounting medium. Stated another way, the refractive index of the mounting medium is lower than 1.5700. By applying the same medium to a glass particle with a lower refractive index, one can find one where the Becke line moves into the mounting medium when focused above the glass. One can then state that the refractive index of the dry mounting medium, for example, lies between 1.5611, which is the next lower refractive index glass powder, and 1.5700.

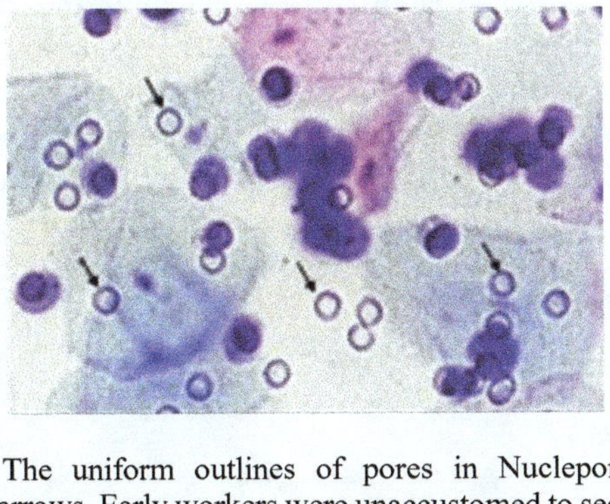

FIG. 16.11 The uniform outlines of pores in Nuclepore filters are identified by arrows. Early workers were unaccustomed to seeing the pores and mistakenly identified them as red cell ghosts, water droplets, or air bubbles. The pores themselves are empty spaces, so the interface between the filter and the space is responsible for the outlines. Dr. Frost often said, “hold your criteria high.”

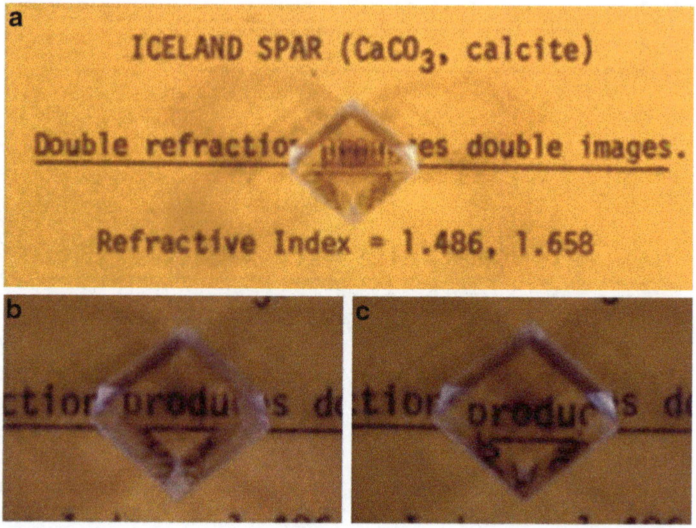

FIG. 16.12. (**a**) Iceland spar's two refractive indexes doubles the image of the word "produces." (**b**) holding the camera one way captures one of the two images; (**c**) holding it at another angle captures the second image of the same word. For this reason, no mounting medium—which can only have 1 refractive index—could ever match both refractive indexes of the Nuclepore filter and make the pore outlines disappear.

For a while, some laboratorians, including me, dissolved the Nuclepore filters in chloroform before mounting to obliterate the pores. Chloroform is a health hazard. Dissolving the filters crazed them and altered cell morphology by air-drying. For these and other reasons, Nuclepore filters in cytology applications went the way of the dodo bird. Under the generic name of polycarbonate filters, they continue to be used in other applications (e.g., TransCyt filters in the ThinPrep Processor).

Conclusion

Mounting media have the potential to positively and negatively impact the quality of cytologic preparations in a variety of ways, including chemical, optical, and physical. Again, I don't use the word quality loosely. As stated previously, quality means useful for its intended purpose. In a well-run cytopreparatory laboratory, there should be no unexpected surprises. Exercise due diligence. Understand the logic behind all products. Ask probing questions relentlessly.

Biological processes provide the raw materials; cytopreparation controls the artifacts on which we base our interpretations cytomorphology.

References

1. Bracegirdle B. A history of microtechnique. Ithaca, NY: Cornell University Press; 1978.
2. Lillie RD, Windle WF, Zirkle C. Interim report of the committee on histologic mounting media: resinous media. Stain Technol. 1950; 25(1):1–9.
3. Lillie RD, Zirkle C, Dempsey EW, Greco JP. Final report of the committee on histologic mounting media. Stain Technol. 1953;28(2): 57–80.
4. Seal SH. A method for concentrating cancer cells suspended in large quantities of fluid. Cancer. 1956;9(5):866–8.
5. Dakin ES. Improved adhesion and visibility of cytologic preparations by use of the frosted glass slide. Science. 1955;121(3144):424–5.
6. Hollander DH. An oil-soluble anti-oxidant in resinous mounting media to inhibit fading of Romanowsky stains. Stain Technol. 1963;38:288–9.
7. Seal SH. A sieve for the isolation of cancer cells and other large cells from the blood. Cancer. 1964;17(5):637–42.
8. Barr WT. Effects of sunlight on stained sections mounted in various media. Stain Technol. 1970;45(1):9–14.
9. Hollander D, Frost JK. Annual bands in synthetic-resin-mounted microscopic slides. Acta Cytol. 1970;14(3):142–3.
10. Hollander DH, Frost JK. Antioxidant inhibition of stain fading and mounting medium crazing. Acta Cytol. 1971;15(5):419.
11. Baker JR. Principles of biological technique: a study of fixation and dyeing. Bungay, Suffolkd, Great Britain: Richard Clay (The Chaucer Press); 1968 reprinting.
12. Billmeyer Jr FW. Textbook of polymer science. 2nd ed. New York: Wiley-Interscience; 1971.

13. Bowden FP. Adhesion and friction. Endeavour. 1956–1957;15–16: 5–18.
14. Davies F. Mounting media and the fading of histological preparations. J Anat. 1939-1940;74:464–70.
15. DeBruyne NA. How glue sticks. Nature. 1957;180:262–6.
16. Greco JP. Refractive indices of currently used mounting media. Stain Technol. 1950;25(1):11–2.
17. Groat RA. Evaluation of isobutyl methacrylate polymer as a mounting medium. Science. 1940;92:268.
18. Jeannot AB. The use of quick drying media for a cytology screening laboratory. Acta Cytol. 1969;13:661.
19. Kirkpatrick J, Lendrum AC. Further observations on the use of synthetic resin as a substitute for Canada balsam. Precipitation of paraffin wax in the medium and an improved plasticiser. J Path Bact. 1941; 53:441–3.
20. Liu W. Further notes on cell refractometry using phase contrast microscopy. Acta Cytol. 1967;11(1):51–3.
21. Marshall W. The influence of refractive index on mounting media. J R Microsc Soc. 1932;52:275–80.
22. Meyrowitz R. A compilation and classification of immersion media of high index of refraction. Am Mineral. 1955;40:398–409.
23. Oettlé AG. "Optical membranes": a common artefact. J R Microsc Soc. 1950;70(3):255–65.
24. Perry JS. A plastic lacquer substitute for coverslips. Q J Microsc Sci. 1954;21:159–61.
25. Ross KFA. The properties of cytological phase objects. In: Phase contrast and interference microscopy for cell biologists. London: Edward Arnold; 1967.
26. Senior W. Polyester resin as a mounting medium for light microscopy. J Microsc. 1970;21(3):207–10.
27. Swartz FJ, Nagy ER. Feulgen stain stability in relation to three mounting media and exposure to light. Stain Technol. 1963;38: 179–85.
28. Wicks LF, Carruthers C, Ritchey MG. The piccolyte resins as microscopic mounting media. Stain Technol. 1946;21(4):121–6.
29. Zimmerman Y. A synthetic medium to replace coverslips. Tech Bull Regist Med Technol. 1963;33:168–9.
30. Baker JR. Cytological technique: the principles underlying routine methods. 5th ed. London, England: Methuen & Co. Ltd.; 1966.

Chapter 17
Cover Glasses

PRINCIPLE NO. 6

Mount to optimize microscope objective's performance.

PRACTICE

Use No. 1 thickness cover glasses.

Historic Milestones

- 1789—Mica coverslip introduced[1].
- 1840—Chance Brothers of Birmingham, England, introduces cover glasses commercially[1].
- 1880s—Otto Schott in Germany makes glass for cover glasses and collaborates with Ernst Abbe and Carl Zeiss in 1886 to make glass for apochromatic microscope objectives.
- 1953—Royal Microscopical Society of London publishes specifications of a standard microscope cover glass[2].
- 1963—American Society for Testing and Materials published standard specification for cover glasses and glass slides for microscopy[3].
- 1967—Thickness of mounting medium between histologic section and underside of cover glass measured with micrometer microscope[4].

G.W. Gill, *Cytopreparation: Principles & Practice*,
Essentials in Cytopathology 12, DOI 10.1007/978-1-4614-4933-1_17,
© Springer Science+Business Media New York 2013

Cover glasses and mounting medium together constitute the de facto front lens of every microscope objective. Even though flat, and not curved as lenses are, their combined thickness and optical properties influence the quality of image formation by objectives and eyepieces. Those who coverslip are, in effect, completing the assembly of a microscope objective with every cover glass they apply. So while coverslipping is a tedious process, its visibly appreciable contribution should not be overlooked.

I first began to pay attention to cover glasses when coverslipping Millipore filters, which we used to collect saponin-treated peripheral blood samples. Millipore filters are highly porous and about as thick as a cover glass. They must be mounted in enough mounting medium to fill-in the pores when the solvent later evaporates. That initial thickness of extra mountant contributed to hazy imaging with 40× objectives and led to our questioning what we thought to be true about mounting slides.

Among other common daily observations that were cover glass, and mounting medium, thickness-related:

- Images of just-mounted cells were sharp when viewed with the 10× objective but not the 40× objective.
- Images were sharp when viewed with both magnification objectives a few weeks after the slides had been mounted.
- Images were sharp when viewed with a 10× objective when the preparation had inadvertently been placed upside down on the microscope stage. Substantial refocusing was required to be sure, but the cause wasn't apparent that I switched to a 40× objective and couldn't focus.

Cover glasses are not simply thin pieces of glass, otherwise they wouldn't cost 15 times more than slides, ounce for ounce. There are scientific reasons on which coverslipping do's and don'ts are based. It is optically and physically impossible for alternatives such as liquid substitutes[5–7] and plastic film[8] to duplicate the specifications and performance of cover glasses. Such substitutes are used occasionally, however, perhaps because little reliable technical information is available to discourage such use. Quality outcomes are based on quality processes, not the other way around.

Royal Microscopical Society Specification of a Standard Microscope Cover Glass

The fog began to clear when I came across a 1953 paper in which the results of a survey of microscope manufacturers were published.[4] The survey was sponsored by the Standards Committee of the Royal Microscopical Society (RMS). It had been brought to their attention that discrepancies in cover glasses occasionally were responsible for an "appreciable falling off in image definition."[2] Of five identified properties of cover glasses, three are quantitative. The author had asked manufacturers of microscope objectives to submit the values they adopted for the variables. The responses were compiled into the following specifications.

The RMS specifications for a standard microscope cover glass are:

1. Thickness: $d = 0.18\text{mm} \pm 0.003\text{mm}$
2. Refractive index: $N_D = 1.524 \pm 0.0007$
3. Dispersion: $V = \dfrac{N_D - 1}{N_F - N_C} = 52.0 \pm 2.75$
4. Homogeneity: N_D consistent within ± 0.0007
5. Surface quality: Machine polished parallel and flat on both faces, to "plate-glass" quality

Refractive index indicates the amount of bending by light as it passes from one transparent medium (e.g., air) into another (e.g., glass). Since light of different wavelengths (i.e., color) bend differently, refractive index is measured using a refractometer and sodium (yellow) light, indicated by subscript D (i.e., $_D$). Dispersion is a measure of chromatic aberration and is responsible for familiar phenomena such as rainbows. Homogeneity simply means the refractive index varies little throughout the glass.

American Society for Testing and Materials Standard Specification E211

In 1963, American Society for Testing and Materials (ASTM) published its first standard specification for cover glasses and glass slides for use in microscopy. It has been republished numerous times essentially unchanged.[3]

According to ASTM Standard Specification E211, "the cover glasses and glass slides covered by this specification shall consist of colorless, transparent glass that is free of pits, nicks, bubbles, striae, scratches, and cloudiness when observed by the unaided eye in a good light in front of a dark surface. The glass for covers shall have refractive indexes as follows:

2.1.1 *Type I* for critical microscopy: 1.523±0.0005
2.1.2 *Type II* for routine use: 1.52±0.02"

The difference in refractive between RMS (1.524) and ASTM (1.523) specifications is inconsequential.

ASTM E211 specifies 4 thickness ranges. See Table 17.1.

No. 1.5 thickness cover glasses are often recommended, as this is the thickness range within which the 0.17 engraved on most microscope objectives appears. The number 0.17 means 0.17 mm, the thickness of cover glass for which the objective is designed to be used.

That recommendation is correct, if and only if, the specimen is in direct contact with the underside of the cover glass, without intervening mounting medium. Examples include cells cultured on cover glasses, blood films spread on cover glasses, Nuclepore filter preparations dissolved on a cover glass, or "cooked" slides

TABLE 17.1. Cover glass thickness is indicated by numbers. Only Nos. 1 and 1.5 are relevant to cytopreparation; Nos. 0 and 2 are not.

No.	Thickness (mm)
0	0.085–0.13
1	0.13–0.17
1.5	0.17–0.19
2	0.19–0.25

(see Chap. 18). A No. 1.5 cover glass is also satisfactory when preparations will be examined *only* by low numerical aperture objectives that are insensitive to cover glass thickness. These scenarios are uncommon. For reasons to be explained, No. 1 thickness cover glasses should be used routinely. It causes me no end of consternation when people who should know better recommend No. 1.5 cover glasses, regardless of application.

Tolerance of Microscope Objectives to Deviations from 0.17-mm Thick Cover Glasses

Setterington concluded his 1953 paper thus: "It should be pointed out that a great deal of routine microcopy can be done without achieving the best possible performance from an objective, since the detail being examined is within the resolution limit of the objective even when some aberration is present.

Several major assumptions are made in the calculations and arguments laid out in this paper and, while the need of a standard cover glass is in no way invalidated by these assumptions, some mention of them should be made in order that all aspects of the problem are appreciated. If we consider each variable in turn, we observe:

(a) The specimen must be mounted in contact with the undersurface of the cover glass, otherwise a layer of mounting fluid is interposed which is equivalent to an increase in cover-glass thickness.

(b) The layer of the specimen must be thinner than 0.003 mm [3 μm], or the lower surface will lie beyond the tolerance on the cover-glass thickness. This is a major problem and some compensation must be provided since it is quite usual, for example, to mount biological sections with thicknesses of the order of 0.02 mm [20 μm]."

It is exceedingly rare that any specimen is routinely "mounted in contact with the under-surface of the cover glass." The tolerance of a microscope objective to deviations from the recommended thickness of cover glass, and by extension, the mounting medium,

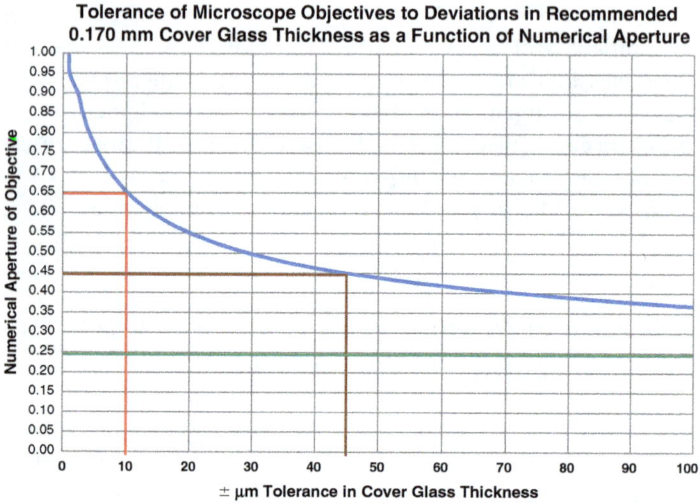

Fig. 17.1. The tolerance of microscope objectives to deviations in the thickness of cover glasses, and added thickness of mounting medium, from the 0.17 mm recommended thickness is a function of numerical aperture. The color of the statements is keyed to the colored lines in the graph.

- Below NA 0.35, objectives are insensitive to cover glass and mounting medium combined thickness. This explains why 10× planachromat objectives, among the most common, image objects sharply even when preparation is upside down on the microscope. Under those circumstances, the 1 mm (1000 µm) thick slide is acting as a cover glass.
- 10× planapochromat objectives, the most highly corrected, have NA 0.4, and can tolerate deviations of ± 45 µm thickness before the resulting spherical aberration degrades image quality perceptibly.
- The NA of 40x planachromat objectives is usually 0.65, and can only tolerate thickness deviations of ±10 µm before image quality is noticeably degraded.
- 0.95 is the highest NA dry objective available. It has virtually zero tolerance for thickness deviations; the working distance is essentially zero. Thus, such an objective is sold with a correction collar that must be adjusted as needed to achieve best image quality. In practice, the user adjusts the correction collar with one hand, while focusing on the image with the other hand (i.e., so-called "dynamic focusing"). It is easy to crack the cover glass; the objective is wildly expensive. For these reasons, I don't recommend these particular objectives for routine use.

depends on its numerical aperture (i.e., NA). Therefore, what deviations from the recommended cover-glass thickness are permissible for different NA objectives?

The following equation calculates the "evaluation of a toleration for cover-glass thickness."[2]

$$\delta d = N'\lambda \mathrm{NA}^2 \frac{N'}{N} - (N2 - \mathrm{NA}^2)1/2(N^2 - \mathrm{NA}^2)$$

δd = permissible change of cover-glass thickness

N' = 1.0 (refractive index of immersion medium [air = 1, immersion oil = 1.515])

λ = wavelength 0.000589 mm (wavelength of spectral line of sodium which is commonly used in measuring refractive index)

N = 1.524 (cover-glass refractive index)

NA = numerical aperture of objective

Figure 17.1 illustrates the tolerance of microscope objectives to deviations in cover glass over the range of possible NA. Cover glasses that are too thick *or* too thin can degrade image quality. From a practical standpoint, however, too thick is usual.

Numerical Aperture Impact on Image Quality

Numerical aperture is the factor that determines thickness sensitivity of objectives. NA is a measure of the angle of the cone of light that enters an objective: the wider the cone, the greater the resolution and the greater the optical challenge of reuniting the marginal image-forming rays into sharply focused points.

Planapochromat objectives with numerical apertures of 0.95 are virtually intolerant of deviations from design specification for cover glass thickness. Such objectives have correction collars that require interactive double focusing by the user. The user focuses on the specimen with 1 hand, while adjusting the correction collar with the other hand until optimal resolution is obtained. This time-consuming task is impractical for routine applications

Table 17.2 lists the three quality levels of objectives and their respective numerical apertures.

TABLE 17.2. **Green NA** objectives are insensitive to cover glass and mounting medium combined thickness, whereas **red NA** objectives are sensitive. See Fig. 17.1.

Objective	Numerical apertures		
Magnification	Planachromat	Planfluorite	Planapochromat
×4	**0.10**	**0.13**	**0.20**
×10	**0.25**	**0.30**	**0.45**
×20	**0.40**	**0.50**	**0.75**
×40	**0.65**	**0.75**	**0.95**
×60	**0.75**	**0.85**	**0.95**

Power for power, achromat objectives—the usual and lowest quality objectives—are less sensitive to cover glass and mounting medium thickness than fluorite and apochromat objectives, since achromats have lower numerical apertures.

Planachromat objectives are perfectly suitable for everyday use. For slightly more money, one can buy planfluorite lenses that image objects that are noticeably sharper and brighter. Fluorite lenses are also known as semi-apochromats. Planapochromats are prohibitively expensive and require nearly ideal preparations to perform best.

Mounting Medium Thickness

The thickness of mounting medium in mounted tissue sections has been measured. Three sets of 4 slides each were broken across the section; the broken edges trued up, polished, and measured with a micrometer microscope.[4] Note the coauthor Setterington also authored the 1953 paper.[2]

Set #1 slides had been mounted routinely, without pressure added to the cover glass. Set #2 slides were mounted with a 1 oz weight applied to each cover glass while the slides were in a drying oven of unspecified temperature. Set #3 slides "had spring clothes-pegs applied for 72 h after mounting."

Figure 17.2 illustrates the thickness of mounting medium that was measured.

The thickness measurements are shown in Table 17.3.

Therefore, the thickness of mounting medium in routinely coverslipped slides is substantial relative to the difference

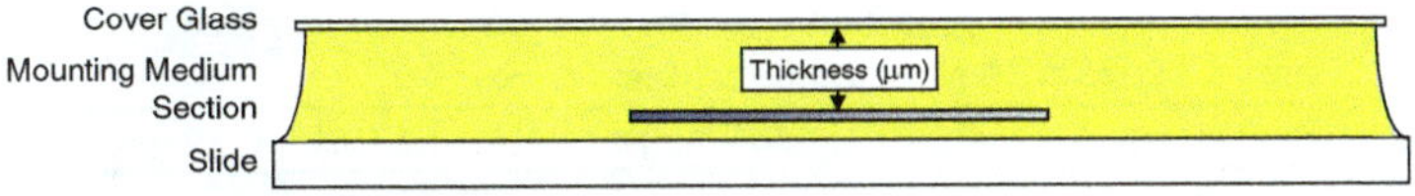

FIG. 17.2. Diagrammatic transverse section of a slide, with the section, mounting medium, and cover glass.

TABLE 17.3. In the thickness of the mounting medium, "it will be noticed that the layer of mountant between the section and the cover glass is never less than the thickness of the section itself, and may be nearly four times this thickness."[4]

Mounting medium thickness under different forces		
Routine (i.e., no added force [gravity])	1 oz weight × 48 h in drying oven	Spring-loaded clothes-peg × 72 h
10 μm	18 μm	5 μm
51 μm	18 μm	10 μm
63 μm	20 μm	10 μm
76 μm	30 μm	20 μm

between the range of thicknesses for No. 1 cover glasses and the tolerance of NA 0.65, 40× achromat objectives to deviations from optimal thickness of 0.17 mm (±10 μm and more). This is the basis for recommending the routine No. 1 cover glasses.

Unitized Pricing

An ounce of cover glasses costs the same regardless of dimensions and thickness. Cover glasses are priced by the ounce. A dollar buys more small thin cover glasses than large thick cover glasses. See Table 17.4 and Fig. 17.3. Save money by purchasing No. 1 cover glasses large enough for the intended preparation and not larger. For example, a 13-mm SurePath preparation can be covered by an 18×18-mm cover glass (233 pieces/oz); 20-mm ThinPrep, 24×30 mm (108 pieces/oz); and 24×50-mm conventional Paps (67 pieces/oz). Automated coverslippers are limited in the range of acceptable sizes and so may not allow the laboratory to use size cover glass sizes that are most economical.

TABLE 17.4. Approximate number of cover glass pieces per ounce varies with size and thickness.

Size (mm)	No. 1 (0.13–0.17 mm)	No. 1.5 (0.16–0.19 mm)
18 × 18	233	202
22 × 22	156	138
22 × 30	116	102
22 × 35	100	87
22 × 40	87	76
22 × 50	70	61
22 × 60	58	51
24 × 30	108	94
24 × 35	94	82
24 × 40	80	70
24 × 50	67	58

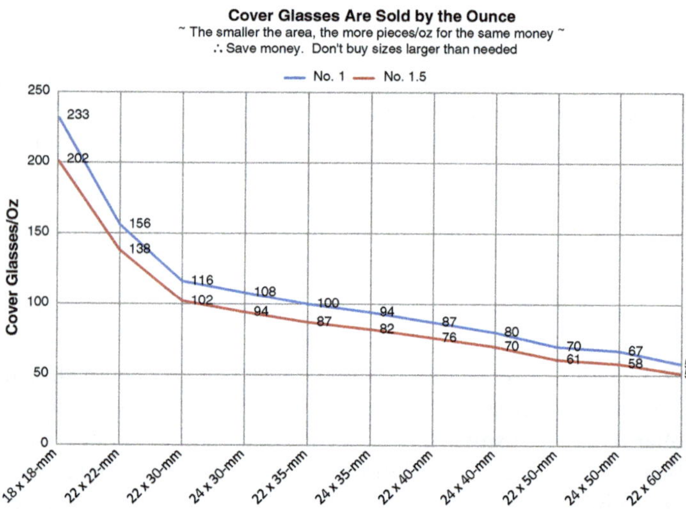

FIG. 17.3. The number of cover glasses/oz varies with cover glass dimensions and thickness. For example, an ounce of No. 1 24 × 30-mm cover glasses has 41 more pieces than an ounce of No. 1 24 × 50-mm cover glassess (i.e., 108 vs. 67) and 50 more pieces than No. 1.5 24 × 50-mm cover glasses (i.e., 108 vs. 58). Buying the smaller No. 1 cover glass saves 38% of the cost of buying the same number of the larger No. 1 cover glass.

Cover Glass Dimensions

For nonliquid-based preparations that may have cells spread over the entire 24 × 50-mm cell display area, the length and width of the cover glass determines the working area of the preparation that can be imaged. The question arises whether there is a standard area cover glass? For conventional Pap tests, the de facto standard size has been the 24 × 50 mm.

In the mid-1990s, the national medical director of a major laboratory increased the standard cover glass size for its Pap tests to 24 × 60 mm. The change was made for risk management purposes. It was believed that abnormal cells might occasionally occur exclusively outside the uncovered area near the label end of the slide and be missed during routine screening. The change in cover glass size was well intended but not supported by any evidence.

Using cover glasses that were 10 mm longer increased the area to be screened by 20%. Cytotechnologists were faced with an unpleasant choice: (1) either overlap less to maintain their screening productivity or (2) overlap the same and watch their output drop. One cannot arbitrarily increase the cover glass size, and at the same time, expect cytotechnologist screening productivity to remain unchanged. Screening is a zero-sum game and something's got to give.

In the late 1990s, I began employment at a large cytology laboratory that was doing overflow screening for the same laboratory that had switched to the larger cover glass size. For convenience, the second laboratory had begun to use the larger cover glass size as well—without appreciating the practical consequences.

I successfully demonstrated that using the larger cover glass was cost-inefficient and ineffective. To do this, I pulled 100 consecutive abnormal conventional Pap tests from the files, placing each dotted slide on a scale drawing of a slide with a 24 × 60-mm cover glass area divided into six 10-mm wide columns. I counted the number of dots that fell within each of the six columns. See Fig. 17.4.

Elsewhere, others had concluded similarly about potential negative consequences possibly associated with using even smaller size coverslips: "Nine hundred and twenty-three smears

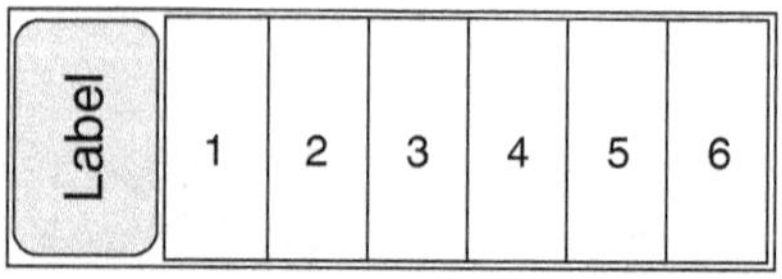

FIG. 17.4. Of 1,230 dots that were counted among 100 abnormal conventional Pap tests, only 68 (5.5%) were in the label-adjacent column. Two-thirds of the time, no dots were in the label-adjacent column. When dots were present at all in that column, they weren't the only dots on the entire slide. Each of the other five columns had more dots on average.[5] These findings convinced my very conservative employer to once again use 24 × 50-mm cover glasses for its conventional Pap tests.

covered by 40 × 22-mm-size coverslips were examined inside and outside the coverslip area to determine whether this coverslip size could be responsible for missed dyskaryotic cells in conventional cervical cancer screening. There was no instance when abnormal cells seen outside the coverslip were not also present within the coverslipped area."[6]

Conclusion

Cytologic preparations are thicker and more problematic than histologic preparations. Conventional Pap smears, for example, can sometimes require up to 12 or more drops of mounting medium to fill in all the valleys of thick preparations. Liquid-based preparations are on average much thinner but still benefit from the use of No. 1 thickness cover glasses.

Cover glass and mounting medium thickness should be controlled within narrow tolerances to avoid unacceptable spherical aberration. Spherical aberration makes object images look washed out, hazy, cloudy, milky, and low in contrast. It looks the same as, and is visually indistinguishable from, that produced by glare and flare. Glare is scattered light and is loosely defined as light that does not contribute to image formation. See Fig. 17.5.

The proactive practice, therefore, is to use No. 1 cover glasses, which are thinner than No. 1.5 cover glasses (0.13–0.17 mm vs.

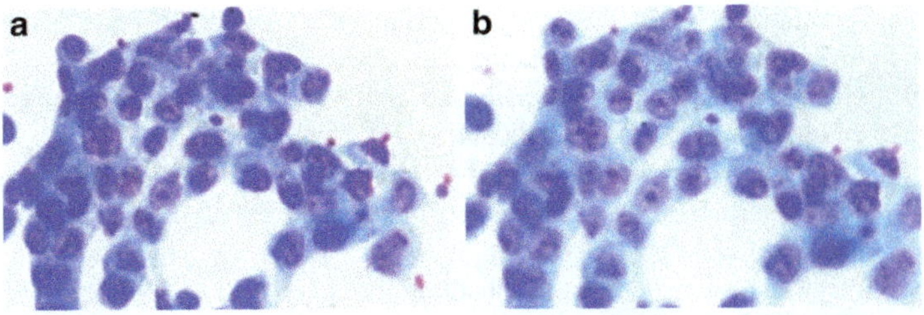

Fig. 17.5. In contrast to the sharply imaged malignant tissue fragment in (**a**), the image of the same fragment in (**b**) appears washed out, hazy, cloudy, milky, and low in contrast. Such features can be caused by excessively thick mounting medium and/or cover glass (i.e., spherical aberration), and/or glare and/or flare (i.e., light scattered by wide open field diaphragm and/or substage condenser aperture diaphragm), and/or dirt on lens surfaces. Since all these possible causes result in similarly degraded images, it is essential to know how to troubleshoot the problem and solve it.

0.17–0.19 mm) and as little mounting medium as is consistent with a permanent mount after the solvent evaporates. Some mounting media contribute to biological dyes fading even when stored in the dark. Since there is no convenient way of knowing in advance which mounting media fade dyes, one must simply be aware of this factor as a possibility. Always store stained slides in the dark.

References [7–23] are included for the historical record. All were published decades ago in journals few of us have ever heard of, much less have access to, or have read. All, however, are timeless.

References

1. Bracegirdle B. A history of microtechnique. Ithaca, NY: Cornell University Press; 1978.
2. Setterington R. VI. – The specification of a standard microscope cover-glass. J R Microsc Soc. 1953;73(2):69–76.
3. ASTM Standard E211 – 82(2010), "Standard Specification for Cover Glasses and Glass Slides for Use in Microscopy." West Conshohocken, PA: ASTM International; 2005. doi: 10.1520/E0211-82R10. http://www.astm.org.

4. Aumonier FJ, Setterington R. Some notes on the mounting of histological sections. Proc R Microsc Soc. 1967;2:428–9.
5. Gill GW. Cover glass dimensions and Pap smears: 24x60-mm vs. 24x50-mm. Acta Cytol. 2000;44(5):925. Abstract 135.
6. Oommen R, Curling M, Adams C, Barrett E, Titmuss E, Widdowson A. Evaluation of 40 × 22 mm coverslip area in cervical cancer screening. Cytopathology. 1998;9:46–9.
7. Perry JS. A plastic lacquer substitute for coverslips. Q J Microsc Sci. 1954;95(Pt 2):159–61.
8. Ransom JP. An acrylic spray used as a substitute for cover glasses. Stain Technol. 1953;28(6):279–80.
9. Zimmerman Y. A synthetic medium to replace coverslips. Am J Clin Pathol. 1963;4:454–5.
10. Richards OW, Small JH, Collyer PW. Microscopy with plastic substitutes for cover glasses. Stain Technol. 1944;119(2):58–62.
11. Anon. Federal specification for slide, microscope. NNN-S-450B, 27 Sep 1976, 11 pages.
12. Anon. Interim federal specification for cover glass, microscope. NNN-C-001434A (GSA-FSS), 8 Jan 1971, 6 pages.
13. Anon. The influence of cover glasses upon microscopic image clarity… Leitz Technol Info Bull. 1961;1(4):1–3.
14. Bracey RJ. I. – The aberrations of microscope objectives and their variations with small departures from optimum working conditions. J R Microsc Soc. 1952;72:1–9.
15. Bracey RJ. III. – A universal tube-length and cover-glass correcting lens system for use with microscope object-glasses. J R Microsc Soc. 1934;51:20–3.
16. Gill GW, Frost JK. In defense of cover glasses. Cytotechnol Bull. 1983;20(4):51–4.
17. Klosevych S. Photomicrography – importance of the cover glass thickness. J Biol Photogr Assoc. 1966;34(1):1–6.
18. Klosevych S. Further notes on the importance of cover glass thickness. J Biol Photogr Assoc. 1966;34(3):101.
19. Norris KP. XXIV. – Some observations on microscope cover-glass. J R Microsc Soc. 1961;79(Pt 4):287–98.
20. Pijper A, Kistner A. VII. – Slides and cover-glasses for dark-ground microscopy. J R Microsc Soc. 1951;71:176–80.
21. Sartory PK. The effects of cover glass thickness on the spherical corrections of objectives. Microscope. 1949;7:122–6.
22. Spinell BM, Loveland RP. V. – Optics of the object space in microscopy. J R Microsc Soc. 1960;79(Pt 1):59–80.
23. Uhlig, von M. Einflusz der Deckglasdicke und des n_d-Wertes der Deckgläser.

Chapter 18
Mounting

PRINCIPLE NO. 6

Mount preparations to optimize microscope objective's performance.

PRACTICE

Apply a thin layer of mounting medium and cover with No. 1 thickness cover glass.

Historic Milestones

- 1953—Mounting medium extends effective cover glass thickness.[1]
- 1963—"Cooking" slides described.[2]
- 1969—Brown artifact described.[3]

Mounting marks the point of departure from the materials and methods that interact with the specimen to those that influence its image. Mounting is also referred to as coverslipping. Mounting completes the glass sandwich: (1) glass slide, (2) cells, (3) mounting medium, and (4) cover glass. The applicable chemical, optical, and physical properties of each of the 4 components are described throughout in this volume.

G.W. Gill, *Cytopreparation: Principles & Practice*,
Essentials in Cytopathology 12, DOI 10.1007/978-1-4614-4933-1_18,
© Springer Science+Business Media New York 2013

This chapter addresses the process of coverslipping and associated events that aren't always obvious. Whether automated or manual, coverslipping constitutes the first lens of microscope objectives. Details matter:

- Apply as little mounting medium to a slide as will remain intact after the solvent has evaporated.
- Coverslip rapidly to minimize the likelihood of air being deposited on the surface of superficial squamous cells (i.e., cornflaking, corn flake cells, the brown artifact).
- Use a mounting medium that promotes an archival quality preparation.
- Select a cover glass size no larger than needed to cover the preparation.

Coverslipping Millipore Filters

Coverslipping highly porous, relatively thick cellulosic membrane filters such as Millipore requires modifications that help ensure satisfactory preparations. Again, I appreciate that few cytology laboratories use Millipore filters today. Nonetheless, Millipore filters are uniquely valuable and in the right hands can answer questions other methods cannot.

Materials for Halving 47-mm Millipore Filters (See Fig. 18.1)

- Small bottle of xylene with Pasteur pipette
- Flat-tip forceps
- Disposable aluminum foil dish, 70 x 17-mm
- Eukitt mounting medium
- Petri dish, top or bottom, 100 x 20-mm
- 41-mm diameter mallet handle die with a diametrically mounted blade; tool steel
- Soft face 16-oz plastic hammer
- 3 x I-in. micro slides, plain, with one side of on end frosted
- No. I thickness, 24 x 50-mm cover glasses

FIG. 18.1. The materials to halve and mount cellulosic filters as shown here simplify the process. They are more than a casual user needs.

Method

1. Remove the filters, one at a time, from xylene and remove from the carrier that was used during staining.
2. Place the filter cell-side-up on a piece of illustration board wet with xylene. Saturate to avoid drying out. See Fig. 18.2.
3. Place the punch eccentrically on the filter, so that (1) the margin in which the accession number is written twice will not be cut off, (2) the halving blade will evenly divide the filter so that each half will retain an accession number, and (3) only acellular margin is removed. If necessary, wet the filter with xylene to keep the cells moist. See Fig. 18.3.
4. Strike the punch once with the hammer. See Fig. 18.4.
5. Discard the trimmed margin.
6. Without touching the cellular area, immerse each filter half in mountant. See Fig. 18.5.
7. Cover the immersed filters. After 5 minutes, remove the filter until the last one or two drops resist dropping off. See Fig. 18.6.
8. Without adding any more mounting medium to the cellulosic filter, carefully center the filter numbered-side-up (i.e., cell-side-up) across the display area of a pre-identified microslide—making certain that all air bubbles are excluded.

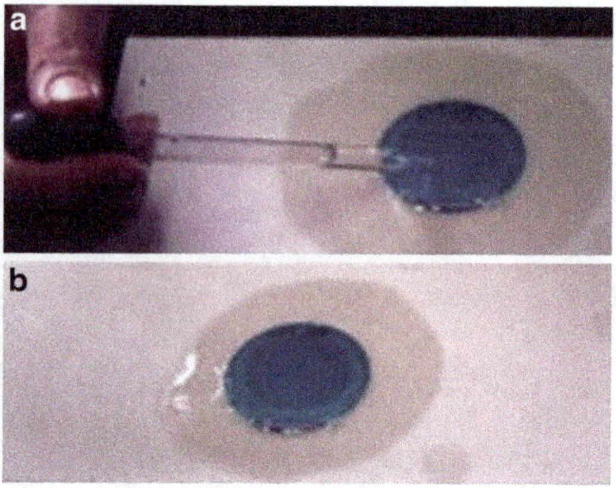

FIG. 18.2. (**a**) Wet filter with xylene. (**b**) Saturate the illustration board that is used as a cutting surface.

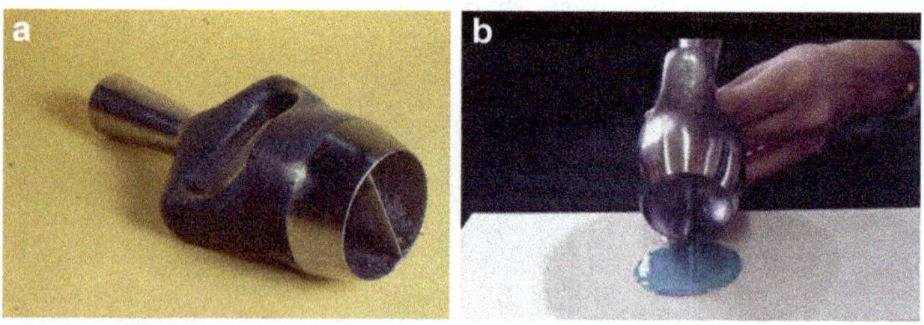

FIG. 18.3. (**a**) Custom-made 41-mm diameter punch with diametrically mounted blade. (**b**) Position the punch as described in step 3.

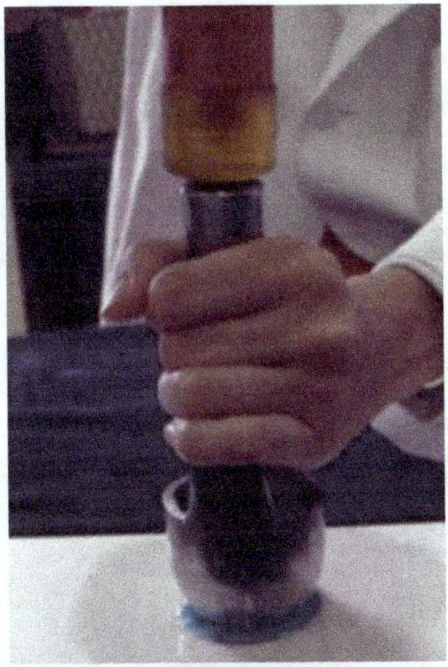

FIG. 18.4. Wipe off the cutting edges of the punch before using it with another specimen.

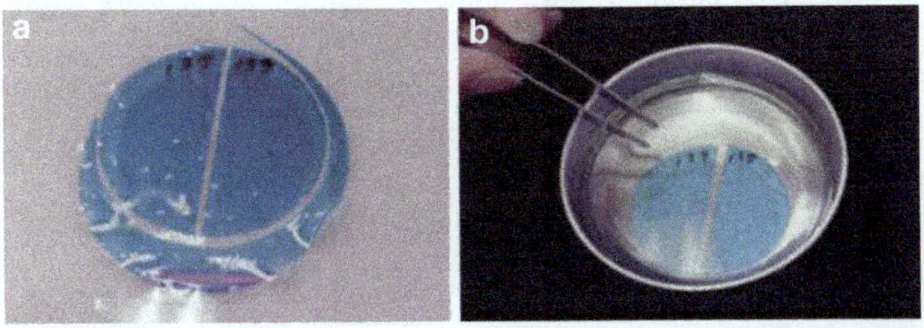

FIG. 18.5. (a) The custom punch neatly halves the filter and trims the acellular margin. (b) Immerse each half of the same filter per preparation in mounting medium.

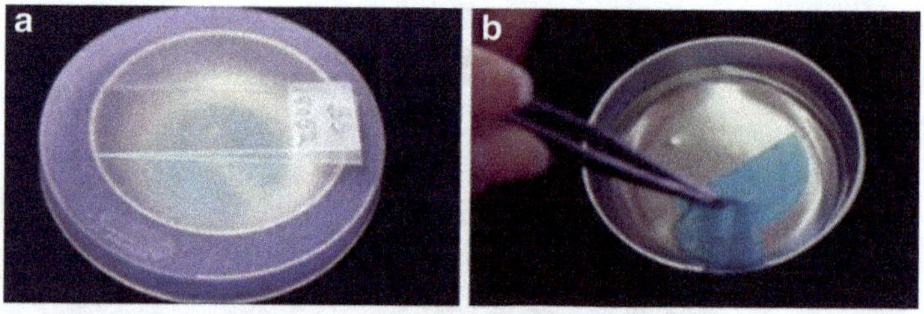

FIG. 18.6. (a) Covering the setup helps maintain the mounting medium for future use after being filtered. (b) Remove each filter half.

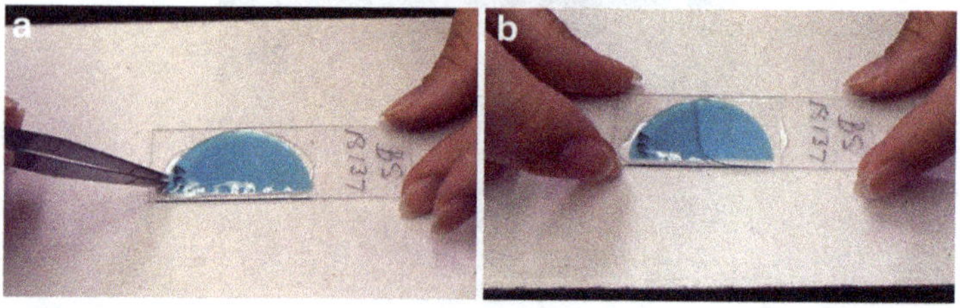

FIG. 18.7. (a) Exclude air bubbles while laying the filter across the slide and (b) while coverslipping.

9. Lower a 24 × 50-mm No. 1 thickness cover glass onto the filter—again excluding all air bubbles. Align the cover glass within the edges of the slide. See Fig. 18.7.
10. If any mounting medium has overrun the boundaries of the cover glass onto the cover glass or slide, wipe it clean with xylene-moistened cheesecloth.

Results

A properly prepared Millipore filter preparation has the potential to exceed all expectations. See Figs. 18.8 and 18.9.

Fig. 18.8. This Millipore filter was prepared in 1977 and photographed 35 years later.

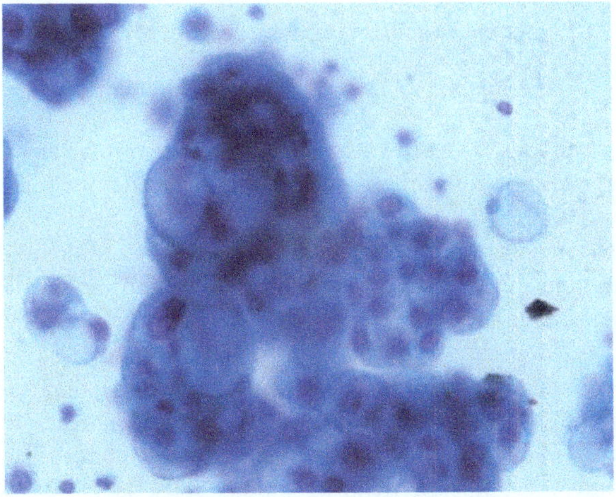

Fig. 18.9. Adenocarcinoma tissue fragment from pleural effusion photomicrographed 35 years after having been prepared in 1977. Papanicolaou stain, original magnification ×400.

Discussion

Brown Artifact (aka "Cornflaking")[4]

The brown artifact is air deposited on superficial squamous cells. It is also known as corn flakes cells, or simply cornflaking, because of the crinkled appearance imparted to the surface of these

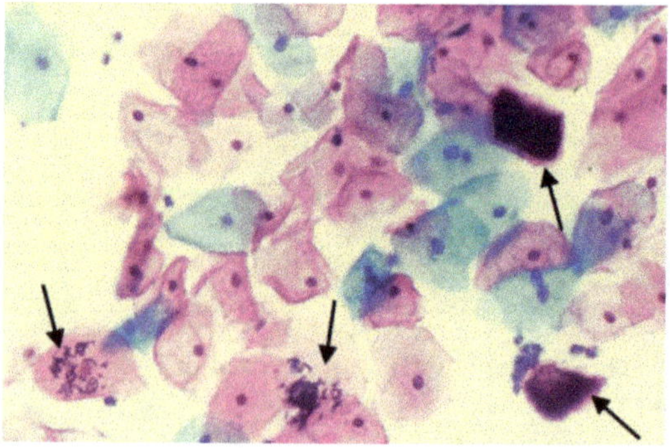

FIG. 18.10. The brown artifact in increasing degrees of coverage.

polygonal cells. The brown artifact is considered to be an extrinsic artifact, meaning it arises from factors outside the cells. See Fig. 18.10.

The brown artifact is observed in slide—not cellulosic filter—preparations in all cytology laboratories. When noticeable on a few cells, as is usual, the artifact is a curiosity. Not uncommonly, however, its presence is sufficiently great to be distracting, and in the worst case, obscuring—literally making it impossible to determine whether abnormal cells are present in affected areas. It occurs most often on conventional Pap test slides due to the preponderance of superficial squamous cells, is more likely to occur on thick preparations as gravity accelerates the drainage of xylene off the topmost cells and exposes them to air sooner, and is almost never seen on non-squamous cells. Liquid-based preparations sometimes exhibit the brown artifact but less often than on conventional Paps, since they are of more uniform thickness.

The brown artifact occurs because preparations dry sufficiently—before being coverslipped—to allow air to come in contact with the surface grooves of superficial cells. As seen clearly in scanning electron micrographs, these grooves, or channels, course in irregular paths across the entire surface and aid in intercellular adhesion. While submicroscopic, their appearance is suggested by the brown artifact seen in progressive formation stages on different cells of the same preparation, occupying small areas on some cells and greater areas on others. It is difficult to recreate the brown artifact experimentally.

The artifact arises at, or after, the first alcohol in the latter half of the Pap stain, after which only alcohol and xylene are encountered. The most likely trigger point is the interval between removing a slide from the last xylene bath and applying mounting medium. Other less plausible possibilities include (1) the interval between successive baths when a slide rack is drained and (2) vigorous dipping. Some staining machines allow slides to drain for 30 seconds between successive baths, which may provide the opportunity for evaporation and consequent air deposition. Vigorous dipping may cause air bubbles to collapse against immersed cells, thus depositing air. The latter possibility seems unlikely. The fact that water removes the brown artifact indicates the air does not occur in the first half of the Pap stain where water is commonly used. Spray fixatives with Carbowax are also ruled out as a potential cause.

Although invisible, air has substance—as is plainly evident to anyone who has tried to purge an air bubble from a liquid-filled syringe. The brown color resembles that of the wall of an air bubble trapped in mounting medium, though the latter is black. The black color is due to destructive interference of light waves by diffraction, which occurs when light bends as it passes through a transparent medium (i.e., mounting medium, air) and encounters an edge (i.e., the interface between air and mounting medium, or perhaps between air and the adjacent mounting medium-infiltrated cytoplasm). Destructive interference occurs when the peaks of one light wave train coincide with the troughs of another and cancel each other out, thus producing darkness. Light is bent into the irregular paths of trapped air by the relatively extreme difference in refractive indexes between air and mounting medium (or fixed protein), 1.0 vs. 1.52, respectively. As microscopic illumination passes through superficial squamous cells, which are stained by eosin (a dye that transmits yellow as well as red light), the entrapped air appears brown. The crinkled appearance is consistent with the submicroscopic topography of superficial squamous cells as described above. The brown artifact is not material deposited by spray fixatives; as will be seen, it would be removed by water.

The following procedure removes the brown artifact, and coincidentally, the counterstain dyes:

1. Remove the cover glass by immersing the preparation in xylene for as long as needed.
2. Remove the mounting medium in xylene, 3 changes × 10 dips each.
3. Remove the xylene in absolute alcohol, 3 changes × 10 dips each.
4. Remove the absolute alcohol in water, 3 changes × 10 dips each.
5. Restain beginning with OG and continue routinely through the Pap stain.
6. Clear and mount.

Removal appears to be related to water's higher surface tension, relative to that of alcohol; scattered residual pockets sometimes remain. Removal might also be due to the slight swelling of cells that occurs upon re-immersion in water, which might loosen the foothold of air in the grooves. If desired, check the completeness of removal microscopically before restaining the slide. Since this entire procedure is time-consuming, prevention is better. Using 50% glycerin instead of water to remove this artifact is unnecessary. No correlation has been observed between its occurrence and the quality of the alcohol that is used.

To prevent the brown artifact, avoid and minimize activities that promote evaporation of alcohol and xylene with subsequent exposure to air, including:

- If staining manually, dip slides at the rate of once per second the distance of the entire slide. Do not submerse a slide rack and rapidly agitate it.
- Limit the "hang" time between baths to no more than 10 s, an arbitrary time.
- Don't remove slides from xylene at one end of the lab and carry the rack to the other end of the lab. Keep the slides wet until ready for coverslipping.
- Avoid draughts (e.g., fans, proximity to open doors, foot traffic). Working under a fume hood promotes the incidence of the brown artifact, as air enters the workspace at the rate of 75–100 linear feet per minute.
- Set up a chemical splash shield at the front edge of the fume hood to divert airflow and create a quiet zone behind it in which

to coverslip. Use a narrow shield, as wide shields can make a technician's shoulder muscles ache.

- Remove slides from xylene one at a time, quickly dry the back and edges, apply mounting medium as needed, and cover. It appears to make no difference whether the mounting medium is applied to the slide or to the cover glass. Applying it to the slide, however, covers the preparation and reduces the exposure time. Speed is essential. Do not drain each slide excessively. Some high-volume laboratories lay out many slides at a time, faceup, and go back to apply mounting medium and cover glasses. Such a practice is not recommended, as it promotes the incidence of this artifact.

Mounting Medium Thickness

As pointed out in Chap. 17, mounting medium increases the effective thickness of the cover glass.[1] Image-forming light waves interact identically with mounting medium and cover glasses. Since the cover glass thickness is manufactured within narrow limits, and mounting medium thickness can exceed those limits to the detriment of image formation, it should be applied sparingly—especially for Millipore filters.

Evaporative Weight Loss of Mounting Medium Solvent

After mounting medium is applied to a slide and covered, its solvent begins to evaporate along the edges. As evaporation continues, an edge seal forms that slows the evaporation of the solvent. Over time, the edge seal becomes wider and thinner, further slowing the evaporation of the remaining interior solvent. At room temperature, complete evaporation and hardening of the mounting medium takes months. See Fig. 18.11.

When ready to be filed upright back-to-face, slides can be placed on aluminum slide trays and put in a 70 °C hot air convection overnight for about 16 h. The metal trays conduct heat

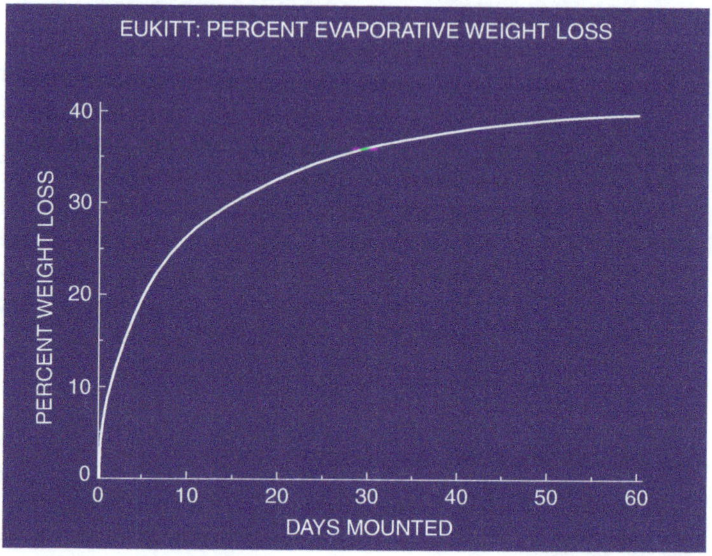

FIG. 18.11. All mounting media are comprised of 1 to 3 solids dissolved in a solvent, which is usually xylene. The weight of solids per 100-mL mounting medium (i.e., w/v) varies among mounting media. After mounting medium is applied to a preparation and coverslipped, its solvent begins to evaporate relatively rapidly at first. As an edge seal forms around the perimeter, it forms a barrier that slows the rate of evaporation. With time, the evaporative weight loss approaches the maximum possible for the particular mounting medium. This particular example graphs solvent weight loss in Eukitt mounting medium for 2 months after being applied to a Millipore filter. If slides filed within a few days after being mounted, the partially hardened mounting medium will be squeezed out and glue adjacent slides together. Most freshly mounted slides are dried in an oven to accelerate hardening.

efficiently and accelerate the drying process. Some mounting media become discolored under those conditions. If that should happen, shorten the time or lower the temperature. Let the slides sit for a day at room temperature after being removed from the oven. File them without risk of the slides sticking together.

Commercially available mounting media are comprised of different concentrations of solids in solvent. When the amount of solids is insufficient to occupy the space between the slide and

FIG. 18.12. "Ferning" in cellulosic filters results from mounting techniques that don't compensate for the filter's 84% porosity.

cover glass, retraction occurs. Retraction means that the mounting medium recedes within the cover glass perimeter, leaving the specimen mounted in air and degrading image formation. Only by its own experience with a given mounting medium can a laboratory learn of a mounting medium's strengths and weaknesses.

An extreme example of retraction is "ferning" or the "fern effect" in Millipore filters, so-called because of its obvious resemblance to the frond of a fern. When the clearing agent xylene is not displaced by mounting medium during the mounting process, its evaporation results in air coursing through the porous filter. See Fig. 18.12. If "ferning" is small in area, its impact may be inconsequential. When extensive, however, it scatters light so badly that cells can't be imaged usefully.

"Cooking" Slides Reduces Mounting Medium Thickness and Hastens Hardening

"Cooking" slides is a term that few know these days, so I'll explain. Slides were being "cooked" in the Cytopreparatory Laboratory when I arrived as a student in August 1963. "Cooking" was discontinued some time later for reasons I can't recall but probably because of safety concerns and alternative safer, effective processes. The cooking process we used was similar to the one described by Graham.[2] She had developed the method to dry slides completely before being screened, which "set" the coverslip.

"Ink dots are valueless if the coverslip moves even slightly. To overcome this difficulty, we developed a 'cooking' process for rapid drying of the slides. After the smears are mounted, they are placed on a porcelain pan that sits on an electric hot plate. The plate is set at 'medium' heat. Two or three slides are placed on the pan at once. While they are heating, the technician presses gently [with forceps] on the coverslip to squeeze out all excess balsam. As soon as the balsam begins to bubble, the slides are immediately removed, pressure is again applied with the forceps to remove all air bubbles, and the slides are allowed to cool. After the slide is cool the coverslip is immovable, even when the slide is immersed in xylol, so that it may then be thoroughly cleaned. No attempt is made to clean the slides until after the 'cooking' process, though they are mounted with as little excess balsam as possible.

We have found this method satisfactory. Not only do the slides for interpretation come to the microscopist ready to be dotted, but the slides may be filed the same day. This simplifies the filing of large numbers of slides. The 'cooking' process does not interfere with the staining properties of cells if the slides are removed as soon as the balsam begins to bubble. All slides used for photographs and drawings in this book have gone through this process.

Not only does the 'cooking' not interfere with the staining quality of the slide, but it helps to preserve the stain. As slides become older, their staining tends to become light and often the stain in old slides is so faint as to make them practically useless. This is because oxidation gradually takes place. By heating the slides until they bubble, all air is driven out. This tends to prevent further oxidation, and the stain remains good. The cells illustrated in Figure 73 are from a slide taken nineteen years ago, and the cells still have a sharp, well stained appearance."[2]

Parenthetically, although not mentioned originally by Graham, cooking slides minimizes the thickness of mounting medium, which reduces the impact of spherical aberration on image quality.

Spherical aberration means that light rays arising from a single point in the object, and passing through different areas of a lens, are not imaged as a single point in the image. The visual effect is a hazy image that is indistinguishable from that produced by glare. Cover glasses that exceed thickness tolerances contribute to spherical aberration. Great deviations cause great spherical aberration. In the absolute worst case, the image quality is so poor that it cannot be interpreted. Spherical aberration is not an inconsequential optical concept. It is real, and it can be resolved if one knows how.

References

1. Setterington R. VI. – The specification of a standard microscope cover-glass. J R Microsc Soc. 1953;73(2):69–76.
2. Graham RM. The cytologic diagnosis of cancer. 2nd ed. Philadelphia: WB Saunders Company; 1963. p. 346.
3. Danos M. An artifact on stained slides (tips on technique). Cytotechnol Bull. 1969;6(1):9.
4. Gill GW. The brown artifact – "cornflaking". ASC Bull. 2001; 38(5):75.

Chapter 19
Köhler Illumination

He that would perfect his work must first sharpen his tools.

Confucius

Principle No. 7

Examine with a clean microscope and Köhler illumination to promote highest resolution.

Practice

Be a microscopist, not just a microscope user.

This chapter could be called "Practical Microscopy," "The Microscope," or simply "Microscopy," but I chose to call it Köhler Illumination to emphasize the role of this essential—yet still not widely known more than a century after its introduction—technique of getting the best performance from a microscope. I have tried to provide in this chapter common, practical microscopy issues that don't require a deep understanding of optical theory, ray diagrams, and the like. As will be seen, Köhler illumination provides a systematic approach to adjusting the illumination system without guessing how best to adjust the field diaphragm, substage condenser height, and opening of the substage condenser aperture diaphragm. In addition to optimizing optical resolution,

G.W. Gill, *Cytopreparation: Principles & Practice*, Essentials in Cytopathology 12, DOI 10.1007/978-1-4614-4933-1_19,

© Springer Science+Business Media New York 2013

Köhler illumination is fundamental to quality photomicrography and digital imaging.

August Köhler published his "new method of illumination for photomicrographical purposes" in 1893 when he was only 27 years old.[1,2] His coworkers included Carl Zeiss, Ernst Abbe, and Otto Schott. The latter three gentlemen collaborated to produce the first apochromatic objectives, which advertising proclaimed "killed all others dead." It wasn't marketing hyperbole; it was true.

Köhler illumination ensures uniform illumination of the object plane (i.e., where the specimen is located) and maximum resolution of detail for the lenses in use. These goals are accomplished by systematically aligning the optical and illumination axes with one another and focusing the conjugate focal planes along each axis so that they alternate with one another from the lamp filament to the retina. Conjugate focal planes are "married" to one another, so that multiple planes appear as inseparable. Köhler illumination focuses the field diaphragm, the specimen, and the field lens of the eyepiece into a single image on the observer's retinas. These you can see.

Simultaneously, Kohler illumination focuses the lamp filament, the substage condenser aperture diaphragm, and the back focal plane of the objective into a single image at the observer's pupils. These you can't see. The illustration shows the optical imaging axis and the illumination axis as separate axes to facilitate visualization of the concept. In reality, of course, they are aligned along the same axis. See Fig. 19.1.

Prior to the 1893 introduction of Köhler's method, there was no generally accepted way of adjusting microscope illumination.

→

FIG. 19.1. (continued) situated below the mechanical stage, and the lenses of the imaging system, situated above it. Köhler illumination centers the illuminating beam relative to the opening of the front lens of each objective and focuses the image of the lamp filament at the plane of the substage condenser aperture diaphragm and simultaneously at the back focal plane of the objective. These simple adjustments make a nonuniform lamp filament illuminate the object uniformly and make it appear as though the object is the source of light (i.e., self-luminous). The latter avors optimal resolution.

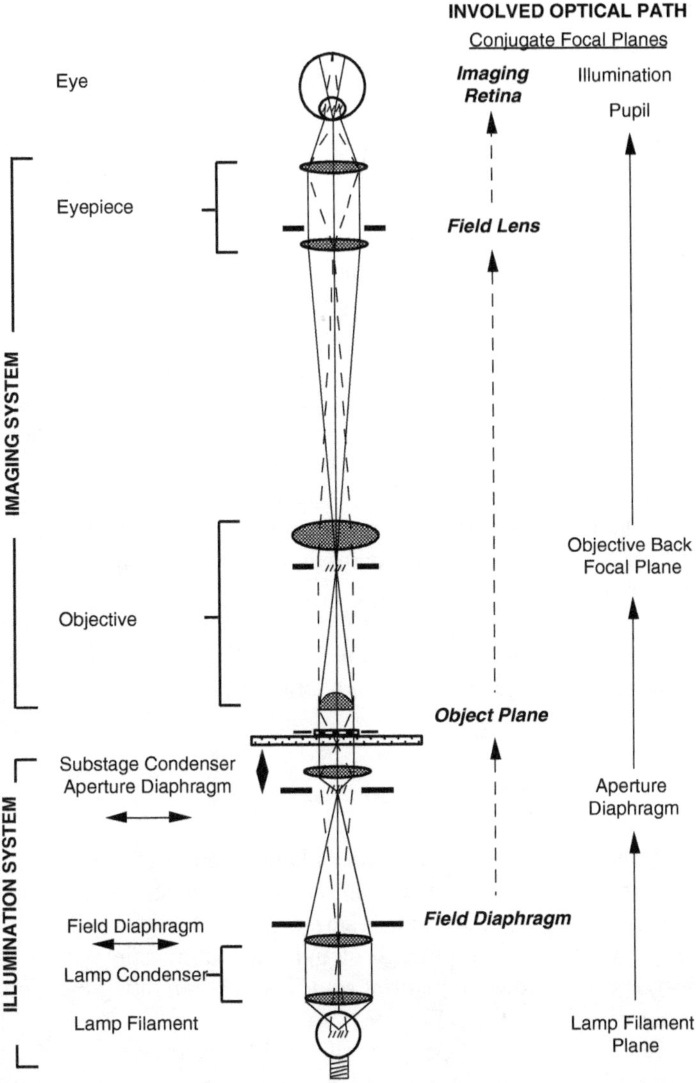

FIG. 19.1. The principle elements of a microscope involved in Köhler illumination. The optical axis of a microscope is "bipolar": the south pole being the lamp filament and the north pole, the microscopist' retinas. Between those two fixed points are the lenses of the illumination system,

Even today, few microscopes are capable of producing strict Köhler illumination in which an image of the lamp filament can be seen by removing an eyepiece and looking down the barrel at the objective's back focal plane. Most microscopes incorporate a frosted lens in the lamp condenser that scatters light, thereby making the light source appear uniform and not allowing an image of the lamp filament to be seen.

A well-prepared and mounted micropreparation has the potential to appear sharp and crisp at all magnifications, to be more signal than noise. Objectively speaking, so to speak, the extent to which it fulfills its potential depends on its being imaged by a clean microscope that is adjusted to produce Köhler illumination. On the other hand, poor quality preparations won't degrade image quality per se. If imaged well, the limitations in these preparations can be seen for what they are. Poor imaging combined with poor preparation, however, can invalidate the results. Whether these degradations are considered problematic is a largely subjective matter. Inexplicably, some microscope users are simply insensitive to poor quality images.

It takes less time to set up Köhler illumination than it does to read about it. See Fig. 19.2. Once established, it is stable and requires slight adjustment throughout the day.

Cleaning the Microscope

A clean microscope[3–5] and Köhler illumination[6] go hand in hand. One without the other makes no sense. Cleaning a microscope means keeping all exposed lens surfaces free of anything that will scatter or impede the passage of light, including things such as dust, eye secretions, fingerprint smudges, ink, mounting medium, and immersion oil. See Fig. 19.3.

To clean a microscope, you will need lens cleaner, lens paper, cotton-tipped swabs, sharpened bamboo sticks, a camel's hair brush, and a can of filtered compressed air. Daily simple housekeeping keeps microscopes top-notch.

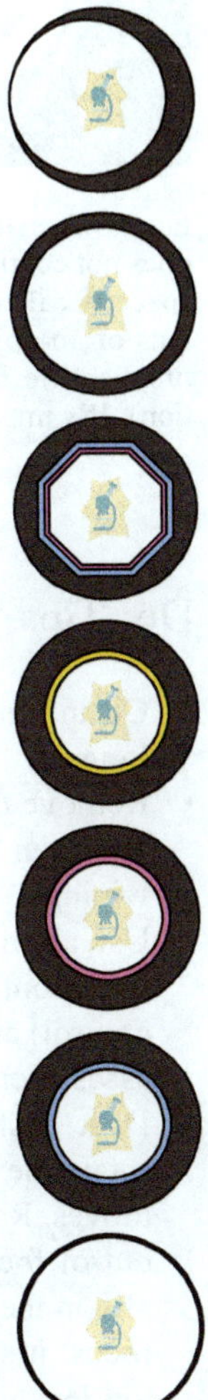

Fig. 19.2. Before beginning, adjust the interpupillary eyepiece distance and focus each eyepiece independently. The entire procedure is completed within seconds. The openings of the field diaphragm and substage condenser diaphragm are different for each magnification. For best results the microscope must be clean

1. At ×10, select a high-contrast object to focus on (e.g., a superficial squamous cell nucleus) and center it within the field of view.
2. **Close** the field diaphragm to see whether its image is centered.
   - Close the substage condenser aperture diaphragm more.
3. **Center** the image of the field diaphragm while using the dark surrounding area as a convenient guide. The cell focus should not change throughout this procedure.
   - Close the field diaphragm more.
4. **Focus** the image of the field diaphragm iris leaves in the object plane by adjusting the height of the substage condenser. The halo will be red and blue; the intensity will vary with the substage condenser aperture opening. After step 6, it is OK to defocus dirt images under ×10.
   - If the image of the field diaphragm iris is ringed by a soft yellow halo, the substage condenser is too high and nearly touching the underside of the slide.
   - Lowering the substage condenser slightly eventually produces a soft magenta halo, which appears more distinctive when using the ×40 objective.
   - Lowering the substage condenser even more produces a soft blue halo. The colors described in these three bullets occur within a few millimeters. To the unaided eye, it is difficult to see that the substage condenser has been moved at all.
5. **Open** the field diaphragm until its image just disappears from view.
6. **Adjust** the substage condenser diaphragm until you see the best contrast. Closing it too far imparts a refractile quality to the cells due to diffraction. Opening it too far creates image-degrading glare. Follow the same 6-step procedure for each objective as needed.

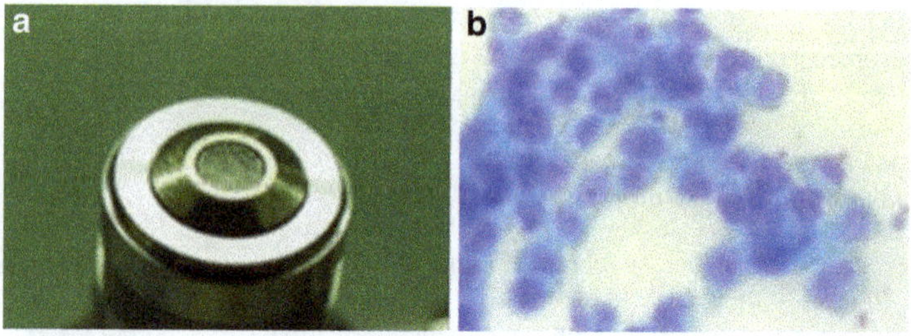

FIG. 19.3. Anything that scatters light causes glare, which is light that does not contribute to image formation. Glare that originates in the image space is called flare. In this example, a greasy fingerprint on the front lens of an objective degraded the image quality. Visually, glare is indistinguishable in appearance from spherical aberration (aka shape distortion). It's important to know the difference and how to fix the problem.

Do: Tips

- Clean microscopes perform best. Lenses cleaned least last longest.
- Remove coarse dirt first and fine dirt last. Brush off, blow off, and then wipe off. Let lens cleaner dissolve dirt film before wiping.
- Use an inverted eyepiece as a jeweler's loupe to check objective front lens surfaces for dirt before cleaning (i.e., quality control) and to confirm dirt removal after cleaning (i.e., quality assessment).
- Dust busting: To locate the source of imaged dirt particles, rotate the eyepieces and move the slide to see whether the dirt moves. Raise and lower the substage condenser. If the dirt goes out-of-focus, it is located below the substage condenser—usually in the neighborhood of the blue filter. If the dirt remains in focus, it's located above the substage condenser, either on the top lens of the substage condenser, or more often, on the eye lens of the ocular. Dirt, dust, and smudges scatter light and degrade image quality.

Do: Techniques

Eyepieces

- Brush or blow off dust.
- Place a few drops of lens cleaner on twice-folded lens paper. Gently wipe lens in a circular motion. Move to a dry area of paper and wipe until dry, taking special effort to remove lens cleaner around the circumference. Use a dry cotton tip applicator if needed. Don't remove the eyepiece from its sleeve if you can avoid it.

Substage Condenser Top Lens

- Lower the condenser to its lowest point and remove from its mount.
- Brush and blow off loose glass particles.
- Apply a drop or two of lens cleaner to one end of a double thickness of lens paper placed on the top lens. Slowly pull the lens paper across the lens until dry. Repeat as needed.
- Replace the condenser, raise, and recenter.

Objectives

- Remove and replace objectives from the nosepiece with extreme care. Dropping them is easy.
- The front lens surfaces are often recessed or concave, thus requiring cleaning techniques that will reach the dirt. Use lens cleaner for most jobs; xylene for mounting medium and immersion oil.
- Work over a table, with a cushioned surface if possible.
- Check the front surface with an inverted eyepiece. If dirty, proceed with cleaning.
- Apply a drop of lens cleaner to a double thickness of lens paper.

- Using a sharpened bamboo stick, move the moistened paper over the surface. Dry it.
- Check for success using an inverted eyepiece. Sometimes first cleaning efforts make matters worse. Replace the objective in the nosepiece; hold it until certain it's secure.

Do: Timing

Daily

- Brush off each eyepiece's eye lens, the substage condenser's top lens, the blue filter, and the microscope overall.
- Cover unused microscopes.

Weekly

- Blow off glass particles that accumulate around the top lens mount of the substage condenser.

Do: As Needed

- Clean each eyepiece's eye lens.
- Clean the front lens of each objective.
- Clean both sides of the blue filter and blow off dust particles on top surface of the supporting lens.
- Enter the day, date, and your initials in a Microscope Maintenance Log whenever you clean all the exposed glass surfaces between the blue filter and the eyepieces.

Don't[8]

- Use saliva to clean a lens surface.
- Apply solvent directly to a cemented lens (e.g., objective front lens).

- Let go of an objective while removing it or replacing it.
- Clean the lens inside an objective.
- Use rough stuff (e.g., paper towels, facial tissue, wipe for eye-glasses, or an unwashed handkerchief) on lens surfaces.
- Wipe the lens excessively, as the antireflection surface coating is easily scratched and promotes glare.
- Disassemble an objective.
- Put any liquid inside an objective.
- Clean first surface mirrors (e.g., in the base). Front surface means the silver coating that makes it a mirror is on the front surface. In that location, it's vulnerable to scratches.

Practical Microscopy (aka "Glass-and-Brass" Tacks)

In the nineteenth century, microscopes were often referred to as glass and brass. Brass tacks, of course, means getting down to business. Hence, glass-and-brass tacks.

Working Köhler illumination

In routine daily use, it is unnecessary to change the opening of the field diaphragm, the height of the substage condenser, and the opening of its aperture diaphragm with every change of objective or slide. Instead, after establishing Köhler illumination, keep the field diaphragm at the diameter for the 10× objective and the aperture diaphragm opening for the 40× objective. Lowering the substage condenser slightly will optically erase any imaged dust particles.

Well-prepared specimens appear sharp and crisp at all magnifications when imaged by a clean microscope adjusted for Köhler illumination. See Fig. 19.4. The background should be uniformly illuminated and largely glare-free.

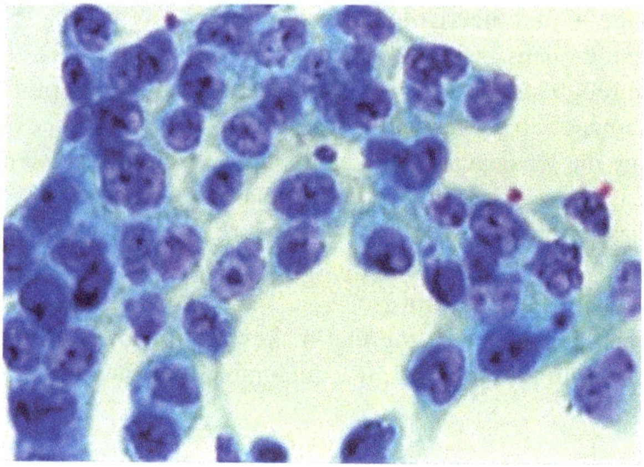

FIG. 19.4. Papanicolaou stained sheet of malignant cells in a pleural fluid. Original magnification—×600. As a visiting pathologist remarked, "if my preparations looked this good, I think I could do cytopathology."

Light show

The opening of the substage condenser aperture diaphragm influences image quality more than any other step. Low power objectives require a wider opening of the field diaphragm and a smaller opening of the substage condenser aperture diaphragm. The opposite is true for high power objectives. See Fig. 19.5.

The glass cube is not available commercially. Mine was a gift from Dr. Haselmann.[7, 8] At one time, small uranium glass blocks were available to demonstrate light paths but no more. The only commercial source today of anything comparable is Bioindustrial Products, who sell an inexpensive smoked 40×35×20-mm acrylic block.[10] A similar teaching tool can be homemade. Add eosin-colored alcohol to a screw top, flat tissue culture flask.[11]

Cleanliness Is Next to Goodliness

Köhler illumination images 4 conjugate image focal planes: (1) field diaphragm, (2) object, (3) the field lens of the eyepiece or ocular

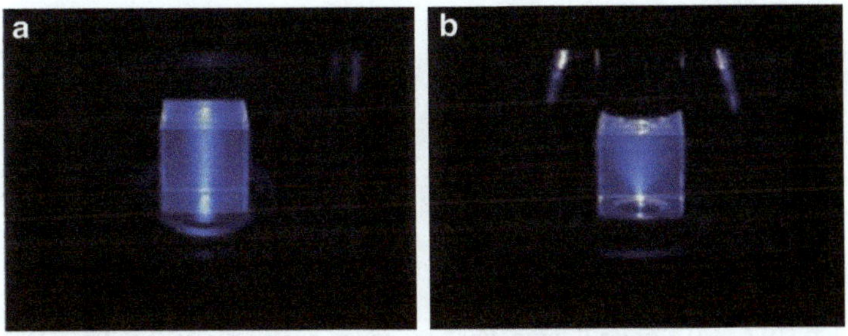

FIG. 19.5. A 20-mm glass cube with fine particles brings light to life.[7, 8] Absent such a tool, it's difficult to teach students what happens. The cube is sitting on the top lens of a substage condenser to illustrate how adjusting the field diaphragm and the aperture diaphragm affects light: (**a**) a ×10 objective images a wide field of view and accepts a narrow cone of light. (**b**) A ×40 objective images a smaller field of view and accepts a wider cone of light. The diameter of the field of view is a function of the eyepiece field number and the magnification of the objective. The angle of the cone of light is a function of each objective's numerical aperture.

(i.e., the lens closer to the field; the eyepiece lens closer to the eye is the eye lens), and (4) retina. Any dust particles on the blue filter that sits above the field diaphragm may be imaged in the field of view of the 10× objective, which has a great depth-of-field. The same dust particles will not be seen in the field of view of the 40× objective, which has a shallow depth-of-field. Lowering the substage condenser until the image of the particles is defocused will not compromise the quality of Köhler illumination. See Fig.19.6.

Thickness of Slide

The thickness is immaterial up to a point—assuming it's always underneath the specimen. However, excessively thick slides can raise the preparation beyond the substage condenser's ability to focus an image of the field diaphragm in the object plane. For example, gluing the pieces of a broken slide to another slide to repair it creates an excessively thick slide pair.

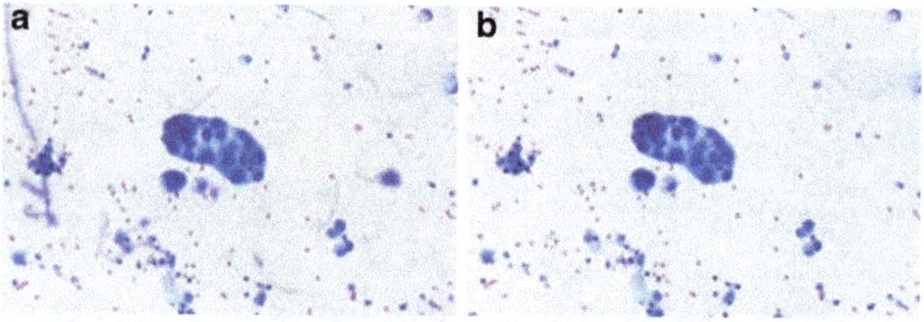

FIG. 19.6. (**a**) Köhler illumination can image dust particles that are on the *blue filter* above the field diaphragm into the same plane as the specimen. The particle images don't move when the slide is moved, which is distracting. (**b**) Slightly lowering the substage condenser defocuses the image of the dust particles without compromising imaging quality. This is seen when using ×10 objectives, which have a great depth-of-field compared to that of ×40 objectives.

Working Numerical Aperture Salvages Image Quality

In my experience, the combined thicknesses of mounting medium and cover glass (i.e., effective cover-glass thickness [ECT]) in freshly mounted preparations virtually always exceed the tolerance limits of high dry objectives to deviations from the recommended thickness (i.e., 170 μm). Consequently, images appear sharp with low NA objectives (e.g., 10×, NA 0.25) but low in contrast with higher NA objectives (e.g., 40×, NA $\geq$0.65). Slightly closing the substage condenser aperture diaphragm reduces the *working* numerical aperture of the objective, thus reducing its intolerance to excessive ECT and restoring image quality.

Using Köhler illumination, focus on a preparation. Now, remove an eyepiece and look through the tube at the lighted circle. That lighted circle is the working NA. The diameter changes as the substage condenser aperture diaphragm is opened and closed. Close it completely and look through the one eyepiece. Dimly lit and highly refractile, the image quality is obviously unacceptable.

Open the diaphragm completely, and look again. The image quality has improved, but it's still not the best. It is somewhat obscured by glare and flare. If the field diaphragm is opened wider than necessary to just illuminate the object field, light reflects off the metal opening in which the objective's front lens is located and creates glare. If the substage condenser diaphragm is opened so it just fills the back focal plane of the objective, it causes flare. Flare lights the interior walls of the space between the objective's back focal plane and the top of the tube. Flare degrades image quality.

Closing the substage condenser slightly until flare disappears, and only dark walls remain, improves contrast and image quality. There is no single magical opening that is best for all specimens. The working numerical aperture is always less than the nominal maximum numerical aperture that is engraved on each objective.

Depth-of-Field Versus Depth-of-Focus

Depth-of-field and depth-of-focus are sometimes used interchangeably, as though synonymous. They are not. Depth-of-field is the total distance, measured in the field, between the nearest point of acceptable focus, when the image is viewed, and the farthest point of acceptable focus along the optical axis or parallel to it.[11] depth-of-field is what your eyes see. It is a function of an objective's numerical aperture (NA). The greater an objective's NA is, the shallower its depth-of-field. Differences in depth-of-field explain why dust particles that are visible in the field-of-view of 10× objectives aren't seen in the field-of-view of 40× objectives.

Depth-of-focus, on the other hand, is in the image space. It is the depth along the axis of acceptable definition. It is also the axial distance the lens can be moved without noticeable degradation of definition in a single plane in the image space.[12] Differences in depth-of-focus among objectives with different NA explain why capturing sharp photomicrographs are less challenging with a 40× objective (NA 0.65) than with a 100× one (NA 1.3).

Magnification Versus Enlargement and the "×" Factor

Magnification and enlargement are also used sometimes as though synonymous with one another, but they're not. Primary lens systems (e.g., objectives and eyepieces) magnify objects and resolve details that could otherwise not be seen. The limits of useful magnification are between 500 and 1,000 times NA. When the image of a magnified object is enlarged beyond the limit of resolution of the lens system, the image loses resolution and can look blurred. The latter circumstances create what's known as empty magnification. On the other hand, an enlargement process makes the image larger without resolving fine details.

By convention, magnification is expressed as multiples of the original object diameter by placing the "×" after the number (e.g., 40× objectives). When referring to number of times enlargement, the "×" is placed before the number (e.g., ×400).

Photomicrograph Versus Microphotograph

Again, the same but different. Photomicrographs enlarge something small; microphotographs ensmall something large (e.g., microfiche).

A Breath of Fresh Air

Exhaled air contains naturally distilled water, which can be applied to selected lens surfaces for cleaning with lens tissue.

Köhler illumination is curiously absent from most cytology textbooks, curricula of cytotechnology schools and medical schools, and pathology residency programs. It isn't among the questions in any certification examinations. Most pathologists who many years ago attended Dr. Frost's annual 2-week Postgraduate Institute for Pathologists in Clinical Cytopathology had never heard of Köhler illumination. I doubt it's any different today. Many times during these continuing education marathons, visiting pathologists would ask me to do a quick-and-dirty cleaning of their microscopes and establish Kohler illumination. The reactions were always the same: Wow, I've never seen cells so clearly!

"Plus ça change, plus c'est la même chose." The more things change, the more they remain the same.

References

1. Köhler A. Ein neues Beleuchtungsverfahren für microphotographische Zwecke [New method of illumination for photomicrographical purposes]. Z wiss Mikr. 1893;10(4):433–40. Original article available at http://www.archive.org/download/cbarchive_38379_1018934page433440einneuesbeleu1893/1018934page433440einneuesbeleu1893.pdf. Accessed 27 Feb 2012.
2. English translation available at http://books.google.com/books/download/Journal_of_the_Royal_Microscopical_Socie.pdf?id=0O4BAAAAYAAJ&output=pdf&sig=ACfU3U1DMogG4ZRyiuGdBqkVMWFzKWz6AQ. Accessed 27 Feb 2012.
3. Gill GW. Microscope cleaning tips, techniques and timing. LabMedicine. 2005;36(8):460.
4. Duke CV. Lens cleaning. Best practices review. Micros Today. 2004;12(2):42, 4.
5. Larson JM. Cleaning a microscope's optical surfaces. Micros Today. 2001;10(4):27–8.
6. Gill GW. Köhler illumination. LabMedicine. 2005;36(9):550.
7. Haselmann H. Teaching microscopy. I. Microscopy as a field of teaching and research. Proc R Microsc Soc. 1967;2:337–43.
8. Haselmann H. Teaching microscopy. II. Didactic methods in teaching microscopy. Proc R Microsc Soc. 1967;2:344–55.
9. Gill GW. Care and feeding of microscopes: cleanliness is next to "goodliness.". ASCT J Cytotechnol. 1997;1(2):73–5.
10. Bioindustrial Products. http://www.bioindustrialproducts.com/january_2012_catalog.pdf. Accessed 1 Mar 2012.
11. BD Biosciences. Cell culture flasks. http://www.bdbiosciences.com/cellculture/flasks/index.jsp. Accessed 1 Mar 2012.
12. Loveland RP. Photomicrography – a comprehensive treatise, vol. I. New York, NY: Wiley; 1970.

Part III
Everything Else

Chapter 20
Screening

Principle No. 8

Screen preparations in ways that facilitate abnormal cell detection.

Practice

Overlap juxtaposed field-of-view diameters by 30% to image every μm^2 of preparation area at least once, slow down, and take frequent breaks to help sustain vigilance. The ability to recognize abnormal cells is insufficient by itself to find them.

The first schools of cytotechnology in the world were established in America in 1947 in New York city and Hartford, Connecticut. The American Society for Clinical Pathology (ASCP) certified Rosalyn S. Yaskin Abrams as the first cytotechnologist in 1957. In 1960, the Department of Health, Education, and Welfare began paying stipends of $225.00 per month for up to 12 months to encourage enrollment of cytotechnology students. That amount is about $20,000 per year in 2012 dollars. The number of cytotechnology schools reached a peak of about 130 in the early 1970s, and today, the number is about 32. As of June 2011, ASCP reports that 15,224 cytotechnologists have been certified. As of 2009, the latest year as of this writing for which CMS has provided data, 6,064 cytotechnologists screen Pap smears. As the outgoing president of the American Society of Cytopathology in 1996, Prabodh Gupta,

G.W. Gill, *Cytopreparation: Principles & Practice*,
Essentials in Cytopathology 12, DOI 10.1007/978-1-4614-4933-1_20,
© Springer Science+Business Media New York 2013

declared, "the Pap test is cytopathology."[1] Cytotechnology: The First Half-Century, a 3-part series by Florence W. Patten, is available online.[2–4]

The brief background serves as a preamble for a remarkable—but generally overlooked—fact about screening Pap smears. Given the potential medicolegal risk and cost of a false-negative test result to everyone, one might expect the practice and process of screening would have long ago been defined. But that is not the case.[5, 6] Prior to 1987,[7–9] The Doctors Company, a medical malpractice insurer of pathologists, had experienced only 7 Pap test malpractice claims in its 12-year history. In 1988, 11 Pap test malpractice claims were filed. Another 184 claims were filed between 1989 and 1995. From 1998 thru 2003, another 42 claims were filed. The Doctors Company represents about 16% of the number of pathologists who examine Pap tests.

Virtually all lawsuits are based on false-negative Pap test results. With rare exception, these are due to "locator" error by the primary screening cytotechnologists. These cases are difficult to defend. In one recent study, 34 of 37 verdicts found in favor of the plaintiff.[10] Recent papers have addressed potential error-related issues such as workload, time of day, day of the week, and unidirectional versus bidirectional screening.[11–13] All appear to assume there is an SOP for screening.

As a cytotechnology student in 1963, I was advised to overlap scan paths slightly so that I could pick up what I missed in the previous scan path. That was state-of-the-art advice then, and little has changed since. That which follows is what I've learned about screening.

The objectives of screening are threefold: (1) find and dot at least 1 abnormal cell, regardless of its location on a slide; (2) find additional examples: and (3) interpret the dotted cells for subsequent review by a pathologist.[14] Abnormal cells can be found in any number, in any morphologic presentation, anywhere on a slide. Given that experiential reality, it makes sense to systematically search for abnormal cells to promote the likelihood of a successful outcome.

I define manual screening as a process by which a Pap smear is moved successively in small steps to increase the probability that at least 1 abnormal cell will fall within the narrow field of vision of an alert professional observer long enough to be recognized and

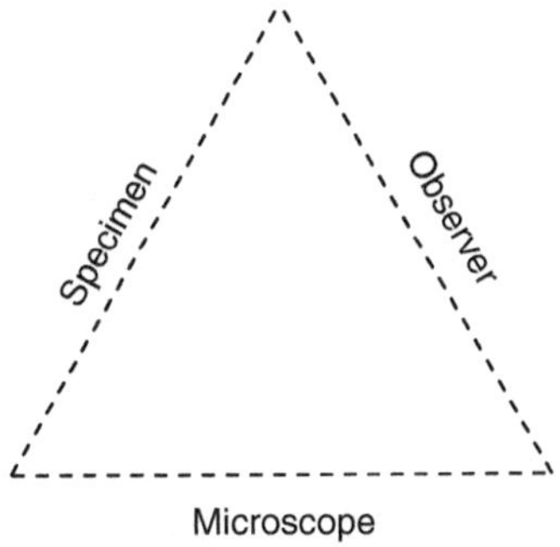

FIG. 20.1. "CytoTect Triangle" is a portmanteau meaning "cytodetection triangle."

interpreted.[15] This definition brings together the three principle elements of the CytoTect Triangle: specimen, microscope, and observer. See Fig. 20.1.

The CytoTect Triangle relates the interdependent roles of the specimen, observer, and microscope usage in the detection of abnormal cells. Relating this model to electronics, the specimen is the source of the signal; the observer, the receiver; and the microscope, the transmitter. The many variables that impact the process introduce noise. By optimizing and standardizing the three processes, the signal is strengthened, and the noise is reduced. Optimized processes increase the probability of abnormal cell detection (i.e., true positives, high sensitivity) and reduce the incidence of missed abnormal cells (i.e., false negatives, high specificity). In Fig. 20.1, the probabilistic nature of the entire process is represented by dashed lines, rather than solid lines, as would be the case for a deterministic process such as the fire triangle.

An expanded illustration of the CytoTect Triangle is presented in Fig. 20.2.

The fire triangle, also known as the combustion triangle, illustrates simply the relationship among three elements essential to starting and sustaining combustion: heat, fuel, and atmospheric oxygen. When present in suitable proportions, these elements will *always* result in combustion. To extinguish a fire, take away any 1 of the elements.

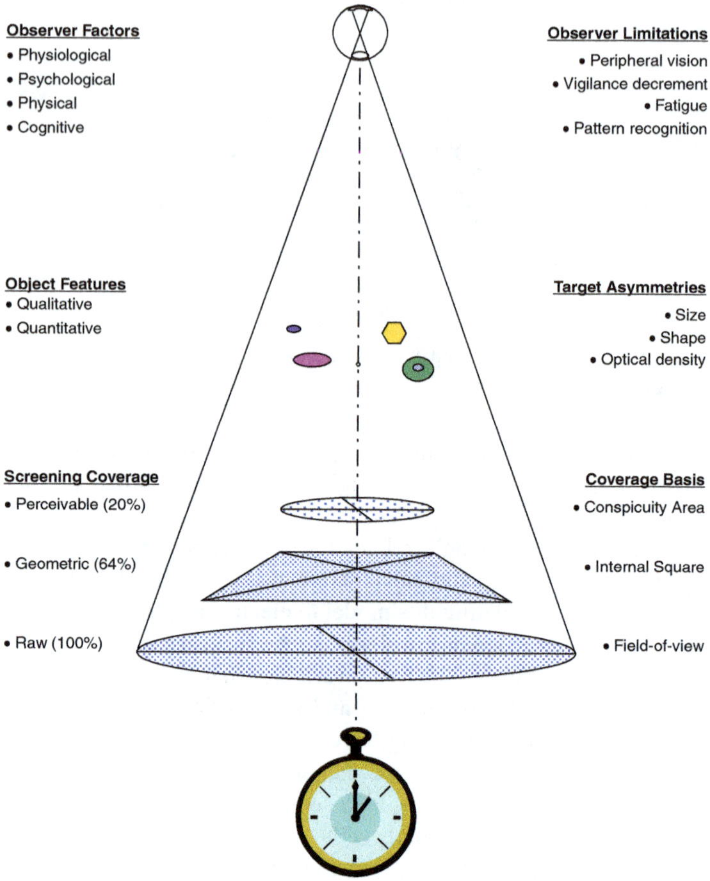

FIG. 20.2. There is more to screening than meets the eye. Detecting the presence of abnormal cells is more likely when they are abundant, recognizable and stand out in relation to the surrounding cluttered displays (i.e., the presence of confusing nontargets in the display leading to competition search), and fall within the observer's conspicuity area. Screening is performed in the context of "beat the clock."

The probabilistic CytoTect Triangle connotes the concept that abnormal cells will usually, but not always, be detected during the complex process of screening. To *not* find abnormal cells, take away any 1 of the elements.

SPADE: Screening Protocol to Assist Detection

SPADE, Screening Protocol to Assist Detection, is a systematic approach to examining a cytological preparation to promote the likelihood of finding at least one abnormal cell.[16] This is the first objective of screening and a prerequisite for avoiding false-negative results. The need for this protocol arose from a realization that schools of cytotechnology do not teach rational screening practices, and as a result, all practicing cytotechnologists screen slides differently. Even if one possessed superhuman cognitive and vigilance capabilities, it would still be easy not to find abnormal cells since they can be present in any number, anywhere on a slide. A conventional preparation magnified ×100 is equivalent to an 8.3 × 16.6-ft mural of Where's Waldo, the colorful illustrated children's rare event search game, which means abnormal cells have an extremely large area in which to hide.

SPADE uses 4× and 10× objectives in combination with plane geometry-based screening overlap practices to promote uniformly thorough examination of slides by individual cytotechnologists. It constitutes quality control of the mechanical aspect of moving a slide under an objective. SPADE is summarized in Table 20.1.

Preview

Screening is a coarse sifting process that identifies specimens as normal, abnormal, or indeterminate. Each finding determines how the Pap smear should be examined next. Screening is performed using a 4× objective, while overlapping field diameters by 50% to bring the most peripheral cells to the center of the field for easier viewing.

Normal cases are readily identified; they contain a single cell type distributed as a monolayer. The appearance is so monotonous that the presence of any other cell type is obvious. The usual

Table 20.1. SPADE: Screening Protocol to Assist Detection.

Step	Level	Objective	Overlap[a]	Purpose	Emphasis	Decision
I. Preview	Screen	4×	50%	R/O abnormal	Specimen	Normal vs. indeterminate vs. abnormal
II. Examine	Scan	10×	30%	R/I abnormal	Whole field	Normal vs. abnormal
	Search	10×	30%	Find more	Cell(s)	Abnormal category
III. Review	Screen	4×	50%	Find best	Obvious cells	Confirmation

[a]Overlap is 2-way (i.e., within the same, and between adjacent, scan paths.)
R/O = rule out.
R/I = rule in.

examples are a Pap smear from a hysterectomized woman being treated with estrogen or a menopausal woman with an atrophic smear. Also included are cellularly inadequate cases that are unsatisfactory. Such infrequent cases do not need additional examination.

Abnormal cases are also readily identified when there is an abundance of abnormal cells or you get lucky and spot an isolated, but obvious, abnormal cell. These cases will be examined in the search mode.

Indeterminate cases comprise about 95+ percent of screening findings. These Pap smears contain a mix of cell types in sufficient numbers to require scanning. You cannot tell whether abnormal cells are present or not. This is the usual finding.

Examine

Scanning is a more accurate term for screening, which is undefined in current practice and performed highly individualistically. It is a *systematic* search of a preparation that ensures every μm^2 of cellular preparation area is imaged at least once. It effectively images the greatest area with the least effort and is the most cost-effective examination approach. Imaging devices scan, not screen, cytological preparations.

Using a 10× objective, begin at one corner of a preparation and advance the slide from field-to-field, overlapping field diameters by 30%, if using 10×/FN 20 eyepieces, and by 36% if using 10×/FN 22 eyepieces. (FN means field number and is explained fully later in this chapter.) Overlap by the same amount when moving the slide to the next scan path. This percent overlap is based on the existence within the circular field of view of a virtual square with corners that just touch the boundaries. Moving the slide, the length of 1 side of the square from field-to-field "lays down" juxtaposing squares that ensure total imaging screening coverage.

Searching is scanning performed more slowly. Upon finding the first abnormal cell, regardless of its location, move the slide in the same fashion as when screening. Simply slow down, overlapping more as necessary, while your eye examines each field of view more thoroughly as you look for more abnormal cells. Don't overlap in acellular areas.

Review

Having identified abnormal cells, rescreen the slide using the 4× objective to find better examples that may have "fallen between the cracks."

Pap smear screening is part of a series of sequential samples of successively diminishing size: the collection device samples the ectocervix and endocervix, the transfer process samples the collection device, the pattern of slide movement samples the slide, and the cytotechnologist visually samples each field—subject to real limitations in specimen presentation, pattern recognition skills, peripheral vision, vigilance, and time. Therefore, SPADE does not and cannot guarantee that all cells will be evaluated or that there will be no false negatives. It is simply an essential step in the series of successive samplings required for accurate results. One can utilize this protocol or not, but at the very least, it represents a logical approach to a microscopically challenging task and underscores the limitations of current methods.

The 4× Objective

According to the Bethesda System: "An adequate endocervical/transformation zone component should consist, at a minimum, of two clusters of well-preserved endocervical glandular and/or squamous metaplastic cells, with each cluster composed of at least five cells." Given such a low numerical threshold, these cells can easily be overlooked when using a 10× objective as too much of the slide remains unexamined. Diligent searching can still yield an EC false negative—not to mention its being time-consuming and cost-ineffective.

TABLE 20.2. Object field diameter and area are a function of magnification and field number.

	Object field diameter			
	Eyepiece magnification/field number			
Objective	×10/20	×10/22	×12.5/16	×15/18
×**4** diameter	**5.0 mm**	**5.5 mm**	**4.0 mm**	**4.5 mm**
Object field area	19.6 mm^2	23.7 mm^2	12.6 mm^2	15.9 mm^2
×**10** diameter	**2.0 mm**	**2.2 mm**	**1.6 mm**	**1.8 mm**
Object field area	3.1 mm^2	3.8 mm^2	2.0 mm^2	2.5 mm^2

Instead, use a 4× objective to look for endocervical cells when their presence is not obvious. This low power objective has many other practical applications:

Working distance. The longer working distance allows greater clearance for a dotting pen.

Depth-of-field. Its great depth-of-field allows you to locate the plane of focus in sparsely populated specimens faster than with a 10× objective. The presence or absence of cells can be determined quickly and confidently. (Don't confuse depth-of-field with depth-of-focus. Depth-of-field is the vertical distance along the optical axis where the specimen appears to be in focus at the object plane, while depth-of-focus is the distance the image appears to be in focus at the image plane [e.g., retina, film]).

Field-of-view. A 4× objective images an object field that is 2–1/2 times wider and 6–1/4 times greater in area than that imaged by a 10× objective used with any eyepiece. See Table 20.2.

The 4× objective provides a bird's eye view that can be used to:

- Provide a snapshot of the composition and distribution of the overall preparation to help you to anticipate the unique screening challenges provided by each
- Detect rare events in addition to endocervical cells, such as hyperkeratotic plaques and HPV-infected cells
- Find best examples of abnormal cells and tissue fragments
- Quickly and more reproducibly estimate proportions of cell types and numbers that determine whether the degree of obscuration by blood and/or inflammation constitutes an unsatisfactory result

- Locate the starting point for screening slides covered partially with cells
- Alternate with a 10× objective in screening irregularly distributed specimens
- Trace the trail of abnormal cells to examine
- Identify the remnants of incompletely removed ink dots from test slides—a handy tip for those engaged in giving, or taking, slide-based examinations, as the case may be

If unaccustomed to using a 4× objective, or you feel imaged objects appear to be too small to be useful, practice looking at benign cells at 10× and then at 4× to become familiar with their appearance. Overlap adjacent field diameters by half. You can cover the entire slide in less time than possible when using a 10× objective and still find more endocervical cells. You'll soon consider the 4× objective to be a valuable search tool.

Ink Dots

Each laboratory should have a written policy about its ink dotting practices that includes: (1) who uses what color ink, (2) whether different colors are used for different entities (e.g., abnormal cells, microorganisms, endocervical cells), (3) how the dots are applied (e.g., 1 dot at 9 o'clock with the cells-of-interest centered, 2 dots [1 above and 1 below the centered cells], a C-shaped parenthesis, or a circle), and (4) whether the slide label is routinely on the left or right.[17] Most laboratories keep the dots intact,[18] but at least 1 defense attorney recommends removing the dots from all slides before being filed.[19] Removing ink dots might be considered spoliation.[20]

Who put what color dots in what location on the slide is a common issue in Pap test-related litigation. I have seen blue, black, green, and red dots on a set of slides in litigation, and no one appeared to know who put them there and when. That has the potential to create problems, especially for the defense. If a cytology laboratory finds itself in litigation about alleged misread Pap tests, it should memorialize the original slides and dots by making color electronic copies of the slides—among other things.

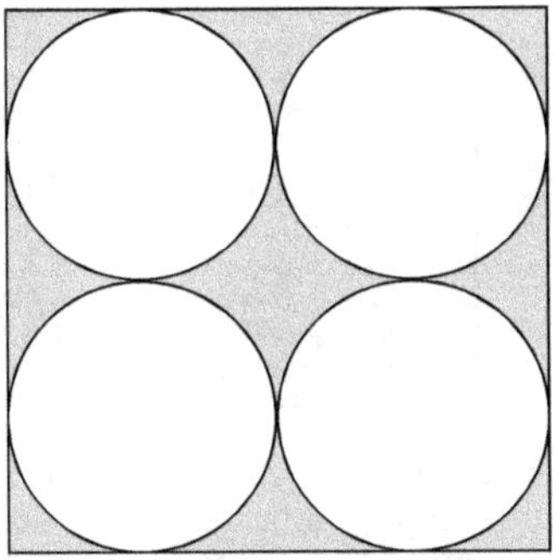

FIG. 20.3. Just-touching fields-of-view (juxtapositioned FOV) leaves 21.5% of the immediate surrounding area unseen. The areas outside of the fields of view are imaged as the slide is moved, but any objects of potential interest cannot be seen in those areas. The images move too fast and fall outside the observer's peripheral vision and conspicuity area.

Percent Overlap

In the context of screening slides, overlap is moving a slide less than the diameter of 1 field of view so that a portion of the same field is visible twice. Not overlapping at all is known as underlap.[21] Knowing how much overlap will ensure that each square micrometer of the preparation is imaged at least once is not readily apparent with circular fields of view. See Fig. 20.3.

Figure 20.4 illustrates an entire preparation covered by a 24 × 50-mm cover glass that has been screened with juxtapositioned FOV.

Screening without overlapping FOV may contribute to false-negative results. After all, abnormal cells *may* be present only in the areas that aren't imaged or near the periphery of the field of view. Knowing how much to overlap fields of view is problematical, because they are circular. How much should adjacent FOV overlap to ensure 100% geometric coverage of the area?

Imagine that within each FOV is a virtual square with corners that touch the circumference. That fact provides the basis for calculating how much adjacent fields of view must overlap to ensure total imaging screening coverage. The diameter of the field

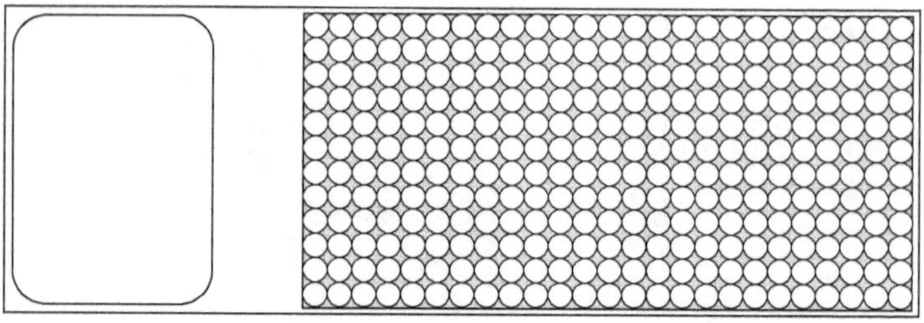

FIG. 20.4. The area under a cover glass defines the maximum area to be screened in the case of nonliquid-based preparations. Historically, the usual example is conventional Pap tests. In America, the de facto standard cover glass dimensions are 24 × 50-mm (i.e., 1,200 mm^2). Assuming for the purposes of this figure that the specimen is spread over the entire area under the cover glass, and the slide is screened with a 10× objective and 10× eyepieces with a field number of 20, 300 juxtapositioned 2-mm diameter FOV of 3.14 mm^2 each are required to cover the entire area. That means 21.5% of the preparation is not imaged (i.e., $1.200\,\text{mm}^2 - \left[300 \times 3.14\ \text{mm}^2\right] / 1{,}200\ \text{mm}^2 = \left[1{,}200\ \text{mm}^2 - 942\ \text{mm}^2\right] / 1{,}200\ \text{mm}^2 = 258\ \text{mm}^2 / 1{,}200\ \text{mm}^2 = 0.215$). AND THAT WHICH IS NOT IMAGED CANNOT BE SEEN.

of view is the same as the diagonal of the interior square and so determines the size of the square. Again, assuming the same objective/eyepiece combination described in Fig. 20.4's legend, and applying the Pythagorean theorem, the sides of the interior square are 1.414 mm each (i.e., the square root of 2).

Successively moving a slide along its x- and y-axes 1.414 mm per stop images the entire preparation area (i.e., total imaging screening coverage). See Fig. 20.5.

Figure 20.6 illustrates an entire preparation covered by a 24 × 50-mm cover glass that has been screened with 30% overlapping FOV diameters.

Figure 20.7 illustrates the underlying checkerboard-like pattern of juxtaposed interior squares.

To estimate one's own mean percent overlap, one must first know the field number of the eyepieces (e.g., 20 or 22). Use the 24-mm side of a cover glass as a ruler. Next, position one corner of a coverslipped side at the 6 o'clock position within the field of

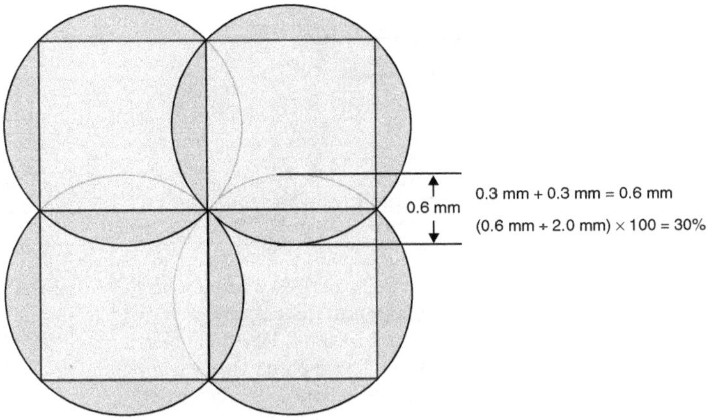

FIG. 20.5. Overlapping field-of-view diameters along the same scan path and between adjacent scan paths by 30% ensures that every square micrometer of the preparation is imaged at least once. Such overlapping provides complete geometric screening coverage with the least amount of effort. It increases the likelihood—but does not guarantee—that an abnormal cell will be detected. Nothing is that simple.

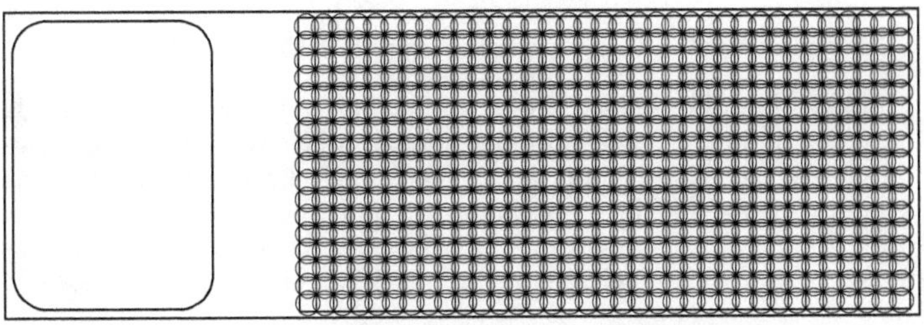

FIG. 20.6. To image the entire 24/50-mm area, 600 FOV with 2 mm^2 interior squares are required. This is twice as much work as screening without overlapping.

view of a 10× objective. Then, count the number of fields needed to move the 24-mm distance to the 12 o'clock position of the field of view at the nearest corner, moving the slide as you ordinarily would. See Fig. 20.8 and the accompanying instructions.

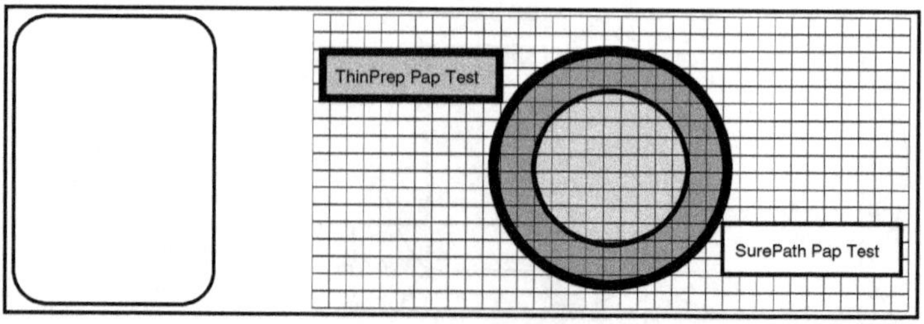

FIG. 20.7. The underlying simplicity of 30% overlapping FOV diameters is revealed when the FOV circumferences are removed. Hologic's ThinPrep is 20-mm diameter and 314 mm^2. Becton Dickinson's SurePath is 13-mm diameter and 133 mm^2. These preparations can be imaged with about 157 and 67 juxtaposed 2-mm^2 areas each, respectively.

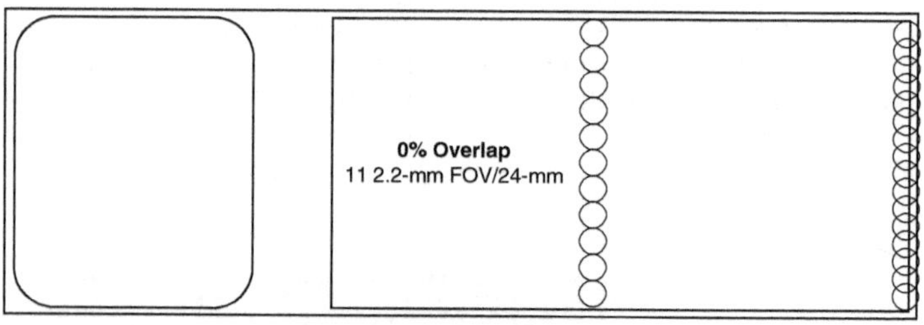

FIG. 20.8. In this example, 16 overlapping fields of view were required to cover the 24-mm distance from one corner of the cover glass to the nearest corner. Dividing 24 mm by 16 FOV equals 1.5 mm/FOV, the mean distance moved per FOV. Subtract 1.5 mm from 2.2 mm, divide the difference by 2.2 mm, and multiply the quotient by 100, which equals 31.8% mean overlap ([{2.2 mm – 1.5 mm}/2.2 mm] × 100 = 31.8%.).

Gill Screening Reticle

In 2000, I developed a reticle that was designed to assist cytotechnologists: (1) find abnormal cells by promoting systematically thorough screening coverage when using 10× objectives and 10× eyepieces and (2) reproducibly compare nuclear areas of suspect ASC-US nuclei imaged by a 40× objective and 10× eyepieces to assist cytomorphological interpretations as needed.

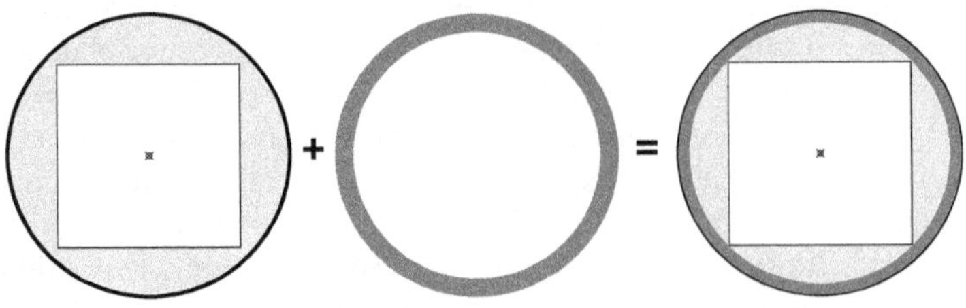

FIG. 20.9. The Gill screening reticle is intended for use with ×10 and ×40 objectives. Under a ×10 objective, the interior square allows the user to overlap adjacent field-of-view diameters by 30% to ensure complete imaging coverage. Under a ×40 objective, the two size comparators allow the user to compare the area of a suspected ASC-US nucleus with that of a "standard" intermediate squamous cell nucleus. The reticle, metallic-side down, fits easily into an eyepiece. The eyepiece with reticle is placed in the focusable eyepiece tube. The diameter of the reticle must match that of the eyepiece in which it will be installed, and therefore must be specified by the potential purchaser. When used with a ×10 objective and ×10/FN 20 eyepieces, the interior square images a 2-mm^2 object plane. One reticle is required per eyepiece pair, and two reduction rings are required.

The reticle reduces to practice the concept that an interior square could be outlined by depositing a neutral density filter-like metallic coating on a glass round between the square and the four segments between it and the boundary. The diameter of the glass varies with the intended eyepiece and depending on the microscope manufacturer ranges from 20.4 to 27 mm. Only 1 reticle is needed per microscope. See Fig. 20.9.

The reticle is manufactured by Klarmann Rulings.[22] The catalog number is KR-27401-XX. XX is the diameter in millimeters of the eyepiece for which the reticle will be installed, and must be specified by the customer (e.g., KR-27401-24). Your microscope distributor can provide the information if you are unsure. These reticles are made as needed and cannot be returned if you order the wrong size.

By moving a slide while in cellular areas so that cells at one side of the "interior square" go to the opposite side, cytotechnologists will image every μm^2 of the preparation at least once. Moving a slide so systematically does not guarantee that an abnormal cell will be detected, as it may fall outside the observer's field of vision, which is smaller than the microscope's field of view, or the cell may go unnoticed due to cellular limitations, or the observer may experience lapses in observer vigilance, or natural limitations of peripheral vision. Nonetheless, this device introduces a measure of mechanical quality control into the screening process. The square is designed to just touch the boundaries of eyepieces' field of view that has a field number of 20. Field number 22 in today's eyepieces is too large for screening purposes. Such eyepieces image the most peripheral cells up to 10% farther from the center of the field of view and image 21% more area. Both factors increase the probability of not finding any abnormal cells, thus resulting in a false negative.

By aligning the nucleus of a morphologically problematic cell with the cross-hair bull's eye *under high dry magnification*, cytotechnologists may more uniformly differentiate cell areas. For example, nuclei of ASC-US cells are described as being 2–1/2 to 3 times larger in area than the nuclei of intermediate squamous cells. The bull's eye circle equals 6.7-μm diameter (the diameter of an average intermediate squamous cell nucleus [i.e., 35 μm^2][23]), when viewed at 400×. Each of the two crossed lines equals 11 μm when viewed at the same magnification, which represents the diameter of an ASC-US nucleus that is 2.7× greater in area.

The circle/crossed-line aid is uniquely novel. Humans visually estimate area poorly, especially when comparing the area of one nucleus with another.[24] See Fig. 20.10. The areas of nuclei of similar cell types vary appreciably in the same preparation and more so among different preparation types (i.e., conventional Pap tests vs. ThinPrep Pap tests vs. SurePath Pap tests). An objective area reference is needed.

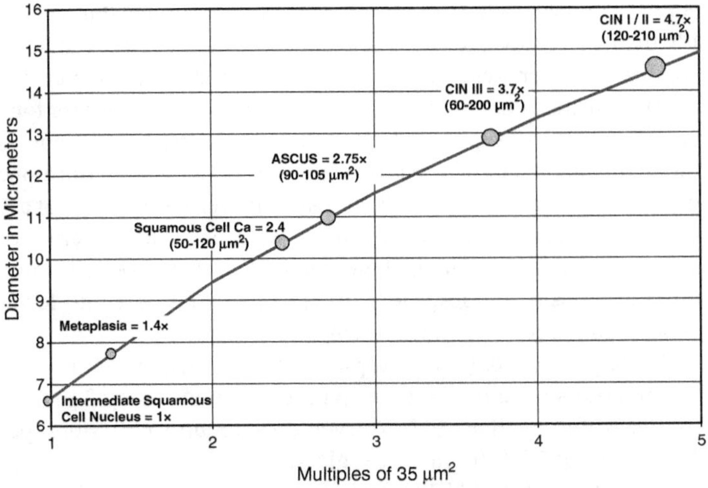

FIG. 20.10. Increasing the 6.7-mm diameter of a 35-mm² reference intermediate squamous cell by only 42% doubles the nuclear area; thus, the human eye will consistently underestimate increases in nuclear area without benefit of an objective size comparator (i.e., Gill screening reticle.)

Eyepiece Field Number[25]

Eyepieces have two related features that impact the search for abnormal cells: magnification and field number. Magnification is straightforward; standard is 10×. Field number (FN), or field-of-view index, equals in millimeters the diameter of the microscopic field of view (also known as object field) when divided by an objective's magnification. For example, FN 22 eyepieces used with a 10× objective image a 2.2-mm diameter-field of view (i.e., 22/10 = 2.2). An eyepiece's FN is usually engraved on the mount.

Most cytotechnologists use 10× eyepieces when screening slides. Some, however, use 15× eyepieces. While the higher magnification may help us see small HSIL cells that account for some false negatives, these eyepieces necessarily have a smaller FN than do 10× eyepieces. Consequently, these eyepieces image a smaller area per field of view, and compared to the same total number of fields as imaged by FN 20 eyepieces, image less total area. Using 15× eyepieces doesn't have any downside as long as there are abundant abnormal cells present on the slide. Since one doesn't know in advance whether abnormal cells are present at all, much less how many, using 10× eyepieces is a conservative approach that is likely in the long run to allow the observer to find more true positive cases. More true positives translate into fewer false-negative results, which is why we screen.

In the good old days, the eyepiece FN was 18, and later FN, 20. The interior square of FN 20 eyepieces is 2 mm^2. Using 2 mm^2 as the basis for overlap, FN 20 eyepieces require 600 stops per 1,200 mm^2 of 24 × 50-mm cover glass.

In the late 1990s, FN 22 eyepieces were introduced with advertisements that extolled the virtues of the wider field of view. Since 10× eyepieces with FN 22 image 2.42-mm^2 squares, a 1,200-mm^2 area can be imaged in 496 fields or stops, which is an attractive prospect. While visually impressive and potentially beneficial for stationary viewing or photomicrography, however, large field-of-view eyepieces insidiously increase the challenge of finding abnormal cells during the rapid stop-and-go pace of production rate screening.

Compared to FN 20 eyepieces, FN 22 eyepieces image an additional 21% area that is up to 10% farther from the center of field of view where an abnormal cell is more likely to be seen. Proportionally, the area difference is approximately the same as that between 14- and 16-in.-diameter pizzas. To image the same area as that by FN 20 eyepieces, FN 22 eyepieces require 36% overlapping diameters.

On the other hand, eyepieces with FNs less than 20 image smaller internal squares and for the same number of stops image less of a preparation. To image the same area requires more work, meaning more time per slide unless one is ignorant of the implications. Understand the potential practical consequences of your choice of eyepiece magnification and field numbers.

Using 15× eyepieces, which may have FN as low as 14, can unwittingly contribute to false-negative results. An internal square of FN 14 eyepieces is 0.98 mm^2, which is 49% of the internal square area of FN 20 eyepieces and 40.5% of FN 22 eyepieces. The fewer the abnormal cells on a slide, the less likely they will be detected by someone screening a slide using 15× eyepieces. Since one never knows upon commencing screening whether abnormal cells are present at all, much less the number, it does not make sense—in my view—to use 15× eyepieces. The devil is in the details.

Conspicuity Area

Regardless of the area imaged by an eyepiece, the area that really counts is that which an observer can actually see in the blink-of-an-eye or so that each field is imaged on average during the screening of a normal smear. Physicists call this small field of vision the "conspicuity area." It is the part of the visual field in which an object can be discovered in a background during a short presentation while the eye is fixed on the center of the field of view.[26] It is roughly a rectangular area equal to approximately 0.5–0.75 of the FOV diameter in the horizontal direction and 0.33–0.4 in the vertical direction. Our conspicuity area is wider than it is high in humans because of evolutionary development. Our prehistoric ancestors had to be more alert to threats at ground level than those from above.

The conspicuity area is about 20% of the area of a 2-mm diameter field and 16% that of a 2.2-mm diameter field. For comparison, the blink-of-an-eye lasts about 25–400 ms. An individual's conspicuity area is constant, regardless of eyepiece field number. If it were possible to make eyepieces with FN 500, it would be possible to image an entire slide in 1 field of view, but the observer's conspicuity area would remain unchanged. See Fig. 20.11.

In summary, there is a direct connection between an eyepiece's magnification and field number and the probability of

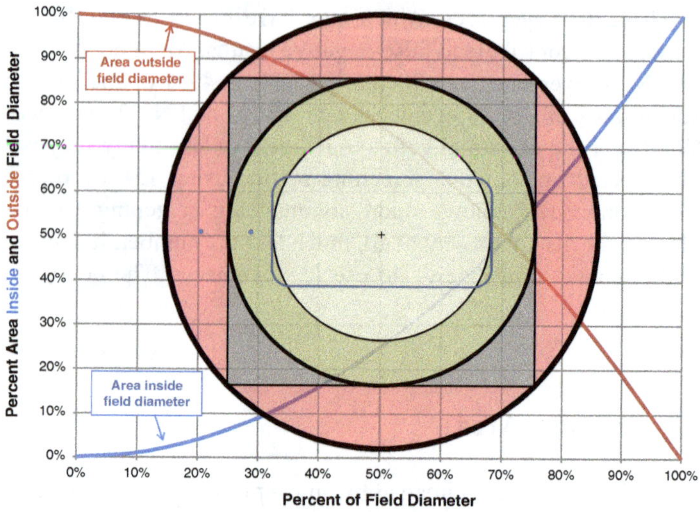

FIG. 20.11. The likelihood of seeing an abnormal cell decreases the farther it is located from the center of the field of view. In the ordinary course of screening, one's eyes are fixated on the FOV center, and one's peripheral vision does not extend to the circumference. The likelihood of seeing HSIL cells is a function of their location within the field of view: (1) The likelihood is least when an abnormal cell falls within the pink zone. (2) The likelihood improves when an abnormal cell falls within the BEIGE zone. (3) The likelihood is best when an abnormal cell falls within the WHITE zone. The rectangular area represents an individual's conspicuity area, which is the visual "sweet spot." The plus sign marks the "bombcyte." HSIL cells "hide out in the open," and their presence may be seen but not perceived even when in the field's crosshairs. The square is the interior area on which 30% overlapping is based. As indicated by the curves, at 1/2 the full FOV diameter, 25% of the area falls within the virtual diameter, and 75% falls outside. The areas inside and outside are equal at 78% of the diameter of the full FOV. An obvious implication is that observers must examine the entire FOV of imaging devices designed to locate objects of interest. An HSIL cell 20 mm in diameter has an area of 314 mm^2 and occupies only 0.008% of the 3.8 mm^2 FOV of ×10 objective and ×10/FN 22 eyepieces. The square area is the basis for the Gill screening reticle, and the rectangle represents an observer's fixed conspicuity area. Both are discussed elsewhere in this chapter.

finding abnormal cells. Higher eyepiece magnifications may increase Pap smear sensitivity for infrequent HSIL cases with small cells but decrease the sensitivity for other abnormalities, thereby increasing the false-negative proportion overall. Higher field number eyepieces image larger areas per FOV, but decrease the probability of finding at least one abnormal cell, thus increasing the false-negative proportion for all abnormalities. It is not clear whether any combination of eyepiece magnification and field number is best, though 10×/FN 20 appears to work well for most cytotechnologists. Nikon, Olympus, and Zeiss sell inexpensive field number reduction rings that reduce FN 22 eyepieces to FN 20.

Vigilance and the Vigilance Decrement

Last, but not least, the human element must be addressed. Screening slides is an extreme example of a vigilance-intensive task. Vigilance means sustained attention.[27] Vigilance is the ability of observers to remain alert; it is a state of maximum physiological and psychological readiness to react and readiness to detect and respond to certain specified small changes occurring at random time intervals in the environment.

Until recently, psychologists and human factor researchers have viewed vigilance tasks as benign work that does not require much effort. Such tasks are seen as "no-brainers," but no longer. "Vigilance requires hard mental work and is stressful."[28]

Vigilance decrement is the progressive decline in performance of a task over time. Attention cannot be sustained indefinitely without errors of omission and commission. Lapses of attention come and go randomly, without warning, or the observer even being aware of the lapses when they occur.

Such lapses, therefore, are insidious. Insidious is rooted in a word that means to ambush. One possible way to maintain attention is to mount eyepiece guards with spikes mounted toward the observer's eyes. Such devices could be called *vigilantes*.[29]

Vigilance as a field of psychological study began around the time of World War II when the Royal Air Force asked psychologist Norman Mackworth to study the problem of U-boat contacts missed by radar observers on antisubmarine patrol. He devised a

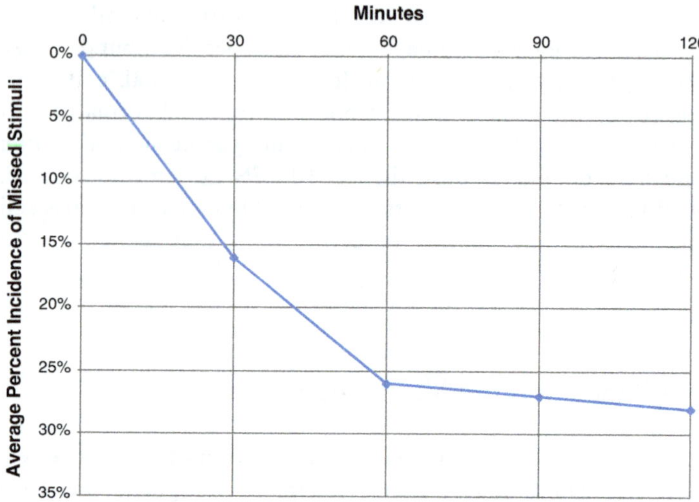

FIG. 20.12. Mackworth demonstrated that observers became more inefficient at detecting radar signals as time on watch progressed. Note that the incidence of missed signals increased sharply from the first 30 min to the second 30 min and then declined more slowly for the remainder of the 120-min period. What do cytotechnologists do when they fall behind? Sit longer.

simulated radar display that enabled him to chart the course of subjects' performance over time. His findings confirmed suspicions that the quality of sustained attention in monitoring tasks declines rapidly. Over a 2-h watch period, detection declined rapidly within the first half-hour and then continued to decline, though at a slower rate over the remaining 1–1/2 h.[30] See Fig. 20.12. The progressive decline in performance is known as the decrement function or the vigilance decrement. It is the most common finding in vigilance research.[27]

Cytotechnologists are notoriously bad about taking breaks from screening. "The mind can absorb no longer than the fanny can endure," which likely contributes to screening errors. Some air traffic flight controller unions require that its members take 30-min breaks for every 2 h on duty, operational circumstances

TABLE 20.3. Factors that may impact screening vigilance.

	Vigilance performance	
Factor	Helps	Hurts
Target	One cell type only	Many cell types
Signal conspicuity	Large and dark objects	Small or pale objects
Signal density	High	Low
Display quality	Impoverished	Cluttered
Background event rate	Low	High
Color of different cell types	Polychromatic	Monochromatic
Spatial uncertainty	Monolayer prep devices	Conventional Pap smears
Display movement	Slow (e.g., 45–60 slides/day)	Fast (e.g., 100 slides/day)
Observer age	Younger	Older
IQ	No relationship	No relationship
Personality	Introvert	Extrovert
Experience	Offsets age differences	–
Visual acuity/peripheral vision	Helps	–
Eye movement in field of view	Multiple fixations	Few fixations
Perceptual field cues	Field-independent	Field-dependent
Mental state	Challenge expectancy	Daydreaming
Habituation	Slow event rate	Fast event rate
Exercise	Mild, frequent breaks	None, screen until you drop
Drugs	Stimulants (e.g., coffee)	Depressants (e.g., alcohol)
Environmental background	Quiet	Noisy
Temperature	Comfortable	Too warm or too cool
Time of day	Afternoon	Morning

permitting. "Break" means doing something other than what you're doing in the tower. Break does not mean do nothing. I believe that cytotechnologists and flight controllers have in common the grim reality that a mistake can be fatal, the major difference being how many people die and how soon one learns about it

Table 20.3 summarizes some of the factors that vigilance research suggests could help or hurt vigilance in the context of screening.[31]

Screening: Search and Attention—Misunderstood and Undervalued

Abundant relevant research into screening-like tasks exists that has not been incorporated into screening performance. Consider, for a moment, the jargon: search trial, search time, target, fixations, saccades, scan path, display, impoverished displays, cluttered displays, nontargets, competition, free search, irregular array, and display contents.[32] Of particular interest in the context of screening and productivity are the lengths of time consumed by individual actions that are usually overlooked and simply subsumed into a mean time per slide. Examples include:[26, 33–35]

- Intercase interval: 1–2 min.
- Moving the slide field-to-field: 180 ms.
- Mean viewing time per field of view (i.e., fixations): 230 ms (i.e., the blink-of-an-eye). A cytotechnologist covers each field of view with a sequence of fixations. Depending on the contents, the number of fixations may be as few as 1 or more.
- Viewing time per fields of view with conspicuous objects: ≥350 ms.
- Saccades last about 50–60 ms each. Saccades are the eye movements from one fixation to the next.
- Mean percent time per slide spent using the 4× objective: 4.97%.
- Mean percent time per slide spent using the 10× objective: 91.0%.
- Mean percent time per slide spent using the 20× objective: 2.32%.
- Mean percent time per slide spent using the 40× objective: 0.53%.
- Mean percent time per slide spent switching from one objective to another: 1.18%.

In other words, cytotechnologists have precious little "eyes-on-cells" time to locate abnormal cells. Screening slides is a zero-sum game. In essence, this means quality is compromised when more slides per hour are screened rather than fewer. The zero-sum game aspect of screening Pap tests is expressed in anti-Parkinson's Law: Work per Pap test contracts as the difference between the number of Pap tests you planned to screen and the number you've actually completed becomes greater as the day wears on. This is the opposite of the more familiar Parkinson's Law, which states: Work expands so as to fill the time available for its completion.

The length of time required to perform each of the many tasks associated with screening a slide is inelastic. To examine many slides per hour translates into skipping details (e.g., reduced screening coverage). If we knew in advance which of the 90% or so of Pap tests were normal, we could afford to cut corners. However, we don't know, and every Pap test deserves to be examined thoroughly. Nearly all Pap tests will only be examined once.

> The only reason for time is so that everything doesn't happen at once.
> Albert Einstein

Teaching search concepts. My fellow cytotechnology students and I were taught by tachyscopy exercises (i.e., literally "quick look"). This exercise was conducted in the darkened old hematology laboratory that had been used by Dr. Maxwell Wintrobe, author of Wintrobe's *Clinical Hematology*. Images of photomicrographs were projected onto a screen for a fraction of a second. We were then asked to describe what we saw and where, when abnormal cells were present. Figures 20.13, 20.14, and 20.15 attempt to illustrate some key concepts.

All cytotechnologists who are shown abnormal cells they have missed react internally the same: (1) I recognize those cells, (2) I don't know how I missed them, and (3) I don't know what I can to do that will guarantee I'll never miss them again.[14] In truth, there is little one do proactively that will ensure 100% screening accuracy. It is for this reason that I believe that CLIA '88-mandated gynecological proficiency testing has little demonstrable value. Passing an artificial test does not require the same skills as screening in real life and probably does not prevent fatal false negatives

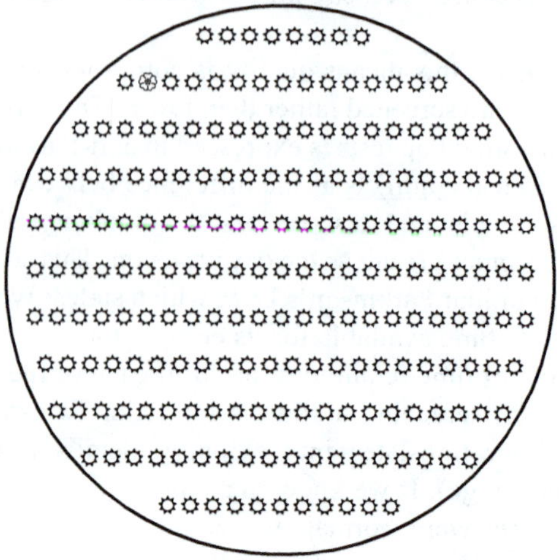

FIG. 20.13. This is a highly regular and artificial display of 1 abnormal cell (❀) among a field of 204 normal cells (✿). It is intended to show that peripherally located targets that are similar in size, shape, and optical density fall outside the observer's conspicuity area are unlikely to be seen. The abnormal cells fall inside the microscope's field of view, but outside the observer's much smaller field of vision (i.e., conspicuity area).

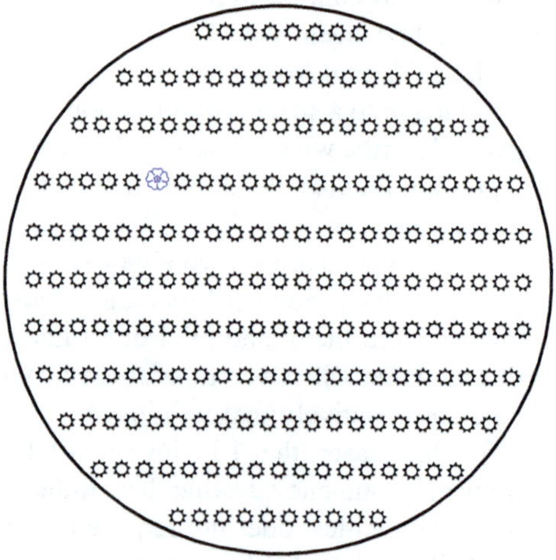

FIG. 20.14. When the abnormal cell is located closer to the observer's conspicuity area and is noticeably larger, more colorful and distinct from its normal neighbors, it is likely to be identified.

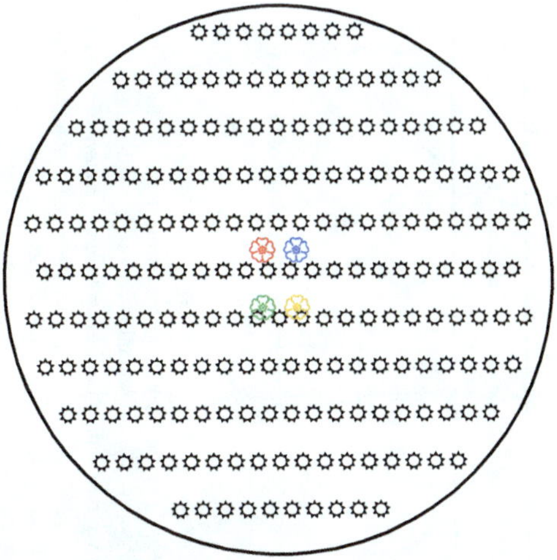

FIG. 20.15. The probability of locating abnormal cells is nearly assured, but not guaranteed, when the abnormal cells stand out in the observer's visual "sweet spot." Numerous large cells with distinctive abnormal cytomorphology that set them apart from their benign cohorts, located in the center of the observer's conspicuity area will probably be perceived, though not necessarily. I have seen false-negative Pap tests in the context of medical malpractice litigation in which the obviously malignant cells were "wall-to-wall."

False-negative Pap tests result from a multiplicity of factors as shown in Table 20.4.

Conclusion

Numerous elements must converge in time and space to satisfy the first objective of screening: find one abnormal cell. Therefore, chance plays a not inconsiderable role. "Good pickup" translates into "you got lucky." Just as an abnormal case with few abnormal

TABLE 20.4. Factors that may contribute to false-negative results.[36]

Abnormal cells present on slide…	Location and interpretation limited by…		
	Specimen	Microscopy	Observer
But not imaged		Inadequate coverage Stage drift	
Imaged, but not looked at		Cells outside field of vision	Incomplete examination of field of view
Imaged and looked at, but not seen	Few abnormal cells Many normal cells	Screening too fast	Vigilance decrement Mental fatigue
Imaged, looked at, and seen, but not recognized	Small cells, inconspicuous low contrast; air-dried; hypo-, hyper-, or monochromatic; cells from well-differentiated lesion	Spherical aberration (i.e., cover glass and mountant too thick) Glare/flare (i.e., no Köhler illumination) Light scatter by dirty microscope lenses	Lack of proficiency: pattern or cytomorphology not recognized as abnormal

cells may be found by chance alone, an abnormal case with many abnormal cells may be missed—but not by chance alone.

A man's got to know his limitations.

Clint Eastwood character detective Harry Callahan in Magnum Force, a 1973 movie.

References

1. Gupta PK. Cytopathology today: challenges and opportunities. Acta Cytol. 1997;41(1):1–5.
2. Patten FW. Cytotechnology: the first half-century. Part I. ASC Bull. 2002;39(3):37, 40–43, 46–47. Available at http://www.cytopathology.org/website/article.asp?id=2355. Accessed 15 Mar 2012.
3. Patten FW. Cytotechnology: the first half-century. Part II. ASC Bull. 2002;39(4):53, 56–57, 59, 61–62. Available at: http://www.cytopathology.org/website/article.asp?id=2355. Accessed 15 Mar 2012.
4. Patten FW. Cytotechnology: the first half-century. Parts III. ASC Bull. 2002;39(7):101–103, 109–111, 120–122. Available at: http://www.cytopathology.org/website/article.asp?id=2355. Accessed 15 Mar 2012.
5. Gill GW. Is here a standard of practice for screening Pap smears, and if so, what is its significance? LabMedicine. 2006;37(1):40.
6. Gill GW. Automated cytology workload: global considerations (manual and automated). Public Comment. Presented at CLIAC Meeting, Atlanta GA, 02.14-15.2012. Tab_26_CLIAC_2012Feb_Public_Comment_Cytology_Gill.pdf. Available at http://wwwn.cdc.gov/cliac/cliac_meeting_view_documents.aspx?MeetingDocumentID=196. Accessed 22 Mar 2012.
7. Bogdanich W. False negative. Medical labs, trusted as largely error-free, are far from infallible. Haste, misuse of equipment, specimen mix-ups afflict even best labs at times. Regulation; weak and spotty. Wall St J. 1987.
8. Bogdanich W. Lax laboratories. The Pap test misses much cervical cancer through labs' errors. Cut-rate 'Pap mills' process slides with incentives to rush. Misplaced sense of security? Wall St J. 1987.
9. Bogdanich W. Physicians' careless with Pap tests is cited in procedure's high failure rate. Wall St J. 1987.

10. Kornstein MJ, Byrne SP. The medicolegal aspects of error in pathology. A search of jury verdicts and settlements. Arch Pathol Lab Med. 2007;131(4):615–8.
11. Elsheikh TM, Kirkpatrick JL, Fischer D, et al. Does the time of day or weekday affect screening accuracy? A pilot correlation study with cytotechnologist workload and abnormal rate detection using the ThinPrep Imaging System. Cancer Cytopathol. 2010;118(1):41–6.
12. Renshaw AA, Elsheikh TM. Predicting screening sensitivity from workload in gynecologic cytology: a review. Diagn Cytopathol. 2011;39(11):832–6.
13. Holbrook M, Roebuck J, Rana DN, et al. Unidirectional versus bidirectional screening of SurePath® liquid based cytology slides – pitfalls & perils of study design. Cytopathology. 2006;17 Suppl 1:16.
14. Gill GW. Unaddressed issues in cytotechnology I: how did I miss those cells? Adv Med Lab Prof. 1995;7(23):5–7, 24.
15. Birdsong G. Panel discusses impact of new technologies on workload limits. ASC Bull. 1999;36(7):96.
16. Gill GW. Optimising and standardising microscopic screening coverage. First find one abnormal cell. SCAN. 2001;12(1):10–2, 5.
17. Hollander DH, Frost JK. Retrieval of located cells in screened cytologic material. Acta Cytol. 1969;13(11):603–4.
18. Bierig JR. Removing dots from cytology slides—liability issues. CAP Today. 2005;19(1). Available at http://www.cap.org/apps/portlets/contentViewer/show.do?printFriendly=true&contentReference=cap_today%2Fpap_ngc%2F0105NGC_CytoDots.html. Accessed 15 Mar 2012.
19. McCoy DR. Defending the Pap smear: a proactive approach to the litigation threat in gynecologic cytology. Pathology Patterns Reviews. 2000;114(Suppl 1):S52–8. Available at http://ajcp.ascpjournals.org/content/supplements/114/Suppl_1/S52.full.pdf. Accessed 15 Mar 2012.
20. Gerstein MD. It's a matter of record for the laboratory. LabMedicine. 2001;32(5):235–8.
21. Baker RW, Brugal G, Coleman DV. Assessing slide coverage by cytoscreeners during the primary screening of cervical smears, using the AxioHOME Microscope system. Analyt Cell Pathol. 1997;13(1):29–37.
22. Klarmann Rulings, Inc., 480 Charles Bancroft Highway, Litchfield NH 03052-1088; in NH: (603) 424-2401; outside NH toll-free: (800) 252-2401, sales@reticles.com.Tutorials of Cytology http://www.reticles.com/. Accessed 26 Mar 2012.

23. Patten Jr SF. Diagnostic cytopathology of the uterine cervix. 2nd ed. revised. Basel, Switzerland: S. Karger AG; 1978.
24. Schmidt JL, Henriksen JC, McKeon DM, et al. Visual estimates of nucleus-to-nucleus ratios. Can we trust our eyes to use the Bethesda ASCUS and LSIL size criteria? Can Cytopathol. 2008;114(5): 287–93.
25. Gill GW. Eyepieces – bigger isn't better. ASCT J Cytotechnol. 1997;1(1):29–33.
26. Reuter B, Schenck U. Investigation of the visual cytoscreening of conventional gynecologic smears. II. Analysis of eye movement. In: Wied GL, Bartels PH, Rosenthal DL, et al., editors. Compendium on the computerized cytology and histology laboratory. Chicago, IL: The Tutorials of Cytology; 1995. p. 49–56.
27. Warm JS, editor. Sustained attention in human performance. New York, NY: Wiley; 1984.
28. Warm JS, Parasuraman R, Matthews G. Vigilance requires hard mental work and is stressful. Human Factors. 2008;50(3):433–41. Available at http://uc.academia.edu/GeraldMatthews/Papers/927907/Vigilance_Requires_Hard_Mental_Work_and_Is_Stressful. Accessed 25 Mar 2012.
29. Gill GW. Vigilantes – eyepiece guards for maximum attention. ASC Bull. 1998;35(4):58.
30. Mackworth NH. The breakdown of vigilance during prolonged visual search. J Exp Psychol. 1948;1:6–21.
31. Gill GW. Unaddressed issues in cytotechnology III. Vigilance in cytoscreening. Looking without seeing. Adv Med Lab Prof. 1996;8(15):14–5,21.
32. Monk TH. Search, chapter 9. In: Warm JS, editor. Sustained attention in human performance. New York, NY: Wiley; 1984.
33. Schenck U, Reuter B. Analysis of the cytoscreening technique as a method of quality control. In: Wied GL, Keebler CM, Koss LG, Patten SF, Rosenthal DL, editors. Compendium on diagnostic cytology. 7th ed. Chicago, IL: The Tutorials of Cytology; 1992/3. p. 437–40.
34. Schenck U, Planding W. Quality assurance by continuous recording of the microscope status. Acta Cytol. 1996;40(1):73–80.
35. Schenck U, Reuter B, Vohringer P. Investigation of the visual cytoscreening of conventional gynecologic smears—I. Analysis of slide movement. Anal Quant Cytol Histol. 1986;8(1):35–45.
36. Gill GW. Pap smear risk management by process control. Can Cytopathol. 1997;81(4):198–211.

Chapter 21
Bethesda System 2001, CLIA '88, and Data Analysis

Not everything that counts can be counted, and not everything that can be counted counts.

Einstein

PRINCIPLE NO. 8

Screen preparations in ways that facilitate abnormal cell detection.

PRACTICE

Be realistic about your screening performance. Errors are the norm; no one's perfect.

Historic Milestones

- 1967—Clinical Laboratory Improvement Act of 1967 implemented by Public Health Service Act.
- 1987—Bogdanich's Wall Street Journal articles set the stage for Clinical Laboratory.[1–3] Improvement Amendments of 1988.
- 1988—President Reagan signs CLIA '88 into law on October 31st.
- 1988—National Cancer Institute sponsors workshop that results in the Bethesda System.
- 1991—NCI sponsors second TBS workshop.

G.W. Gill, *Cytopreparation: Principles & Practice*,
Essentials in Cytopathology 12, DOI 10.1007/978-1-4614-4933-1_21,
© Springer Science+Business Media New York 2013

- 1994—Krieger and Naryshkin introduce false-negative fraction metric for Pap test screening.[4]
- 1996—College of American Pathologists sponsors Conference XXX: Quality and Liability Issues with the Papanicolaou Smear.
- 1997—Conference XXX proceedings published in Archives of Pathology & Laboratory Medicine.
- 1997—Nagy explains that false-negative proportion is mathematically correct terminology.[5]
- 2000—Naryshkin and Davey publish *Incredibly low false-negative proportion: Watch out!*[6]
- 2000—Gill illustrates what 10% cellular coverage looks like.[7]
- 2000—Gill asks and answers whether very low false-negative proportions make sense.[8]
- 2000—Gill describes how to calculate false-negative proportion.[9]
- 2001—NCI sponsors third TBS workshop.
- 2002—Rowe et al. rescreen all NILM ThinPrep Pap tests, which allows for the first time the calculation of a laboratory's true—not estimated—false-negative proportion.[10]
- 2003—CLIA '88 finalized on April 23rd, 5,287 days after becoming law, which includes for the first time this language: "*Automated and semi-automated screening devices.* When performing evaluations using automated and semiautomated screening devices, the laboratory must follow manufacturer's instructions for preanalytic, analytic, and postanalytic phases of testing, as applicable, and meet the applicable requirements of this subpart K."
- 2010—FDA issues "How Laboratorians Can Safely Calculate Workload for FDA-Approved Semi-Automated Gynecologic Cytology Screening Devices."[11]
- 2012—American Society of Cytopathology publishes its workload recommendations for automated Pap test screening.[12]
- 2013—October 31 marks the 25th anniversary of CLIA '88 being signed into law.

This chapter concludes a cytologic specimen's quality journey from specimen collection to specimen processing and screening to data analysis. The 18 historic milestones span 45 years, which is a long time not to get it right. Not to get it right means that the fundamentals of screening and data analysis aren't widely understood, and as a result, expectations about screening productivity and screening performance are unrealistically high. It's not unlike trying to nail Jell-O to a wall.

In a perfect Pap test screening universe, all Pap tests are adequate samples, abnormal cells are clearly abundant and abundantly clear, cytotechnologists are always alert, and the skies are not cloudy all day. Would that it were so, but alas it is not. Real life can be really messy.

Consider the following cytoliths, my word for small cytological truths that are tough to swallow:

1. Quality means useful for its intended purpose. Once the intended purpose has been defined, contributing factors can be identified and measured. Anything that does not add value to a product or service from the standpoint of the external customer is waste.
2. All cytotechnologists screen Pap test slides differently.[13]
3. There is no standardized approach to screening slides that is in print, is taught, or is practiced.[14]
4. Even if there were a screening SOP, errors would still occur as it's a human activity.
5. False-negative Pap test results can be fatal. There are no definitive data regarding annual numbers. Whatever numbers exist are likely to be understated. Clinicians tend not to tell women who develop cervical cancer that their Pap tests may have been misread. Lawsuits that are settled before the verdict of a judge or jury are signed, sealed, and not disclosed.
6. Screening a Pap test is a process by which a slide is moved successively in small steps to promote the likelihood that at least one recognizable abnormal cell will fall within the narrow field of vision of an alert professional observer long enough to be perceived, recognized, and interpreted.[15]
7. Screening time is interpreted differently by different labs, often in patient adverse ways—an hour is not an hour.[16] "Instead of screening workload, the term 'analytic productivity workload' was used to refer to the analytic component of the test process. This consists of review of the clinical information,

evaluation of the case, and recording of the results. This definition specifically excludes pre- and postanalytic processes such as specimen preparation, resolving clinical history discrepancies, QA review, and batch data entry into a laboratory information system."[15]

8. Another way to determine what counts as screening time and what doesn't is to refer to CPT coding language, for example, CPT 88142: cytopathology, cervical, or vaginal (any reporting system, collected in preservative fluid, automated thin-layer preparation; manual screening under physician supervision). In other words, if it can't be billed, it's not screening.
9. Not everyone understands the difference between QC and QA, or that QA now means quality assessment in the lexicon of CLIA '88.
10. Error detection relies on the same process that caused the error in the first place: (re)screening by humans. Underfinding errors is the rule, not the exception.
11. The relationship between screening and rescreening can be characterized as a self-reinforcing negative feedback loop. The worse the rescreening, the better the screening appears to be.
12. To achieve perfection in screening, don't detect errors. If rescreening identifies few errors, is it because the primary screening was that good or the rescreening was that bad?[17] Labs prefer the former possibility, I prefer the latter. There is little incentive to rescreen well, protestations to the contrary.[18]
13. True negatives will always be negative, regardless of how poorly a slide was screened or whether it was screened at all. This realization introduces the concept of the false true negative: you got the right answer not because you're good, but because you can't be wrong.
14. In laboratories with a culture of high screening productivity, the false true negative can be self-delusional. For example, one laboratory reported finding zero screening errors by 33 of 65 cytotechnologists among 200,000 Pap tests rescreened over a 3-month period.[19]
15. Publishing one's false-negative proportion (FNP) without also publishing the accompanying sensitivity is an exercise in futility. Ideally, FNP + sensitivity = 1.[7] Therefore, FNP 0.05

implies a screening sensitivity of 0.95. If a lab's screening sensitivity were truly 0.95, why rescreen?[4]

16. Rescreening sensitivity is likely to be lower than that for primary screening. Requiring "qualified" individuals to review slides doesn't guarantee quality. When a lab thinks it knows its screening sensitivity, that value must be divided into the rescreening findings to account for the fact that rescreening sensitivity is not 100%.[20]
17. It is probable that most labs don't know how to calculate their screening sensitivity.[21]
18. FNP 0.05 can be achieved when (1) screening sensitivity is 90% and rescreening sensitivity is 50%, (2) screening sensitivity is 80% and rescreening sensitivity is 25%, and (3) screening sensitivity is 70% and rescreening sensitivity is 12.5%. These combinations are calculated using simple arithmetic. Using real data without understanding its limitations bolsters confidence falsely.
19. The FNP so-called floor of performance (0.05) is potentially dangerously misleading.[6] It implies a screening sensitivity of 0.95. If that were true, all false negatives would be identified if, and only if, all initially NILM Pap tests were rescreened with 100% sensitivity.
20. Properly calculated FNP can be used to estimate the number of unidentified ASC-US, LSIL, and HSIL cases screened once and categorized initially as NILM.[8,9]
21. Labs don't know how to compare one cytotechnologist's rescreening findings against those of the entire laboratory. As a result, cytotechnologists can be disciplined unfairly. Use chi-square.
22. Labs don't know how to *simply* determine whether cytotechnologist productivity is within CLIA limits.
23. Laboratories don't know how to assess or reassess each individual's workload limit and adjust when necessary. They may do it, but it's not reliable. Corrective actions fail because they do not address the root cause(s).
24. Errors can be acceptable or unacceptable. The former can occur in any laboratory (e.g., few abnormal cells, well-differentiated

cells that do not stand out). Unacceptable errors are blunders: well-preserved obviously abnormal cells everywhere.

25. Few cytoprofessionals know what vigilance (i.e., sustained attention) means, what's required to maintain it (e.g., frequent breaks), and vigilance decrement's role in screening errors.[22] Vigilance requires hard mental work and is stressful.[23]
26. High screening productivity means uninterrupted screening for long periods, which increases errors. As a result of such "good soldier" behaviors, cytotechnologists develop hyperchronokathistophobia, an occupational hazard that means fear of sitting for extended periods. Also known as fanny fatigue, it's another way of saying the mind can absorb no more than the bottom can endure.
27. Rapid anything (e.g., prescreening, review[24]) doesn't improve screening quality measurably, reproducibly, and sustainably. There's no free lunch!
28. Proficiency testing is an artificial exercise unrelated to real world screening circumstances and errors, which are random.
29. CLIA expects better interobserver agreement between cytotechnologists and pathologists on cell samples than the literature has demonstrated is possible among pathologists on tissue.
30. Discrepant true positives don't kill women; unidentified false negatives do. CLIA's emphasis in cytology is misplaced.[25]
31. Education is key to understanding the complexity of the screening process.[26–31]

Now that those indigestibles are under our belts, I want to discuss cellular adequacy, define screening and screening time, workload calculations, workload limits, evaluating a cytotechnologist's rescreening errors with that of the laboratory, and false-negative proportion.

Cellular Adequacy

In 1991, the Bethesda System's criterion for specimen adequacy stated that "well-preserved and well-visualized squamous cells should cover more than 10% of the slide surface." That criterion

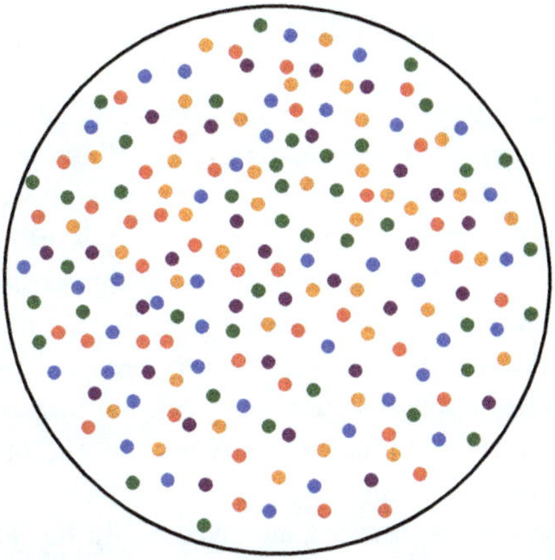

FIG. 21.1. There are 190 "cells" in this field of view or 60,000 on the entire slide surface under a 24 × 50-mm cover glass. Their combined areas equal 10% of the slide surface area of a conventional Pap smear. Clearly, the original Bethesda System adequacy criterion was set too high. Each circle represents a superficial squamous cell that is 1/50th the diameter of this representative ×10 objective field of view as seen through a ×10 eyepiece with a field number of 20.

was not based on evidence; it just sounded about right. No one knew what that coverage looked like in terms of numbers of cells, but it turned out to be so many cells that everyone would classify such a Pap test as adequate. It occurred to me that I could draw a representation of such coverage using MS Publisher and data about areas of squamous cells and field of view. See Fig. 21.1.

In response to this objective information, Bethesda 2001 changed the criterion for adequacy from 10% coverage to 8,000 to 12,000 cells per conventional Pap smear and a minimum of 5,000 cells on a liquid-based Pap test. For a conventional Pap test, those numbers are 87% to 80% less than 60,000 cells and represent about 1.3% to 2% coverage instead of 10% coverage.

Definitions of Screening and Screening Time

On July 22–23, 1999, I participated in the Centers for Disease Control (CDC) panel to discuss the impact of new technologies for gynecologic cytology on workload limitations. Instead of screening workload, the term "analytic productivity workload" was used to refer to the analytic component of the test process. This consists of review of the clinical information, evaluation of the case, and recording of the results. This definition specifically excludes pre- and postanalytic processes such as specimen preparation, resolving clinical history discrepancies, QA review, and batch data entry into a laboratory information system.

I defined manual screening as: "Screening the Pap smear is a process by which a slide is moved successively in small steps to increase the probability that at least one abnormal cell will fall within the narrow field of vision of an alert professional observer long enough to be recognized and interpreted."

These definitions are key to understanding the complexity of the screening process and accounting fairly for screening time. One cannot arbitrarily ramp up expectations for screening productivity without compromising screening sensitivity. Nonetheless, there are laboratories that count the minute a cytotechnologist clocks-in to the minute s/he clocks-out as screening time, regardless if the cytotechnologist is screening or not. The net effect is cytotechnologists who work under those circumstances must compress screening productivity into fewer screening hours to look good at the end of the day. Screening sensitivity must diminish, but it's difficult—if not impossible–to measure. A lawsuit representing the family of a woman who died because of an alleged false-negative-based Pap test is the ultimate quality assessment monitor.

Workload Calculations

On July 27, 2010, FDA published How Laboratorians Can Safely Calculate Workload for FDA-Approved Semi-Automated Gyn Cytology Screening Devices, which included:

"The purpose of this communication is to clarify for laboratories how workload should be calculated when using currently FDA-approved semiautomated gynecologic cytology screening devices. This communication is intended for cytotechnologists, technical supervisors, and laboratory managers using these systems and addresses how to count fields of view (FOV) and full manual slide reviews (FMR), as well as establishing maximum workload limits. Exceeding the designated maximum workload jeopardizes the ability of device users to detect precancerous and cancerous lesions of the cervix and is a public health risk.

It has been brought to our attention that the current product labeling regarding workload recording for these devices has been difficult to interpret, resulting in variability and lack of standardization in counting methods.

How Laboratorians Can Safely Calculate Workload for FDA-Approved Semi-automated Gynecologic Cytology Screening Devices

What Are the Current Issues with Workload Recording and Maximum Workload Limits?

In addition, individual maximum daily workload limits are not being established by the technical supervisor as mandated by CLIA'88. *The maximum daily limit specified in each of the device product labeling is only an upper limit and should never be used as an expectation for daily productivity or as a performance target.*

How Can Laboratorians Safely Calculate Workload for FDA-Approved Semi-automated Cytology Screening Device [sic]?

To ensure the safety and effectiveness of these devices, given their importance as women's health screening tests, the FDA has determined that laboratorians should use the following method when calculating workload. The calculation method applies to both semiautomated cytology screening systems currently on the market (Hologic's ThinPrep® imaging system and Becton Dickinson's Focal Point™-guided screening system):

- All slides with full manual review (FMR) count as 1 slide (as mandated by CLIA'88 for manual screening).
- All slides with field of view (FOV) only review count as 0.5 or ½ slide.
- Then, slides with **both** FOV and FMR count as 1.5 or 1½ slides.
- Use these values to count workload, not exceeding the CLIA maximum limit of 100 slides in no less than an 8-h day.

FMR = 1 slide
FOV = 0.5 slide
FMR + FOV = 1.5 slides
Upper Limit = 100 slides"

The Division of Laboratory Science and Standards, Centers for Disease Control and Prevention (CDC), will convene a cytology workgroup in Baltimore, MD, on August 15–16, 2012. The workgroup is charged with providing input to the CDC, the Centers for Medicare & Medicaid Services (CMS), and the Food and Drug Administration (FDA) on (a) the challenges associated with establishing a maximum workload limit for individuals who screen Pap smears utilizing semi-automated screening devices and (b) suggestions for how to obtain data to determine the maximum workload limit. Therefore, the workload limits may change, as well as how to "safely calculate workload."

Workload Limits

Without appreciating that there is no consensus standard for screening slides, without understanding the definitions of screening and screening time, without realizing that screening and rescreening constitute a self-reinforcing negative loop, without knowing how to compare screening performance among cytotechnologists, without knowing the cytopathologist's interpretation is the gold standard against which the cytotechnologist's interpretation is compared is not always correct, and without knowing how to account for discrepancies or errors and attempt solutions in ways that make a measurable difference, how can one implement any of the following?

"CLIA '88 § **493.1274: Standard: Cytology:** (d) *Workload limits.* The laboratory must establish and follow written policies and procedures that ensure the following:

(1) The technical supervisor establishes a maximum workload limit for each individual who performs primary screening.

(i) The workload limit is based on the individual's performance using evaluations of the following:

(A) Review of 10% of the cases interpreted as negative for the conditions defined in paragraph (e)(1) of this section.

(B) Comparison of the individual's interpretation with the technical supervisor's confirmation of patient smears specified in paragraphs (e)(1) and (e)(3) of this section.

(ii) Each individual's workload limit is reassessed at least every 6 months and adjusted when necessary.

(2) The maximum number of slides examined by an individual in each 24-hour period does not exceed 100 slides (one patient specimen per slide; gynecologic, nongynecologic, or both) irrespective of the site or laboratory. This limit represents an absolute maximum number of slides and must not be employed as an individual's performance target. In addition—

(i) The maximum number of 100 slides is examined in no less than an 8-hour workday."

Evaluating a Cytotechnologist's Rescreening Errors with that of the Laboratory

Workload is always divided unequally among cytotechnologists, and so it is essential to know how to evaluate different numbers of errors detected among unequal numbers of rescreened Pap tests among cytotechnologists. If observed differences appear relatively large, but are nonetheless sufficiently small, differences in errors detected could be due to chance alone. Visual inspection of the data would not reveal whether one cytotechnologist's errors could be due to chance or not.

To be completely fair to all its cytotechnologists, laboratories should use chi-square (χ^2) analysis. Online χ^2 calculators simplify analysis (e.g., http://www.physics.csbsju.edu/cgi-bin/stats/contingency_form.sh?nrow=2&ncolumn=2, http://faculty.vassar.edu/lowry/tab2x2.html). Probabilities less than 0.05 (e.g., $p=0.01$) mean the differences are statistically significant. P values equal to, or greater than, 0.05 mean the observed differences are likely due to chance alone.

False Negative Proportion

False negative proportion is the proportion of true positives missed on the first screening. Stated another way, false negative proportion is the proportion of false negatives (i.e., true positives once removed) identified by rescreening to the total number of true positives identified by screening and false negatives identified by rescreening. Since laboratories do not rescreen 100% of eligible cases, the "true" false negative proportion based on total rescreening findings cannot be calculated. Instead, the number of false negative cases identified by rescreening some of the eligible cases is extrapolated to a number expected to be found if all eligible cases were rescreened. Under the latter circumstances, the correct term is false negative proportion (estimated).

Krieger and Naryshkin identified a "floor" of performance quality near the 5% level, but without providing an estimate of the screening sensitivity, and introduced FNF for individual cytotechnologists based on error quotients, but without providing a relevant statistical test.[4] Ever since, many published reports try to equal or go below the 5% floor without showing the underlying calculations. However, a very low value may result from suboptimal rescreening or inaccurate calculations. I believe it's likely both.

We have never been as good at identifying true positive Pap tests by primary screening as we like to think we are. For example, one cytotechnologist stated in a deposition that s/he had never missed anything in 30 years of screening. That's impossible, of

TABLE 21.1. Abnormal ThinPrep Pap test cases identified by screening and 100% rescreening[10]

	Primary screening		100% QC rescreening	
Categories	Numbers	Percent	Numbers	Percent
Total TPPT cases	53,419	100.0	47,247	88.4
Abnormality	5,368	10.0	804	1.7
ASC-US	3,619	6.77	678	1.44
LSIL	1,496	2.8	116	0.25
HSIL	244	0.46	10	0.02
Carcinoma	9	0.17	0	0

These data permit calculation of the true, not estimated, false negative proportion:
FNP = 804/(5,368 + 804).
FNP = 0.13.

course. It never occurred to any of us that we could miss abnormal cells in conventional Pap smears, but we did. When the first liquid-based Pap test came on the market in 1996, FDA allowed the manufacturer to claim that it was significantly better than the conventional Pap test. The same manufacturer was later allowed to claim increased HSIL pickup. And later still, imaging devices identified even more abnormal cases. Where does it end?

I know of only 1 published paper in which 100% of a laboratory's entire annual eligible workload was rescreened.[10] See Table 21.1.

Since the rescreening sensitivity cannot be 100%, the FNP must be corrected by dividing it by the 86.97% estimate of screening sensitivity (i.e., 5,368/6,172). Thus, 0.13 becomes 0.15 (i.e., 0.13/0.8697).

Even this corrected FNP is likely an underestimate, as there are always unidentified false negative cases. No laboratory's screening sensitivity is 100%. Nonetheless, this "true" false negative proportion is 3 times greater than the putative 5% "floor" of performance quality. FNP 0.15 is closer to the "truth" and validates the rescreening quality. A value of 0.15 means that 15 of every 100 true positive cases—including ASC-US, LSIL, HSIL, and carcinoma—are missed on the first screening.

An FNP value of 0.05 reported by any laboratory—whether estimated or true—is too good to be true. It implies that the primary screening sensitivity is 95%. There is no evidence that is

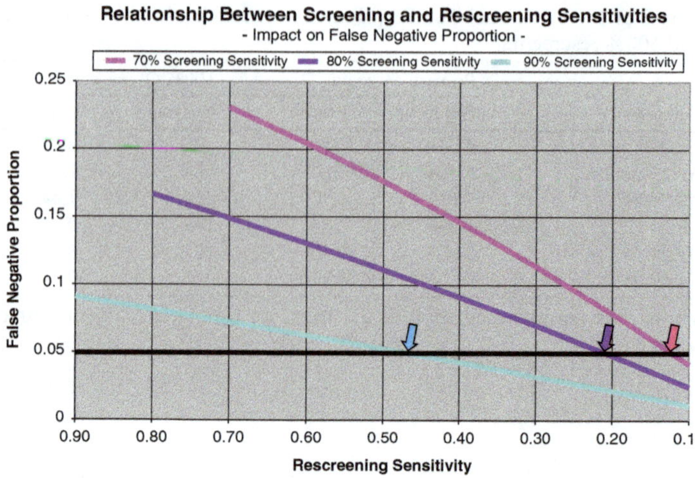

FIG. 21.2. These curves illustrate the relation between primary screening sensitivity and rescreening sensitivity and their combined impact on FNP. Any FNP by itself is meaningless without an accompanying estimate of primary screening sensitivity. FNP 0.05, indicated by the bold horizontal line, can result when rescreening sensitivities are 47.4%, 21%, and 12.3% for screening sensitivities of 90%, 80%, and 70%, respectively. In no case does the rescreening sensitivity approach that of primary screening sensitivity. If FNP satisfied the equation 1 = sensitivity – FNP, then the observed FNPs should be 0.1, 0.2, and 0.3, respectively, assuming rescreening sensitivity is 100%.

ever the case. It is reasonable to ask: what rescreening sensitivity will result in FNP of 0.05 at screening sensitivities of 70%, 80%, and 90%?

Given a workload that includes 1,000 true positives, screening sensitivities of 70%, 80%, and 90% will not identify 300, 200, and 100 true positives (i.e., false negatives) after the first screening. See Fig. 21.2.

Calculating FNP (Estimated)

The simple formula is: $\text{FNP (estimated)} = \dfrac{\text{FN (estimated)}}{\text{TP} + \text{FN (estimated)}}$

The data that go in the numerator and/or the denominator:

- FN includes all false negatives identified by rescreening.
- "Estimated" is a multiplier. It is calculated by dividing the total number of cases eligible to be rescreened by the total number actually rescreened. For example, if 20,000 cases are eligible for rescreening and 2,000 cases were rescreened, the multiplier is 10.
- TP includes all true positives identified by screening at ASC-US and higher.
- Unsatisfactory results are ineligible for rescreening and are excluded.

FNP has potentially far-reaching implications in terms of:

- A laboratory's realistic appreciation of its screening and rescreening performance, and its awareness of the need to improve screening and rescreening performance
- The number of false negative cases that remain unidentified among those cases that have been screened only once
- Whether the Pap test is interpreted as a screening test or a diagnostic test
- The performance level of Pap tests expected by external observers (e.g., attorneys)

A laboratory's quality assessment program of its screening performance is only as good as the quality of its rescreening. Using the same method to detect errors that causes errors—screening by humans—is inherently flawed. "…only a fool trusts a critic who is paid by the restaurant. Why should we accept quality assessment results determined by those most likely to benefit from them?"[32]

The relationship between screening and rescreening can be characterized as a self-reinforcing negative feedback loop by which poor rescreening performance gives the mistaken impression of good screening performance. Taken to the limit, a false-negative proportion of zero is achieved when rescreening identifies no abnormal case. *Ergo*, "perfection." See Table 21.2.

Numerous possible reasons have been suggested to account for diminished rescreening performance: (1) cytotechnologists have not been educated to expect more abnormal cases, (2) rescreening may be put off until day's end and screening time is compressed, (3) cytotechnologists are overworked and fatigued, (4) cytotechnologists

TABLE 21.2. FNP decreases as the rescreening sensitivity decreases relative to various levels of assumed screening sensitivity, and the number of unidentified abnormal cases increases in a workload that includes 1,000 true positives, of which 10% are rescreened.

True positives (TP)	Screening sensitivity (SS)	False negatives (FN)	True FNP	FN left after rescreening 10% of negative cases at various rescreening sensitivities (RS)							
				RS = SS		RS = 0.5		RS = 0.25		RS = 0.125	
				FNP	Left	FNP	Left	FNP	Left	FNP	Left
1000	0.7	300	0.3	0.23	279	0.176	285	0.097	292	0.05	297
1000	0.8	200	0.2	0.167	184	0.111	190	0.059	195	0.03	197
1000	0.9	100	0.1	0.09	91	0.05	95	0.027	97	0.014	99

Especially note that FNP 0.05, 0.059, and 0.05 can result when the rescreening sensitivities are 12.5%, 25%, and 50% when the screening sensitivities are 70%, 80%, and 90%, respectively. Especially note that the number of unidentified false negative Pap tests that remain increases as the FNP decreases, which is the exact opposite of what a casual observer might expect.

know someone else has already screened the case once and believe the likelihood of it being abnormal is slim, (5) some may not want to identify an error that may be small or subjective, (6) others may be concerned that any error they find could have legal repercussions, (7) most of the obvious abnormal cases have already been identified so the ones that are left are more difficult, and (8) there is no way to check on the quality of rescreening.

If FNP 0.05 were applied to 60 million annual Pap test workload in the United States, it can be shown that false negative Pap tests remaining after 100% screening at 85% sensitivity and 10% rescreening at 30% sensitivity are 873158. In other words, 14.5% of the total true positive cases remain undetected (i.e., 873,158/6,000,000), and only 3% of undetected true positives (i.e., false negatives) were detected (i.e., 2,6842/900,000). Sobering, but true.

I believe the untapped value of FNP (estimated) lies in estimating the number of unidentified false negative cases that remain among the Pap tests that have been screened only once. By constructing an Excel spreadsheet with embedded formulas and entering the right data in the right cells, one can quickly calculate FNP (estimated), screening sensitivity, FNP (estimated [and corrected for screening sensitivity]), remaining unidentified false negative Pap tests, and remaining unidentified false negative Pap tests (corrected for screening sensitivity) globally, and by ASC-US, LSIL, and HSIL.

100% Rapid Review Versus 10% Slow Review (i.e., CLIA '88 10% Review)

I'm not a big fan, or even a small one, of rapid rescreening or rapid prescreening for any purpose. That's my tweet and I'm sticking to it.

Conclusion

I recommend meaningful education for cytotechnologists and pathologists about the screening process (e.g., SPADE [Screening Protocol to Assist Detection]), quality assessment, performance measurement, and monitoring. Cytotechnologists are not taught about the preparation, microscopy, physical, psychological, and pattern recognition factors that impact screening and interpretation. Laboratory management should learn how to calculate performance measurements that will provide useful insights into the performance of individuals and the lab as a whole. The education process should include information such as: screening overlap, conspicuity area, limitations in peripheral vision, vigilance, vigilance decrement, signal vs. noise, self-reinforcing negative feedback loop, ASC-US: hrHPV+, ASC-US: SIL ratios (i.e., global, primary screening, QA rescreening), ASC-US(CT): ASC-US (MD), primary screening sensitivity (estimated), FNP (estimated), and chi-square analysis.

When a false negative is bought to the attention of the responsible cytotechnologist, these thoughts occur:

1. I recognize those cells.
2. I don't know how I missed them.
3. I don't know what I can do to guarantee I'll never miss them again.

Most false negatives are the result of random, not systemic or systematic, errors that are largely uncontrollable and cannot be prevented proactively. For this reason, errors are categorized as acceptable and non-acceptable. Acceptable errors are those that can occur in any laboratory; unacceptable errors are those that fall beneath the standard of practice, which is generally defined as practice exercised with the degree of care used by a reasonably careful individual of like qualifications in the community in which he or she practices under the same or similar circumstances. Saying it is easier than proving it.

Cytology laboratories should be concerned more with the number and kinds of abnormal cases that remain unidentified in their files and less with those that have been identified. Emphasis on quality must be prospective, not retrospective. Assess, train, educate, evaluate, monitor, and provide constructive feedback. There is no safety in understated false-negative numbers.

References

1. Bogdanich W. False negative. Medical labs, trusted as largely error-free, are far from infallible. Haste, misuse of equipment, specimen mix-ups afflict even best labs at times. Regulation; weak and spotty. Wall St J. 1987.
2. Bogdanich W. Lax laboratories. The Pap test misses much cervical cancer through labs' errors. Cut-rate 'Pap mills' process slides with incentives to rush. Misplaced sense of security? Wall St J. 1987.
3. Bogdanich W. Physicians' careless with Pap tests is cited in procedure's high failure rate. Wall St J. 1987.
4. Krieger P, Naryshkin S. Random rescreening of cytologic smears: a practical and effective component of quality assurance programs in both small and large cytology laboratories. Acta Cytol. 1994;38(3):291–8.
5. Nagy GK. False negative rate – a misnomer, misunderstood and misused. Acta Cytol. 1997;41(3):778–80.
6. Naryshkin S, Davey DD. Incredibly low false-negative proportion, watch out. Diagn Cytopathol. 2000;22(2):63–4.
7. Gill GW. Pap smear cellular adequacy: what does 10% coverage look like? What does it mean? Acta Cytol. 2000;44(5):873. Abstract No. 32.
8. Gill GW. False negative proportion: do very low values make sense? Acta Cytol. 2000;44(5):851. Abstract No. 8.
9. Gill GW. How to calculate false negative proportion. Acta Cytol. 2001;45(5):883. Abstract 130.
10. Rowe LR, Marshall CJ, Bentz JS. One hundred percent thorough quality control rescreening of liquid-based monolayers in cervicovaginal cytopathology. Cancer Cytopathol. 2002;96(6):325–9.
11. FDA. How laboratorians can safely calculate workload for fda-approved semi-automated gynecologic cytology screening devices. Available at http://www.fda.gov/MedicalDevices/Safety/AlertsandNotices/TipsandArticlesonDeviceSafety/ucm220292.htm. Accessed 24 May 2012.
12. Elsheikh TM, Austin RM, Chhieng DF, Miller FS, Moriarty AT, Renshaw AA. American Society of Cytopathology workload recommendations for automated Pap test screening: Developed by the productivity and quality assurance in the era of automated screening task force. Diagn Cytopathol. 2012. doi: 10.1002/dc.22817. [Epub ahead of print].
13. Gill GW. Optimising and standardizing microscopic screening coverage – first find one abnormal cell. SCAN. 2001;12(1):10–12,15.
14. Gill GW. Q&A: is there a standard of practice for screening Pap smears, and if so, what is its significance? LabMedicine. 2006;37(1):40.

15. Birdsong G. Panel discusses impact of new technologies on workload limits. ASC Bull. 1999;36(7):96.
16. Gill GW. A rogue's gallery of screening policies and practices. Focus. 1998;5(1):4–6. Newsletter of Papanicolaou Society of Cytopathology. Invited article.
17. Gill GW. New metrics set to assess Pap test cytologic screening performance. Acta Cytol. 2003;47(5):825. Platform presentation abstract 8.
18. Renshaw AA. Experts in wonderland – in search of the right test and the scientific method. Diagn Cytol. 2000;23(5):297–8.
19. Krieger PA. Strategies for reducing Papanicolaou smear screening errors: principles derived from data and experience with quality control. Arch Pathol Lab Med. 1997;121(3):277–81.
20. Renshaw AA, DiNisco SA, Minter LJ, Cibas ES. A more accurate measure of the false-negative rate of Papanicolaou smear screening is obtained by determining the false-negative rate of the rescreening process. Cancer Cytopathol. 1997;81:272–6.
21. Gill GW. Self-referenced approach to estimating primary cytological screening sensitivity. Acta Cytol. 2001;45(5):884. Poster abstract No. 132.
22. Gill GW. Unaddressed issues in cytotechnology: vigilance in cytoscreening. Adv Med Lab Prof. 1996;8(15):14–15,21.
23. Warm JS, Parasuraman R, Matthews G. Vigilance requires hard mental work and is stressful. Hum Factors. 2008;50(3):433–41.
24. Gill GW. Rapid review – is it effective? Is it efficient? SCAN. 2002;13(1):8–11.
25. Gill GW. CLIA '88: are we there yet? LabMedicine. 2005;36(11):694, 732. Quality Qorner [sic].
26. Gill GW. Unaddressed issues in cytotechnology: "How did I miss those cells?". Adv Med Lab Prof. 1995;7(23):5–7,24.
27. Gill GW. Pap smear risk management by process control. Can Cytopathol. 1997;81(4):198–211.
28. Gill GW. Viewpoint: on CT workload overload and doing the job right the first time. ASCT News. 1997;18(4):57–61.
29. Gill GW. Quality control – quality assurance in diagnostic cytopathology: what's the difference? Diagn Cytopathol. 1995;13(3):270–2.
30. Gill GW. Chapter 28: The laboratory. In: DeMay RM, editor. The art & science of cytopathology, vol. 3. 2nd ed. Chicago: ASCP Press; 2011. p. 1539–92. ISBN 978-089189-6449.
31. Gill GW. Eyepieces–bigger isn't better. ASCT J Cytotechnol. 1997;1(1):29–33.
32. Renshaw AA. Experts in Wonderland: in search of the right test and the scientific method. Diagn Cytopathol. 2000;23(5):297–8.

Appendix A: Word Notes

Language is the skin of living thought.
Oliver Wendell Holmes

Abbreviations, initialisms, and acronyms. Basic abbreviations are pronounced like the full words they represent. For example, "St." is pronounced Street or Saint, depending on context, and "Ct." is pronounced Court, and so on. Basic abbreviations are usually followed by a period. An initialism is a type of abbreviation pronounced one letter at a time (e.g., PGA, AARP, FDA, IRS). An acronym is a type of abbreviation that is pronounced as a word (e.g., SCUBA [self-contained underwater breathing apparatus], LASER [light amplification by stimulated emission of radiation], and RADAR [radiodetection and ranging]).

"*Artifact.* It was presumably a mere slip of the pen that caused the poet Coleridge to write artifact when coining this term, for English words do not include Latin ablatives such as *arte*. *Ars* being a Latin noun of the third declension, its stem in compound words is *arti-*."[1]

"*Basiphil.* Ehrlich was interested in the classics as a young man and frequently used Latin tags in conversation throughout his life. It is strange that he should not have recognized the error in 'basophil.' The Latin noun *basis* is of the third declension, and its stem in compound words is therefore *basi-*. 'Basophil' is as wrong as 'basosphenoid,' 'matroarchal,' and 'regocide' would be. There is no reason for adding *-ic* at the end of the word: it should have the

G.W. Gill, *Cytopreparation: Principles & Practice*,
Essentials in Cytopathology 12, DOI 10.1007/978-1-4614-4933-1,
© Springer Science+Business Media New York 2013

same termination as the English adjective 'Francophil.' Similarly one should write 'acidophil,' not 'acidophilic.'"[1]

Buffer. The etymology is interesting. Buffer comes from the German *puffer*, which some languages translate as tampon, which in turn means absorb. In chemistry, therefore, a buffer absorbs changes in pH. For example, Hanks balanced salt solution is buffered to resist changes in pH.

Cytology. "**Note on the word 'cytology'**—Librarians and booksellers are apt to be confused because the word 'cytology' has taken on two different meanings. The general study of cell biology has been called 'cytology' for at least 85 years, and the word is still used in that sense; examples are the periodical *International Review of Cytology, of Bourne and Danielli, and A History of Cytology by Arthur Hughes. Darlington's* Recent Advances in Cytology (1st edition, 1932) had a more restricted subject, that of the chromosomes, particularly those of plants! With the widespread development of cytodiagnostic laboratories in hospitals, the name 'cytology' has been generally adopted for that specialty. (The adjective 'exfoliative' was dropped because in many case the cells examined have not been exfoliated.) Accordingly, we have *Acta Cytologica* as the principal journal of the specialty, began as the official journal of the International Academy of Gynaecological Cytology (1958), which has since changed its name to the International Academy of Cytology.

Now that the word has become ambiguous, the best that can be done is to qualify it or use another. The study of chromosomes has become 'cytogenetics,' while the expression 'cell biology' has been adopted by the scientists who previously called themselves 'cytologists.' Medical cytodiagnosticians should speak of 'cytopathology,' 'clinical cytology,' 'diagnostic cytology,' or 'cytodiagnosis.'"[2]

Fumes vs. vapors. The word "fumes" is frequently used when "vapors" would be more accurate. The word "fumes" is rooted in the Latin *fumare*, which means to smoke. "Vapors" is rooted in the Latin *vapor*, which means steam. Strictly speaking, therefore, fumes are the result of combustion and typically include solid particles. Vapors, on the other hand, are the gaseous state of a heated liquid. Why are fume hoods called fume hoods and not vapor hoods? Who knows? In England, they are referred to as fume cupboards.

-ic vs. -ical word endings. Word pairs such as cytologic/cytological, histologic/histological, microscopic/microscopical, physiologic/physiological represent differences with a distinction. All are adjectives. Adjectives ending in -ic mean the noun they modify is related directly, whereas -ical means the noun is related indirectly. For example, consider: cytologic changes vs. cytological reports, histologic preparations vs. histological equipment, physiologic solution vs. physiological values, and microscopic changes vs. Journal of the Royal Microscopical Society.

Mounting media. Plural of mounting medium. Mounting media is often used incorrectly as though it were singular.

pH. Potential hydrogen

"*Supervital.* Arnold made a careful study of the dyeing of *überlebender* cells. He chose this very suitable tem because the cells, having been removed from the body, *survived* while being dyed. Instead of leaving well alone, he later coined the expression *Die supravitale Methode*, and the word 'supravital' is generally used. The Latin preposition *super* would have been the proper one to include in the compound word, since it is contained in *supervivo* (I survive). This method of staining should be called the 'survival' or 'supervital' method. The English word 'survive' is derived through French from the Latin *supervivo*."[1]

That vs. which. Use "that" with restrictive clauses. A restrictive clause is one that limits—or restricts—the identity of the subject in some way. When writing a restrictive clause, introduce it with the word "that" and no comma. (However, if the subject is or was a human being, use "who" to introduce the clause.) Use "which" with nonrestrictive clauses.

A nonrestrictive clause may tell us something interesting or incidental about a subject, but it does not define that subject. When writing a nonrestrictive clause, introduce it with "which," and insert commas around the clause. However, if the subject is, or was, a human being, use "who" to introduce the clause, and insert commas around the clause.

To test whether restrictive or nonrestrictive, omit the clause in question. If the omission doesn't change the sense of the sentence, the clause is nonrestrictive; use which. Otherwise, use that.

References

1. Baker JR. Principles of biological microtechnique—a study of fixation and dyeing. London: Methuen; 1958. [1968 reprinting].
2. Spriggs AI. History of cytodiagnosis. J Clin Pathol. 1977;30(12):1091–102. Free full text article available at http://jcp.bmj.com/content/30/12/1091.long. Accessed 27 May 2012.

Appendix B: Arithmetic in the Cytopreparatory Laboratory

Arithmetic in the cytopreparatory laboratory is comprised of simple calculations of relative (i.e., percent) and absolute concentrations of solutes and solutions (i.e., molarity for nonacids and normality for acids), dilutions, temperature conversion, and centrifugal force—all factors that influence the quality and quantity of cells. Performing accurate calculations is, of course, prerequisite to such issues as quality control, reproducibility of results, usable communications, and checking published results.

While such calculations are easy to do, recalling the specific problem-solving steps is often difficult since such learning may have occurred a long time ago, is not taught routinely in schools of cytotechnology, and is used infrequently. Nowadays, the widespread availability of off-the-shelf products that are ready to use has virtually eliminated the need to know how to perform such calculations. The downside, in my view, is that such convenience comes at the cost of diminishing our problem-solving skills and has dumbed down our critical thinking skills.

Even though the trend in laboratory medicine is the use of kits and other devices that minimize the intellectual component of technologists' contributions, knowing how to do these calculations is an essential skill to possess. After all, cytotechnologists are responsible for preparing the various reagents and solutions implementing new techniques and are held accountable for the success or failure of laboratory procedures. (If not cytotechnologists, who? Pathologists, on-the-job-trained cytopreparatory technicians?) No. Ultimately, we recognize that the qualitative

G.W. Gill, *Cytopreparation: Principles & Practice*,
Essentials in Cytopathology 12, DOI 10.1007/978-1-4614-4933-1,
© Springer Science+Business Media New York 2013

appearance of cells is impacted substantively by the quantitative composition of the materials and methods of cytopreparation.

The following calculations are representative of the kinds encountered in cytopreparation and are intended for reference.

Percent Concentrations

The composition of a solution can be expressed as a percent concentration of solute in solution. The amounts of solute and solution can be given as either weight (*w*) in grams or volume (*v*) in milliliters (e.g., *w*/*v*, *v*/*v*, *w*/*w*, though rarely, if ever, *v*/*w*). Note that two solutions of the *same* percent concentrations of *different* chemicals are *not equal* with respect to their respective numbers of particles or molecules, for the same weights of different chemicals will be unequal with respect to their mole concentrations. Concentrations expressed as a percent are primarily a convenient *way* to instruct potential users as to how to prepare the solution. The choice of the percent concentration of a solution is often determined empirically, though it may occasionally represent a specific mole concentration related to some stoichiometric chemical reaction.

Common Examples of Relative Concentration Expressions

w/*v*—normal saline is 0.9% NaCI (i.e., 0.9 g NaCI in water up to a volume of 100 mL)

w/*v*—Saccomanno's preservative is 2% (*w*/*v*) Carbowax 1450 in 50% ethyl alcohol (i.e., 2 g Carbowax 1450 in 50% ethyl alcohol up to a volume of 100 mL)

v/*v*—50% ethyl alcohol (i.e., 50 mL alcohol in water up to a volume of 100 mL)

w/*w*—the classical solvent for the Giemsa stain is 50% (*w*/*w*) glycerin in 50% methyl alcohol (i.e., 50 g glycerin in 50 g methyl alcohol)

Alcohols

To determine how many mL of an alcohol of a given percentage must be used to produce a given volume of less concentration, use:

$(a \cdot x)/b = c$	Known:
	a=% concentration, as decimal equivalent of starting alcohol
	b=desired volume of less % concentration
	c=desired % concentration as decimal equivalent
	Unknown:
	x=volume of starting % alcohol needed

Calculation Example

Given 95% as the starting concentration of ethyl alcohol, how many mL of alcohol and water are required to make 1 L of 50% ethyl alcohol:

$$(0.95 \cdot x)/1{,}000\text{ mL} = 0.50$$
$$x = (1{,}000\text{mL} \cdot 0.50)/0.95$$
$$x = 500\text{mL}/0.95$$
$$x \approx 526\text{ mL}$$

∴ Mix 526 mL 95% ethanol with 474 mL water to equal 1 L 50% ethanol.

To prepare a liter of Saccomanno's preservative (i.e., 2% Carbowax 1450 (*w*/*v*) in 50% ethyl alcohol) using 95% ethyl alcohol as the starting concentration of alcohol, simply make up a liter of 50% ethyl alcohol as described above, but substitute 40 mL of 50% (*v*/*v*) Carbowax for 40 mL of the water. Thus:

526 mL 95% ethyl alcohol + 434 mL water + 40 mL 50% Carbowax.

Biological Dyes

To determine how many grams of impure dye are needed to provide the required grams of pure dye, use:

$a / b = x$	Known: a = grams pure dye needed b = percent dye content, as decimal equivalent, of impure dye Unknown: x = grams of impure dye to provide needed grams of pure dye

Calculation Example

How many grams of orange G (80% dye content) are required to produce 100 mL of a 10% (*w*/*v*) total dye content aqueous stock solution?

$$10\text{g} / 0.80 = 12.5\text{g}$$

The 12.5 g will contain 10 g of orange G and 2.5 g non-dye impurities (e.g., NaCI and dextrin).

Molarity

To determine the molarity of a substance, divide its weight in g/L by its molecular (MW) or formula weight (FW). Unlike equal percent concentrations of different solutions, equal molar concentrations of different solutions contain equal numbers of molecules.

Calculation Example

0.9% (*w*/*v*) NaCI = 9 g/L
FW NaCI = 58.5 g

$$9\text{g} / 58.5\text{g} = 0.154\text{M}$$

Normality

It is not uncommon to prepare an acid of a given normality by diluting the concentrated form of the acid. If the normality is specified on the label of the bottle, the required dilution is straightforward.

Usually, however, only the specific gravity is given. Therefore, to calculate the normality of an acid of a given specific gravity:

(a) Determine the weight of 1 L of solution.
(b) Determine the weight of the solute (i.e., the acid).
(c) Determine the number of moles of acid that are present.
(d) Multiply the number of moles by the number of hydrogen atoms per molecule.

Calculation Example

Given 1 L of H_2S0_4, SG = 1.8337, FW = 98.082 g, and 95% (*w*/*w*) concentration. What is its normality?

(a) Weight, 1 L solution: 1,000 mL × 1.8337 g/mL = 1,833.7 g
(b) Weight, H_2S0_4: 1,833.7 g × 0.95 = 1,742 g
(c) Moles, H_2S0_4: 1,742 g + 98.082 g/L M = 17.76 moles
(d) Normality: 17.76 moles × 2 H/H_2S0_4 molecule = 35.52 N

When the normality of an acid is known, diluting the acid to produce a required normality is simple. To answer the question "How many milliliters of an acid of known normality must be used to prepare a given volume of a lesser normality?", use the following equation:

$$(N_2 / N_1) \times V_{\Sigma} = V_{xN1}$$

Known:
N_1 = starting normality of acid
N_2 = diluted normality of acid 30
V_{Σ} = total volume of diluted acid
Unknown:
V_{xN1} = volume of acid of known normality to be diluted

Calculation Example

How many milliliters of 35 N H_2S0_4 must be used to prepare 1 L of 1 N H_2S0_4?

(1 N/35 N) × 1,000 mL = 28.57 mL 35 N

Therefore, 29 mL of 35 H_2S0_4 must be mixed with 971 mL water to prepare 1 L of 1 N H_2S0_4.

Acid Safety Notes

When handling strong acids, wear rubber gloves, goggles, and a chemical splash apron. Work near a sink with running cold water. Remember, when mixing acid and water, add the acid to the water for your own safety—*not* water to acid.

Acid mixing with water generates heat, that is, it is an exothermic process. When substantial heat concentrates quickly at an interface, one or both of two things can happen: (1) the container can break, spilling the contents unexpectedly and in unpredictable directions, and (2) the contents can be propelled suddenly through the top of the container.

Acids consist of water plus something else and, therefore, are heavier than an equal volume of water. If acid is added first to a container, followed by the less heavy water, the water would float on the acid surface. As slight mixing begins to occur at the acid water interface, heat is generated with potentially dangerous consequences.

When added to water, the acid, which is heavier, falls through it. In the process, the heat so generated is safely dissipated as soon as it is produced. As a result, no hazardous buildup of heat occurs. Remembering the reason for the rule is easier than trying to remember only the rule: Do I add water to acid or acid to water? Which is it? Why? Darn, I can't remember.

Temperature Conversion

A simple formula that permits conversion of Celsius and Fahrenheit temperatures in either direction is:

$$\mathrm{C}/(F-32^{\circ})=5/9$$

Calculation example

What is 25 °C on the Fahrenheit scale?

$$25^{\circ}/(F-32^{\circ})=5/95 \qquad F=225^{\circ}+160^{\circ}$$

$$F=385^{\circ}/5$$

$$F=77^{\circ}$$

Centrifugal Force

Relative centrifugal force (rcf) is the force that acts on a given particle in a centrifugal field and is expressed in terms of multiples of its weight in the earth's gravitational field (i.e., × G).

The relative centrifugal force experienced by an object is determined by how fast it is spinning and its distance from the center of rotation. For a particle (e.g., a cell) in suspension, the magnitude of that force varies, ranging from least at the top surface of the suspension to greatest at the bottom of the centrifuge tube.

In centrifugation intended to sediment all the particles, as is practiced in the cytopreparatory laboratory, only the relative centrifugal force at the bottom is given, which is taken to be the tip of the centrifuge tube:

$$\text{rcf} = 1.117 \times 10^{-5} \times r \times N^2$$

rcf = relative centrifugal force
1.117×10^{-5} = gravitational constant
r = radius in cm
N = revolutions per minute

Note that the rcf increases linearly with increasing radius but exponentially with speed. That is, doubling the radius will double the RCF, but doubling the RPMs quadruples it. Since the radius of a given centrifuge head and buckets is fixed, however, only the RPMs can be varied to change the RCF. To double a given RCF, increase the RPMs by a factor of 1.414 (i.e., the square root of 2). To halve the RCF, decrease the RPMs by a factor of 0.707 (i.e., the reciprocal of 1.414). In cytopreparatory procedures that require conventional centrifugation, a combination of time and speed should be used that will effect differential sedimentation. That is, sediment the cells while leaving debris suspended in the supernatant. Too little force leaves unsedimented cells; too much force sediments everything, which compacts cells. Compacted cells may require excessive force to resuspend them, which may damage them. Furthermore, space-occupying debris will reduce the proportion of cell spread that will contain cells of interest, thereby producing a less useful preparation.

How can one determine in his own lab what rpm works? In my case, it was easy. Being familiar with Millipore filters, I simply filtered the supernatants following centrifugation at various RPMs, fixed, stained, and mounted them, and examined them microscopically.

For fresh cell suspensions of body cavity fluids (i.e., collected without added preservative), 10 min at 3,000 rpm cleared the supernatant of most cells. I used a swinging bucket centrifuge with a maximum radius of 16.4 cm. Under those conditions, the rcf is 1,650.

For preserved cell suspensions (e.g., homogenized sputum suspended in 50% ethanol), 5 min at 1,500 rpm is sufficient; rcf is 412. Note that halving the rpm (i.e., 1,500 instead of 3,000) reduced the rcf fourfold (i.e., 1,650/412=4).

Why the difference between the centrifugation times and speeds? In this particular comparison, fresh mesothelial cells weigh less than preserved squamous cells. Hence, they require longer centrifugation time and greater rcf to be sedimented.

Online rpm–rcf converters (e.g., http://www.centrifuges.co.uk/rcf.htm) simplify the arithmetic:

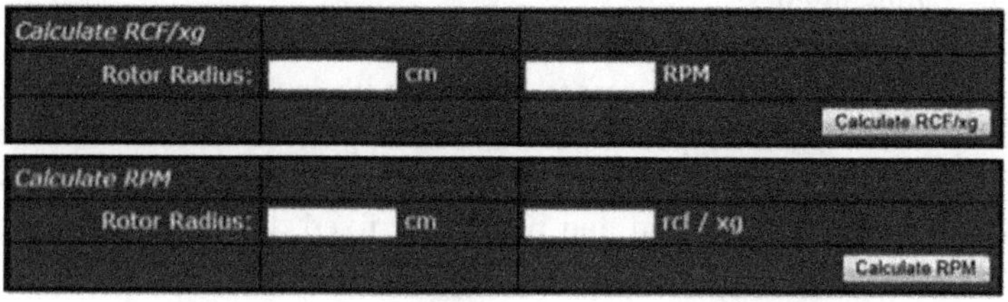

Formaldehyde Versus Formalin

Confusion arises sometimes over differences between formaldehyde and formalin (e.g., definitions, concentration calculations). Formaldehyde is a gas, while formalin is formaldehyde in water. Concentrated formalin is 37–40% (*w*/*v*) formaldehyde in water. The 10% (*v*/*v*) concentration of formalin commonly used in fixatives is prepared by mixing 1 part of concentrated formalin with 9 parts of water. Such a solution will contain a 3.7–4.0% (*w*/*v*) concentration of formaldehyde.

Appendix C: Standard Precautions

The concept of universal blood and body fluid precautions was introduced in 1985 by the Centers for Disease Control and Prevention (CDC). These precautions were identified as "universal" when the CDC published a detailed description of these broad-based guidelines in 1987. Universal Precautions were designed to reduce healthcare workers' risks for exposure to blood and body fluids implicated in the transmission of blood-borne pathogens. The Occupational Safety and Health Administration (OSHA) issued regulations in 1991 that were designed to ensure employer compliance with full implementation of Universal Precautions by mid-1992. The CDC has now included Universal Precautions in a newer approach called "Standard Precautions." Standard Precautions, which are designed for the care of all patients in hospitals regardless of diagnosis or presumed infection status, now replace Universal Precautions.

The CDC recommends **Standard Precautions** for the care of all patients, regardless of their diagnosis or presumed infection status. **Standard Precautions** apply to (1) blood; (2) all body fluids, secretions, and excretions, *except sweat*, regardless of whether or not they contain visible blood; (3) nonintact skin; and (4) mucous membranes. Standard precautions are designed to reduce the risk of transmission of microorganisms from both recognized and unrecognized sources of infection in hospitals. Standard precautions include the use of hand washing and appropriate personal protective equipment such as gloves, gowns, and masks, whenever touching or exposure to patients' body fluids is anticipated.[1]

G.W. Gill, *Cytopreparation: Principles & Practice*,
Essentials in Cytopathology 12, DOI 10.1007/978-1-4614-4933-1,

© Springer Science+Business Media New York 2013

Historical Milestone

- The familiar biohazard warning symbol was created by a Dow Biohazards Research and Development team as part of a contract with the National Cancer Institute in 1966. Charles Baldwin was an Environmental Health Engineer on the team. "Chuck" was stepfather to cytotechnologist Douglas E. King.[2]

References

1. CDC. Guidelines for safe work practices in human and animal medical diagnostic laboratories. MMWR Surveill Summ. 2012;61(Suppl):105 pages. Available at www.cdc.gov/mmwr/pdf/other/su6101.pdf. Accessed 16 Mar 2012.
2. King DE. The biohazard warning symbol: who, what, why, when? Personal tribute. ASCT J Cytotechnol. 1997;1(2):66–7.

Appendix D: Cell Transfer Technique

Purpose and Function

This money-saving technique salvages cytologic or histologic specimens from slides broken in several major pieces. It can also be used to transfer selected areas of cytologic specimen from an intact slide to several others for multiple special staining techniques. Histologic applications are included. It replaces less satisfactory alternative techniques such as taping or gluing pieces together like a jigsaw puzzle or making a kluge-like glass sandwich.

Cells are embedded in situ in mounting medium, peeled off intact, transferred to another slide, reglued, mounting medium dissolved, and remounted. While the elapsed time for the entire procedure is approximately 4 h, actual technician time is minutes.

Materials

• Xylene in Coplin jar	• Single-edge razor blade
• Mount-Quick (see attachment)	• Alcohol-cleaned slide
• 60 °C hot air oven	• Absorbent paper (e.g., Kimwipe)

G.W. Gill, *Cytopreparation: Principles & Practice*,
Essentials in Cytopathology 12, DOI 10.1007/978-1-4614-4933-1,
© Springer Science+Business Media New York 2013

Method

Removing the Cover Glass

1. Remove the cover glass by immersing broken fragments in xylene for as long as needed.
2. Remove any residual mounting medium by dipping in a second xylene bath.

Embedding the Preparation

3. Using a metal slide tray as a work surface, realign the fragments on an intact slide as a template.
4. Cover the pieces completely and uniformly with Mount-Quick.
5. Allow the mounting medium to dry at room temperature until stable (e.g., 15–30 min).
6. Harden the mounting medium in a 60 °C oven for 2 h.

Removing the Preparation

7. Soften the embedded preparation in warm water for as long as needed (e.g., 1 h).
8. Using a razor blade, separate the embedded preparation in a single piece from the slide.

Transferring the Preparation

9. Moisten the preparation in water and apply it to a moistened clean labeled slide.
10. Blot the excess water.
11. Re-adhere the preparation to the slide by returning it to the oven for an hour.
12. Remove the preparation from the oven and bathe it in xylene to remove the mounting medium.
13. Rehydrate the preparation and proceed as needed to restore the stain.

Notes

- Other mounting media may perform as well as Mount-Quick. The prerequisite appears to be that it remains pliable after drying. Mount-Quick is available from Newcomer Supply: http://www.newcomersupply.com/products/mounting-media-lab?page=M#1150 .

Mount-Quick tissue (or cell) transfer technique is available: http://www.newcomersupply.com/documents/product-flyers/Mount%20Quick%20Tissue%20Transfer%20Technique.pdf. Accessed January 17, 2012.

- 60 °C is sufficient to evaporate the solvent. Extremely high temperatures might make the mounting medium brittle and defeat its purpose.
- Superfrost slides were recommended originally but probably are not necessary as long as the slide is cleaned.
- Thick preparations sometimes do not detach completely intact. To promote detachment, first try using several applications of mounting medium to embed the entire thickness, and second, use warm water at step 7 [how warm is arbitrary at this point (e.g., warm to the touch?, not boiling—keep the procedure simple)].
- The water facilitates positioning the piece by lubricating the interface.

References

1. Brown GG, Tao LC. Restoration of broken cytology slides and creation of multiple slides from a single smear preparation. Acta Cytol. 1992;36(2):259–63.
2. Jiminez-Joseph D, Gangi MD. Application of Diatex compound in cytology: use in preparing multiple slides from a single routine smear. Acta Cytol. 1986;30(4):446–7.
3. Sherman M, Jiminez-Joseph D, Gangi MD, Rojas-Corona R. Immunostaining of small cytologic specimens. Facilitation with cell transfer. Acta Cytol. 1994;38(1):18–22.
4. Verbeek DH, Smedts F, Wijnen-Dubbers CW, Mravunac M. Histologic processing of thick tissue specimens from cytology slides. A novel technique. Acta Cytol. 1996;40(6):1198–204.

Appendix E: Lagniappe

Lagniappe answers the recurring question "where can I buy?" for the items listed. Please contact the vendor in advance of placing an order to confirm availability and current pricing.

No.	Item	Catalog no.	Vendor
1.	FNA syringe holder About $300 each	938, 920, and 930 for 10 cc, 20 cc, and 30 cc syringes	Belpro Medical 10450 Rue Secant Anjou, QC H1J 1S3 Canada (888) 230-1010 info@belpro.ca http://www.belpro.ca/about-us.html
2.	Paraffin block transport holds 4 blocks each for mailing. Designed at Mayo Clinic	S.SPBT4.001 (1 transport) to S.SPBT4.500 (case of 500)	Source Medical Products, Inc. 1274 Telegraph Road Lake Forest, IL 60045-3728 (847) 735-9965 customerservice@sourcemp.com http://sourcemp.com/
3.	Slide repair tape 3×1-in. strips $60.00/roll of 500	SRT-700 Item is not listed on website, but is available. Call for details	Label Arts LLC PO Box 727 Kemp, TX 75143 (800) 634-9943 labels@labelarts.com http://www.labelarts.com/

(continued)

G.W. Gill, *Cytopreparation: Principles & Practice*, Essentials in Cytopathology 12, DOI 10.1007/978-1-4614-4933-1,

© Springer Science+Business Media New York 2013

(continued)

No.	Item	Catalog no.	Vendor
4.	Pilot permanent marker pens with extra fine points (aka "dotting pens") for applying ink dots to cover glasses[1]	SCAUFBLK SCAUFBLU SCAUFGRN SCAUFRED $19.32/12	Pilot Pen Attn.: Consumer Service 3855 Regent Blvd Jacksonville, FL 32224 (904) 565-7600 http://www.pilotpen.us/ProductGroup/116-Extra-Fine-Permanent-Marker.aspx
5.	Nikon object marker (cell dotter for cytology) $189.00 each Inquire about purple ink refills	MBW10020 Call to confirm compatibility with your microscope	SEO Enterprises Inc. Microscope Sales and Service 5804 Brannen Road South Lakeland, FL 33813 (800) 330-7654 http://www.seoenterprises.com/shop/home.php
6.	Slide index markers (3.5 × 1-in.). Place-holder for slides removed from files $49.95/1000	SIM-100*n* (*n* is 1 to 6 for each of 6 available colors)	Lab Storage Systems, Inc. PO Box 968 St. Peters, MO 63376 (800) 345-4167 http://www.labstore.com/
7.	Teaching slide holder. Holds four 3 × 1-in slides and a 5.5 × 4-in. index card. About $25/100	SL-3-264[2]	Rochester 100, Inc. 40 Jefferson Road Rochester, NY 14623-2132 (800) 498-1463 http://www.rochester100.com

(continued)

[1]Hollander DH, Frost JK. Retrieval of located cells in screened cytologic material. Acta Cytol. 1969;13(11)603–4.

[2]These "teaching packets" were designed by Mrs. Arline K. Howdon, Chief Cytotechnologist, and Ms. Marianne M. Emery, Secretary. These ladies worked for John K. Frost, MD, Head of the Cytopathology Division, The Johns Hopkins Hospital, Baltimore, MD. Ms. Emery would later become Mrs. Gill.

(continued)

No.	Item	Catalog no.	Vendor
8	Gill screening reticle One per pair of eyepieces $166.00 each Customer must provide eyepiece diameter	KR24701-xxmm (xx are a 2-digit number that equals the eyepiece diameter in which the reticle will be installed)	Klarmann Rulings, Inc. Attn: Chris Wilmot 480 Charles Bancroft Highway Litchfield, NH 03052-1088 (800) 252-2401 sales@reticles.com http://www.reticles.com/
9	Smoked acrylic block (40 × 35 × 20-mm) Demonstrate Köhler illumination. See light rays as they pass through this "smoked" acrylic block	NA $29.00 each	Bioindustrial Products 902 Kitty Hawk Rd, #170-401 Universal City, TX 78148-3825 (210) 655-6403 http://www.bioindustrialproducts.com

Appendix F: Use of the Word "Chromatin"

We incorporate the word "chromatin" in our professional conversations daily. We know its meaning in the context of cytomorphology, but its historical origins are different. For this reason, I reprint verbatim Baker's account:[1]

In general, it is best to avoid words that are supposed to denote chemical composition but stand outside the system of nomenclature adopted by chemists. It seems unlikely that anything is gained by using such words as "linin" and "plastin." A case can, however, be made for the retention of the word "chromatin."

This word was introduced by Flemming in 1880. It is commonly stated to have been introduced by him in 1879, presumably because E. B. Wilson[2] [540a] said so; but it does not occur in the papers by Flemming quoted by Wilson in support of his contention.

Flemming introduced it as follows:

"For further study of the phenomena of division, there is the question of a shorter word for what I have hitherto called the 'colourable substance of the nucleus.' Since the expression 'nuclear substance' is obviously exposed to many misunderstandings, I shall for the time being coin the word chromatin for it. From this name no preconception ought to arise that this substance must be a definitely constituted chemical substance, remaining unchanged in all nuclei. Although this is indeed possible, we do not yet know enough about the nuclear substance to assume it. This only should be denoted by the word chromatin: *that sub-*

G.W. Gill, *Cytopreparation: Principles & Practice*,
Essentials in Cytopathology 12, DOI 10.1007/978-1-4614-4933-1,
© Springer Science+Business Media New York 2013

stance in the cell nucleus which takes up the dye in the treatments with dyes known as nuclear colouring." [3] [171a]

Two years later he added these remarks:

"The [nuclear] network owes its refractivity, the nature of its reactions, and its remarkable affinity for dyes, to a substance which, in consideration of the latter character, I have provisionally named chromatin." He goes on to say that it may be the same substance as nuclein. "I retain the name chromatin," he continues, "until decision on this shall be given by chemistry, and I denote by it, *wholly empirically*, 'the substance in the cell nucleus that takes up the colour in nuclear dyeing.'" He remarks, "As soon as anyone is able to say exactly what the colourable substance in the nucleus is, in terms of *chemistry*, such a name as chromatin will perhaps become useless, unless even then it still commends itself on account of its brevity." [4] [172]

In this book [i.e., Baker's], the word is used precisely in Flemming's sense. Despite all that has been done since his time to enlarge our knowledge of nucleoproteins and DNA, we still do not know exactly to what substance or substances in nuclei and chromosomes the special affinity for particular dyes, to which Flemming refers, is due. It may perhaps be nucleoprotein, but DNA split off from protein by the action of a fixative seems more likely; possibly, in some cases, the protein may itself hold basic dyes after detachment of DNA. It seems safest for the present to retain Flemming's word when describing what we see in microscopical preparations colored with the usual dyes.

References

1. Baker JR. Principles of biological microtechnique—a study of fixation and dyeing. London: Methuen; 1958. [1968 reprinting].
2. Wilson EB. The cell in development and inheritance. New York: Macmillan; 1925.
3. Flemming W. Arch mikr Anat. 1880;18:151.
4. Flemming W. Zellsubstanz Kern und Zelltheilung. Leipzig: Vogel; 1882.

Appendix G: Useful URL[3]

In todays' world of PCs, the Internet, and search engines such as Google, you can find answers to virtually all your questions. First, however, one must be curious enough to ask relevant questions. Ignorance of answers is defensible no longer. My first exposure to the Internet was on August 17, 1998, which was my first day on a new job in Indianapolis, Indiana. Since then, I have accumulated nearly 50,000 files in 4,500 folders.

At a 2005 Program Faculty Seminar in San Diego, I delivered an invited 30-min lecture entitled "Managing cytology information overload: a glimpse into Gary Gill's brain." The organizer, who was responsible for the title, wanted to know: "How do you do that?" "That" being able to answer so many questions posted on professional listserves so quickly with full documentation.

Here are some tips I shared:

- If you're going to act on information, get is straight. Don't rely on someone's word, interpretation, recollection, etc.
- Ask: How do you know that?
- Go to the source. It's part of doing due diligence.
- Hyperlink: everything's connected.
- Keywords are *key* to successful search. An employer wanted to know how long the lab should keep breast implants that had been removed surgically. An initial search that included the

[3] URL means Uniform Resource Locator. In computing, it is a specific character string that constitutes a reference to an Internet resource (http://en.wikipedia.org/wiki/Uniform_Resource_Locator).

G.W. Gill, *Cytopreparation: Principles & Practice*, Essentials in Cytopathology 12, DOI 10.1007/978-1-4614-4933-1,

© Springer Science+Business Media New York 2013

word "implant" was fruitless. Noodling around in PubMed revealed the missing key word: explant.

- Using Google's Advanced Search helps find good information. The Advanced Search page, http://www.google.com/advanced_search?hl=en, allows you to more quickly find what you're looking for by inserting judiciously chosen key words:

- Invest in Adobe Acrobat so you can make PDF copies of everything of interest.
- Invest time up-front to save time down-back.
- Save it now!
- Backup your files in real time. They are your personal professional library.
- Make time to manage files.
- Pick a file-naming convention that works for you.
- Identify files explicitly; you won't remember an incomplete or cryptic file name.
- Use folders within folders, within folders…
- Group-related folders by common prefix. For example, I have 592 folders that begin with CC., which stands for Corporate Compliance: CC. ABN through CC. Waiving Co-pays. "CC" group-related topics. Otherwise, ABN and Waiving Co-pays folders would be far apart. Such a naming device works for me, but it might not work for you. Choose whatever file-naming convention works for you.
- Drop e-copies in multiple folders. You may not remember where you filed a document.
- Use the Search feature of your operating system to locate documents you know you have, but have been unable to locate.

The following URL link to useful professional resources:

No.	Site	URL
1.	Gill GW. Cytopreparation quizzes	http://www.cytopathology.org/website/article.asp?id=559
2.	Patten FW. Cytotechnology: the First Half-century	http://www.cytopathology.org/website/article.asp?id=2355
3.	CDC Centers for Disease Control and Prevention	http://www.cdc.gov/
4.	ASC American Society of Cytopathology	http://www.cytopathology.org/
5.	ASCT American Society for Cytotechnology	http://www.asct.com/
6.	ASCP American Society for Clinical Pathology	http://www.ascp.org/
7.	CAP College of American Pathologists	http://www.cap.org/
8	OIG Office of Inspector General	http://oig.hhs.gov/
9.	CMS Centers for Medicare and Medicaid Services Regulations and Guidance	http://www.cms.gov/Regulations-and-Guidance/Regulations-and-Guidance.html
10.	Publication of OIG Compliance Program Guidance for Clinical Laboratories	http://oig.hhs.gov/authorities/docs/cpglab.pdf
11.	CLIA '88 Clinical Laboratory Improvement Amendments of 1988	http://wwwn.cdc.gov/clia/pdf/42cfr493_2004.pdf
12.	CFR Code of Federal Regulations	http://www.gpo.gov/fdsys/browse/collectionCfr.action?collectionCode=CFR
13.	FDA US Food and Drug Administration	http://www.fda.gov/
14.	FDA Premarket Approval (PMA) Search	http://www.accessdata.fda.gov/scripts/cdrh/cfdocs/cfPMA/pma.cfm

(continued)

(continued)

No.	Site	URL
15.	FDA Premarket Notification (510 K) Search	http://www.accessdata.fda.gov/scripts/cdrh/cfdocs/cfPMN/pmn.cfm
16.	MAUDE Manufacturer and User Facility Device Experience	http://www.accessdata.fda.gov/scripts/cdrh/cfdocs/cfMAUDE/Search.cfm (MAUDE data represents reports of adverse events involving FDA-approved medical devices)
17.	PubMed US National Library of Medicine	http://www.ncbi.nlm.nih.gov/pubmed
18.	Loansome Doc Allows users to order full-text copies of articles from a medical library (local fees and delivery methods may vary)	http://www.nlm.nih.gov/pubs/factsheets/loansome_doc.html
19.	US Trademark search	http://tess2.uspto.gov/bin/gate.exe?f=tess&state=4009:lhvg28.1.1
20.	US Patent search	http://patft.uspto.gov/netahtml/PTO/search-bool.html

Appendix H: Selected Milestones in Microtechnique

Study the past if you would define the future.

Confucius

Cytopreparation preceded histopreparation historically. Given their natural thinness, cells are more readily examined microscopically than is tissue. Tissue requires hardening by fixation and embedding in paraffin before it can be sectioned sufficiently thin and stained to be examined microscopically. From the 1600s when cells were first examined by Leeuwenhoek's microscope, nearly 200 years would pass before chromic acid was used as a hardening agent. Listed below are dates of pertinent events in microtechnique history. Despite advances in molecular biology-based tests, microscopic evaluation remains the most widely used tool for discerning the health or disease status of cells and tissues—primarily because so much can be learned from so little. Light microscopy also serves as the triage platform for additional defining studies.

Selected milestones in microtechnique[1] and cytopreparation.

No.	Year	Contributor	Introduced
1.	1590	Hans and Zacharias Jansen	Compound microscope
2.	1600	Academia dei Lincei	"Microscope" as a term
3.	1665	Hooke	"Cell" as a term
4.	1691	Bonanni	Slider (forerunner of slides)
5.	1743	Baker	Alcohol as preservative

(continued)

G.W. Gill, *Cytopreparation: Principles & Practice*,
Essentials in Cytopathology 12, DOI 10.1007/978-1-4614-4933-1,
© Springer Science+Business Media New York 2013

(continued)

No.	Year	Contributor	Introduced
6.	1789	Ingen-Housz	Cover glass for temporary mounts
7.	1800	Bichat	Beginning of histology
8.	1819	AFJK Mayer	Histology as term
9.	1825	Chevalier	Claims on very early use of coverslips
10.	1827	Gould	Slide
11.	1830	Bowerbank	Canada balsam as permanent mounting medium
12.	1830	Pritchard	Microtome
13.	1833	Jacobson	Chromic acid as first hardening fixative
14.	1835	Ross	Cover glass for permanent mount
15.	1839	Chevalier	"Microtome" first used in print
16.	1840	Microscopical Society of London	Set 3 × 1 in. as standard dimensions for slide
17.	1840	Chance Brothers	First commercial cover glass
18.	1851	Clarke	Alcohol and acetic acid ("Carnoy's fluid")
19.	1853	Bunsen	Bunsen burner
20.	1856	Perkins	Aniline dye (synthetic, not natural)
21.	1865	Böhmer	Hematoxylin with mordant (effective nuclear stain)
22.	1870	Abbe	Sine condition, substage condenser, homogeneous oil immersion objective
23.	1875	Fischer	Eosin
24.	1875	Wissowzky	H&E
25.	1883	Carnoy	Carnoy's fixative
26.	1886	Zeiss, Abbe, Schott	Apochromat microscope objectives
27.	1888	Griesbach	Quadruple staining
28.	1879	Ehrlich	Orange G
29.	1886	Griesbach	Light green SF yellowish
30.	1892	Petri	Petri dish
31.	1893	Friedrich Becke	Becke line
32.	1896	Mallory	Phosphotungstic acid
33.	1897	Coplin	Coplin jar
34.	1898	Köhler	Köhler illumination
35.	1898	Mayer	Mayer's hematoxylin
36.	1900	Harris	Harris hematoxylin

(continued)

(continued)

No.	Year	Contributor	Introduced
37.	1917	Papanicolaou	Alcohol:ether as fixative
38.	1928	Papanicolaou	Pap smear
39.	1931	Trotman	Phosphotungstic acid precipitates basic dyes
40.	1936	Fisher Scientific	Permount (synthetic mounting medium)
41.	1943	Papanicolaou	Pap stain with alcohol-based counterstains
42.	1945	Papanicolaou	Pap test began to be used clinically
43.	1949	Ayre	Spatula patented
44.	1957	Dakin	Frosted microscope slide patented
45.	1957	DeWitt et al.	Modified Carnoy's fixative
46.	1958	Seal	Millipore filters
47.	1962	Baker	Hematal 8 hematoxylin
48.	1963	Saccomanno	Saccomanno Preservative
49.	1965	Seal	Nuclepore filters
50.	1965	Watson	Cytocentrifuge
51.	1969	Gill	Exfoliative Cytology (Millipore Filter Application Report AR-24)
52.	1972	Gill	Gill hematoxylin, modified OG, and modified EA
53.	1983	Gill	Cytospin User Manual for Shandon
54.	1990	Cytyc	ThinPrep Processor patented
55.	1996	Cytyc	ThinPrep Pap test FDA-approved
56.	1996	Gill	Enviro-Pap
57.	1999	AutoPap	AutoCyte Pap test FDA-approved
58.	2000	Gill	Screening reticle
59.	2002	Gill & Snyder	exCellerator patented

Reference

1. Bracegirdle B. A history of microtechnique. Ithaca, NY: Cornell University Press; 1978. (The majority of information that precedes 1898 is taken from this stellar reference.)

Appendix I: Screening and CPR

This 2007 paper[1] demonstrates the connection between one's professional and personal lives. Things have changed, as they inevitably do, since the paper was written in 1998. Our cocker spaniel Rusty was put down on October 15, 2011. Nearly 17 years old, he had run out of gas. He was family, and we miss him.

G.W. Gill, *Cytopreparation: Principles & Practice*, Essentials in Cytopathology 12, DOI 10.1007/978-1-4614-4933-1,
© Springer Science+Business Media New York 2013

Reader Discretion Advised: *This section contains strong language in the interest of advancing professional practices that will improve patient care.*

Screening and cardiopulmonary resuscitation…? No. CPR is "canine poop recovery"—allow me to explain. Our cocker spaniel, Rusty, is let outside daily as needed to "do his business." Since the backyard is not fenced in, he is staked (not literally). That is, his collar is attached to a 40-ft anchored lead that allows him to roam within a relatively large area. Despite our three children's prepurchase promises to clean up after Rusty, it has become my job as a responsible adult to be the primary pooper-scooper (sound familiar?).

While on yard duty one day, I was struck by the similarities and differences between patrolling for Rusty's craposites and screening Pap smears (Table I.1). Whether screening or CPR-ing, I want to find the first and last search target, respectively, to avoid the consequences of a false negative. Therefore (stick with me on this), if I wander aimlessly about the lawn (comparable to screening without a mechanical stage), I will find some regions of interest. If I walk systematically back-and-forth and look from side to side as I go (comparable to using a mechanical stage and overlapping slightly), I will find additional examples. If about to mow the lawn, however, I have learned not to throw away the "biomass cleanup kit." Pushing the mower so that no blade of grass goes uncut (comparable to introducing microscopical quality control into the screening process), I always find previously undetected craposites. A fundamental lesson is that the next step is always the most important.

Mother Nature, in her infinite wisdom, colors Rusty's solid waste products to camouflage their presence on grass. Scattered among the fallen leaves of autumn, these "biomasses" challenge detection. Serendipitously, I discovered a way to highlight their presence. Rusty loves to chew a rope comprised of strands colored similarly to the Pap stain. In a biological demonstration of the conservation of mass, the ingested brightly colored strands subsequently exit intact. *Et voilà*, the craposites are "poop stained" and eminently more visible.

Table Appendix I.1 Screening Pap smears and Pup smears

Consideration	Pap smears	Pup smears
Perspective	Microscopic	Macroscopic
Support medium	Glass	Grass
Target nomenclature	Dyskaryocytes	Craposites
Target variety	See Bethesda System ("BS")	Monotype
Coloration	Pap stain	Poop stain
Senses used for detection	One	Three
Screening approaches	No mechanical stage Mechanical stage Total screening coverage	Walk randomly Walk back-and-forth Walk behind lawn mower
Certitude of target presence	No	Yes
Screening objective	Find one first	Find all last
Probability of detection	<1	1
Target size	Biology dependent	Biology dependent
Target number	Sampling dependent	Days since yard last screened
Target area (TA)	1,000 μm^2	3 $in.^2$
Field area (FA)	1,200 mm^2	5,000 ft^2
TA:FA	1:1.2 x 10^6	1:2.4 x 10^5
False negative consequence	Cleanup	Cleanup
Cost to perpetrator	Substantial long term	Substantial short term
False negative prevention	Quality control of screening	Quality control of screening

In his November 4, 1998, presentation at the ASC Program Faculty Seminar, Bill Crabtree talked about lifelong learning. Afterward, I offered him my definition, which is reconstructed as follows: lifelong learning occurs when an individual integrates all personal and professional experiences into an approach to life that seeks to maximize benefits and minimize costs. Lifelong learning is often a matter of making connections between apparently unrelated observations such as screening and CPR. The latter connection is understandable even to non-dog owners.

The point of this analogy is that the same technique required to find at least one abnormal cell in a Pap smear is identical to that required to find the last biomass on the lawn. Without the introduction of microscopical quality control into Pap smear screening, false negatives are inevitable and will not be substantively reduced by proficiency testing. Quality control is defined as any material or method that is incorporated routinely in a process to promote a desired outcome. As demonstrated by plane geometry, the minimum amount of work required to image every μm^2 of a cytologic preparation at least once is to overlap fields-of-view diameters 30% along both the x and y dimensions of a slide. This process is not taught by schools of cytotechnology and not practiced in cytopathology laboratories, which is unfortunate. It is more cost-effective to prevent a problem than to react to the consequences. *Cave canem!*

Reference

1. Gill GW. Screening and CPR. Micro views. 2007;16(1):20–1. Published by Surgipath Medical Industries, which was purchased by Leica Microsystems in 2009.

Appendix J: Author's Awards and Publications

Notes

- The publications are not readily available to most readers, which is why they are listed.
- The subject matter is almost entirely pragmatic and has stood the test of time.
- The topics addressed cover everything from specimen collection to data analysis.
- Cited publications do not include all those listed in the summary table.
- Most publications are authored by a single individual over 50 years.

Awards and Honors

2011	Dedication of Cytopreparatory Laboratory in the Pathology Building of The Johns Hopkins Hospital, Baltimore, MD
2011	Lifetime Achievement Award, The Imagine a Better World Foundation
2008	Certificate of Merit, ASC Executive Board (2004–2008)
2004	Excellence in Education Award, American Society of Cytopathology
2003	Laboratorian of the Year—Third-place tie, Advance for Medical Laboratory Professionals

(continued)

G.W. Gill, *Cytopreparation: Principles & Practice*,
Essentials in Cytopathology 12, DOI 10.1007/978-1-4614-4933-1,
© Springer Science+Business Media New York 2013

(continued)

2002	Marion and Nelson Holmquist Cytotechnologist Achievement Award, ASCT
1994	Quest Challenge Award, SmithKline Beecham Clinical Laboratories
1989	International Cytotechnology Award, International Academy of Cytology
1983	Cytotechnologist-of-the-Year Award, ASC

Publications (269)

• Books	6	• Videotape Literature	4
• Book Chapters	18	• Exhibit Literature	1
• Peer-Reviewed Articles	16	• Letters to the editor	21
• Abstracts	23	• Laboratory Manuals	9
• Articles	81	• Interview	5
• Posters	15	• Vendor Literature	37
• Cytoteleconference Literature	8	• Trade Literature	1
• Webliography	7	• Featured Photomicrographs	8
• Videotapes	8	• Cartoons	1

Books

1. Exfoliative cytology – Application Report 24. Bedford, MA: Millipore Corporation; 1969. (29 pages).
2. Cytopreparation of Millipore filters in diagnostic cytology – Application Report 24. 2nd ed. rev. Bedford, MA: Millipore Corporation; 1977. (60 pages).
3. The Shandon Cytospin 2 in diagnostic cytology – tips, techniques, and troubleshooting. Sewickley, PA: Shandon Southern Instruments, Inc.; 1982. (73 pages).
4. Pressman NK, Guest Editor, Gill GW, Guest Editorial Assistant. Computer-aided microscopy in medicine, special issue. Applied Optics. 1987;26(16):3199–416. Washington, DC: Optical Society of America.
5. McClatchey KD, Dhurander N, Gallo L, Gill GW, Kurtycz DFI, Plowden K, Radcliffe, Trew C. Nongynecologic Cytologic specimens: collection and cytopreparatory techniques; approved guideline. NCCLS document GP23-A [ISBN 1-56238-380-9]. NCCLS, 940 West Valley Road, Suite 1400, Wayne, PA 19087-1898 USA; 1999.

Book Chapters

1. Methods of cell collection on membrane filters. In: Compendium on cytopreparatory techniques. Chicago: International Academy of Cytology; 1972. p. 8–16.
2. Staining. In: Compendium on cytopreparatory techniques. Chicago: International Academy of Cytology; 1972. p. 17–32.
3. Cytopreparation of prefixed sputum specimens. In: Compendium on cytopreparatory techniques. Chicago: International Academy of Cytology; 1972. p. 33–7.
4. Methods of cell collection on membrane filters. In: Preparatory methods in cytology. Chicago: ASCP; 9–13 Jul 1973.
5. Gill GW, Miller KA. Staining. In: Preparatory methods in cytology. Chicago: ASCP; 9–13 Jul 1973.
6. Methods of cell collection on membrane filters. In: Preparatory methods in cytology. Chicago: ASCP; 8–12 Jul 1974.
7. Gill GW, Miller KA. Staining. In: Preparatory methods in cytology. Chicago: ASCP; 8–12 Jul 1974.
8. Staining. In: Compendium on cytopreparatory techniques. 3rd ed. Chicago: International Academy of Cytology; 1974. p. 9–25.
9. Methods of cell collection on membrane filters. In: Compendium on cytopreparatory techniques. 3rd ed. Chicago: International Academy of Cytology; 1974. p. 26–35.
10. Cytopreparation of prefixed sputum specimens. In: Compendium on cytopreparatory techniques. 3rd ed. Chicago: International Academy of Cytology; 1974. p. 65–9.
11. Principles and practice of cytopreparation. In: Altman NH, Melby EC, editors. Handbook of laboratory animal science, vol. III. Cleveland: CRC; 1976, p. 519–51.
12. Staining. In: Compendium on cytopreparatory techniques. 4th ed. Chicago: International Academy of Cytology; 1976. p. 16–33.
13. Methods of cell collection on membrane filters. In: Compendium on cytopreparatory techniques. 4th ed. Chicago: International Academy of Cytology; 1976, p. 34–44.
14. Cytopreparation of prefixed sputum specimens. In: Compendium on cytopreparatory techniques. 4th ed. Chicago: International Academy of Cytology; 1976. p. 70–4.
15. Controlling the Papanicolaou stain, No. C-66 in ASCP CHECK SAMPLE. Chicago: ASCP; 1979.
16. The Papanicolaou stain, chapter 6. In: Allen K, editor. A guide to cytopreparation. Raleigh, NC: American Society for Cytotechnology; 1995, p. 95–111.
17. Chapter 6: Fixation and specimen processing. In: Gupta PK, Baloch ZW, editors. Cytohistology of small tissue samples. New York: Cambridge University Press; 2011:148–61. ISBN 978-0-521-88358-0.

18. Chapter 28: The laboratory. In: DeMay RM, editor. The art & science of cytopathology. vol. 3. 2nd ed. Chicago: ASCP Press; 2011:1539–92. ISBN 978-089189-6449.

Peer-Reviewed Articles

1. Frost JK, Gill GW, Hankins AG, LaCorte FJ, Miller RA, Hollander DH. Cytology filter preparations: factors affecting their quality for study of circulating cancer cells in the blood. Acta Cytol. 1967;11(5):363–73.
2. Gill GW, Frost JK, Miller KA. A new formula for a half-oxidized hematoxylin solution that cannot overstain and does not require differentiation. Acta Cytol. 1974;18(4):300–11.
3. Roffe BD, Wagner FH, Derewicz HJ, Gill GW. Heparinized bottles for the collection of body cavity fluids in cytopathology. Am J Hosp Pharm. 1979;36(2):211–4.
4. Frost JK, Tyrer HW, Pressman NJ, Albright C, Vansickel MH, Gill GW. Automatic cell identification and enrichment in lung cancer: I. Light Scatter and fluorescence parameters. J Histochem Cytochem. 1979;27(1):545–51.
5. Tyrer HW, Golden JF, Vansickel MH, Echols CK, Frost JK, West SS, Pressman NJ, Albright CF, Adams LA, Gill GW. Automatic cell identification and enrichment in lung cancer: II. Acridine orange for cell sorting of sputum. J Histochem Cytochem. 1979;2(1):552–6.
6. Frost JK, Tyrer HW, Pressman NJ, Vansickel MH, Albright CD, Gill GW, Tiffany SM. Automatic cell identification and enrichment in lung cancer: III. Light scatter and two fluorescence. J Histochem Cytochem. 1979;27(1):557–9
7. Frost JK, Pressman NJ, Albright CD, Gill GW, Vansickel MH. Centrifugal separation of cells in sputum. J Histochem Cytochem. 1979;27(1):7–13
8. Boone CW, Sanford KK, Frost JK, Mantel N, Gill GW, Jones GM. Cytomorphologic evaluation of the neoplastic potential of 28 cell culture lines by a panel of diagnostic cytopathologists. Int J Cancer. 1986;38(3):361–7.
9. Traub RK, Pressman NJ, Frost JK, Gupta PK, Showers RL, Gill GW, Cook DI, Frost Jr JK. Detection and localization of quantum limited events from radionuclide labeled materials by computer enhanced video microscopy. In: Topical Meeting on Quantum Limited Imaging and Image Processing Technical Digest. 1986;86(6):63–6. Washington, DC: Optical Society of America.
10. Pressman NJ, Frost JK, Gupta PK, Showers RL, Gill GW, Cook DL, Frost Jr JK, Traub RK. A quantitative method for the detection and localization of quantum limited events from radionuclides in cells and tissue sections

by computer enhanced video microscopy. Analyt Quant Cytol Histol. 1987;9(4):291–302.
11. Traub RK, Warden KE, Cook DL, Frost Jr JK, Gill GW, Gupta PK, Pressman NJ, Frost JK. Detection and measurement of quantum-limited radionuclide events in tissues using video-intensified microscopy. In: Proceedings of the Sixth Medical Chemical Defense Biosciences Review, US Army Medical Research and Defense Command, 4–6 Aug 1987, p. 483–6.
12. Pressman NJ, Frost JK, Gupta PK, Showers RL, Gill GW, Cook DL, Frost Jr JK, Traub RK. The development of a system for enhanced radionuclide detection. Annual Report, August 1, 1984 – July 31, 1985, of the Quantitative Cytopathology Laboratories, Department of Pathology, The Johns Hopkins University School of Medicine, Baltimore MD. Supported by US Army Medical Research and Development Command USAMRICD Contract No. DAMD17-84-C-4222, 1988.
13. Quality control – quality assurance in diagnostic cytopathology: what's the difference? Diagn Cytopathol. 1995;13(3):270–2.
14. Pap smear risk management by process control. Can Cytopathol. 1997;81(4):198–211.
15. Contributor. Cervical cytology practice guidelines – American Society of Cytopathology. Acta Cytol. 2001;45(2):201–26.
16. Gill hematoxylins: first person account. Biotech Histochem. 2010;85(1):8–17.

Abstracts

1. Frost JK, Gill GW, Jenkins DR, Hollander DH, LaCorte FJ. Density gradient centrifugation for separation of cancer cells in blood. Proc Intern Cong Cytol, Rio de Janeiro, 1968.
2. Frost JK, Pressman NJ, Showers RL, Gill GW, Albright CD, Frost JK. Characterization of cellular activity by spectral information. Analyt Quant Cytol. 1982;4:146.
3. Pressman NJ, Frost JK, Showers RL, Frost Jr JK, Gill GW, Albright CD. Measurement of spectral energy absorption in cytologic preparations using scanning transmission optical microscopy. Analyt Quant Cytol. 1982;4:155.
4. Frost JK, Pressman NJ, Gill GW, Showers RL, Bitman WR, Frost Jr JK, Albright CD, Erozan YS, Gupta PK. Quantitative nuclear analysis of developing squamous cell carcinoma of the human lung. Analyt Quant Cytol. 1983;5:207.
5. Pressman NJ, Frost JK, Albright CD, Makowska WK, Showers RL, Bitman WR, Jeffries LH, Frost Jr JK, Gill GW. Quantitative nuclear

analysis of developing squamous cell carcinoma of the hamster cheek pouch. Analyt Quant Cytol. 1983;5:216.

6. Pressman NH, Frost JK, Gupta PK, Showers RL, Gill GW, Cook DL, Frost Jr JK, Traub RK. A quantitative method for the detection and localization of quantum limited events from radionuclides in cells and tissue sections by computer enhanced video microscopy. In: Proceedings of the Fifth International Conference on Automation of Diagnostic Cytology and Histology, Hilton International Hotel, Brussels, Belgium, May 30–June 1, 1986. (Abstract No. 50, p 12).
7. Tockman M, Pressman N, Gill GW, Piantadosi S, Gupta P, Mulshine J. Quantitation of molecular markers for pre-neoplastic lung cancer. Am Rev Respir Dis. 1991;143(4)/Part 2:A203.
8. Screening coverage: proposed protocol, applications, and implications. Acta Cytol. 1995;39(5):1052. Abstract 166.
9. False negative proportion: do very low values make sense? Acta Cytol. 2000;44(5):851. Abstract 8.
10. Pap smear cellular adequacy: what does 10% coverage look like? What does it mean? Acta Cytol. 2000;44(5):873.
11. Using xylene forever: molecular sieves and daily filtration. Acta Cytol. 2000;44(5):925. Abstract 134.
12. Cover glass dimensions and Pap smears: 24x60-mm vs. 24x50-mm. Acta Cytol. 2000;44(5):925. Abstract 135.
13. How to calculate false negative proportion. Acta Cytol. 2001;45(5):883. Abstract 130.
14. Pilot study experience with a screening reticle. Acta Cytol. 2001;45(5):884. Abstract 131.
15. Self-referenced approach to estimating primary cytological screening sensitivity. Acta Cytol. 2001;45(5):884. Abstract 132.
16. Enviro-Pap: an environmentally friendly, economical, effective Pap stain. Acta Cytol. 2002;46(5):1019–20. Abstract 169.
17. Screening performance analysis tool (SPAT): developing the model. Acta Cytol. 2002;46(5):1020–21. Abstract 171.
18. Johnson FL, Gill GW, Erozan YS, Frable WJ, Gupta PK. Improved diagnostic sensitivity for lung cancer from cell block sections plus Cytospin and smear preparations of sputum specimens induced with INS316. Lung Cancer. 2003;41(Suppl 2):S196.
19. New metrics set to assess Pap test cytologic screening performance. Acta Cytol. 2003;47(5):825. Platform presentation, Abstract 8.
20. Wall RW, Metzler RL, Gill GW, Glant MD, Eisenhut CC. Will CLIA gyn cytology proficiency test scoring negatively impact test takers pass rates? Can Cytopathol Suppl. 2005;105(5):7–8. Platform presentation, Abstract 7.
21. Gill GW, Eisenhut CC, Snyder JA. exCellerator: a new tool to improve specimen adequacy in TPPT. Can Cytopathol. Supplement. 2008;114(S5):386. Poster abstract 57.

Articles

1. Miller KA, Gill GW. Teaching cytopreparatory techniques to cytotechnology students. Cytotechnol Bull. 1973;10(1):6.
2. A new formula for a half-oxidized hematoxylin solution that cannot overstain and does not require differentiation. Scanner. 1974;13(2):8–9.
3. Bismar(c)k brown and Papanicolaou EA stains. Scanner. 1975;14(3):2.
4. A cross contamination control and stain storage system. Cytotechnol Bull. 1975;12(3):12–3.
5. Pitfalls in cytopreparation with membrane filters. Cytotechnol Bull. 1975;12(6):8–10.
6. Dissolving Nuclepore filters on micro slides before staining and without allowing air-drying. Cytotechnol Bull. 1980;17(3):50–5.
7. Causes and prevention of fading in Giemsa stained human chromosomes. Karyogram. 1981;7:29–32.
8. Visual microscopy in diagnostic cytology (controlling microscopic illumination and troubleshooting problems microscopical). Am Lab. 1981;13(12):14, 16, 18, 20, 22–5.
9. Gill GW, Frost JK. In defense of cover glasses. Histo-Logic. 1982;12:175–17.
10. Visual microscopy in diagnostic cytology – Part I. Cytotechnol Bull. 1983;20(2):21–5.
11. Visual microscopy in diagnostic cytology – Part II. Cytotechnol Bull. 1983;20:(3)42–4.
12. Gill GW, Frost JK. In defense of cover glasses. Cytotechnol Bull. 1983;20(4):51–5.
13. Gill GW, Frost JK. In defense of cover glasses. Technical Inform Bull. 1983;1(2):16–7 (a customer magazine published by E. Leitz, Inc., Rockleigh NJ).
14. The role of Carbowax in alcohol – CPT Forum. Cytotechnol Bull. 1983;20(4):56–5.
15. Cleaning the front lens of microscopical objectives – CPT Forum. Cytotechnol Bull. 1983;20(5):84–5.
16. Xylene substitutes – CPT Forum. Cytotechnol Bull. 1984;21(2):32–3.
17. Aspiration biopsy cytology – CPT Forum. Cytotechnol Bull. 1984;21(4):61.
18. An introduction to special stains in diagnostic cytology. Cytotechnol Bull. 1986;23(4):48–52.
19. Thank you, Rhonda. Cytotechnol Bull. 1986;23(6):71. (An appreciation).
20. Practical considerations in microscopy – Part I. Cytotechnol Bull. 1987;24(6):69–71, 74–5.
21. Practical considerations in microscopy – Part II. Cytotechnol Bull. 1988;25(2):19–21, 25–6.

22. Arithmetic in the cytopreparatory laboratory. ASC Bull 1992;29(3):29–31.
23. Quality control–quality assurance in diagnostic cytopathology: what's the difference? Diagn Cytopathol. 1995;13(3):270–2.
24. "How did I miss those cells?" Adv Med Lab Prof. 1995;7(23):5–7, 24.
25. Quality in cytopreparation. Adv Med Lab Prof. 1996;8(5):18–20, 25.
26. Vigilance in cytoscreening. Adv Med Lab Prof. 1996;8(15):14–5, 21.
27. The care and feeding of microscopes. Adv Med Lab Prof. 1997;9(4):5–7, 14.
28. Tips to get the best results from your microscope. Adv Med Lab Prof. 1997; 9(4):7, 38.
29. Gill hematoxylin: setting the record straight. CAP Today. 1997;11(3):5–7.
30. Eyepieces–bigger isn't better. ASCT J Cytotechnol. 1997;1(1):29–33.
31. Troubleshooting problems in microscopy. Adv Med Lab Prof. 1997;9(17):24–7.
32. Fix the causes, not the effects. The ASC Bulletin. 1997;34(7):87, 92–3.
33. Correction to Eyepieces–bigger isn't better. ASCT J Cytotechnol. 1997;1(2):44.
34. Care and feeding of microscopes: cleanliness is next to 'goodliness." ASCT J Cytotechnol. 1997;1(2):73–5. Excerpt.
35. A rogue's gallery of screening policies and practices. Focus. 1998;5(1):4–6. Newsletter of Papanicolaou Society of Cytopathology. Invited article.
36. Pap smear screening practices –If it's broke, fix it! Adv Med Lab Prof. 1998;10(9):4. Guest editorial.
37. Screening Pap smears is a public trust. Adv Med Lab Prof. 1998;10(4):8–11.
38. Cornflaking cells. Micros Today. 1999;2:36.
39. Cover glass perspectives. Micros Today. 1999;4:12–3.
40. Phosphotungstic acid, water, and histologic stains. Micros Today 1999;5:20–1.
41. Troubleshooting eosin staining, and a brighter eosin stain. Micros Today 1999;8:27.
42. What is the Papanicolaou stain anyway? SCAN. 1999;10(2):18–9, 21.
43. The brown artifact – 'cornflaking'. SCAN. 2000;11(1):6–7.
44. Stemming wasteful cytology cashflow. Advance Med Lab Prof. 2000;12(8):8–11.
45. Broadstock M. Effectiveness and cost-effectiveness of automated and semi-automated cervical screening devices. NXHTA Report. 2000;3(1). New Zealand Health Technology Clearing House. ISBN 1-877235-13-X. Gill GW, correspondent. Acknowledgements, page ii. Accessed on 26 Feb 2009 at http://nzhta.chmeds.ac.nz/publications/csv3n1.pdf.
46. Optimising and standardizing microscopic screening coverage – first find one abnormal cell. SCAN. 2001;12(1):10–2, 15.

47. Der Goldstaubeffekt (cornflaking). Cyto-Info (Zeitschrift des VDCA: Verband Deutscher Cytologisch Tätiger Assistenten (roughly: Journal of Active German Cytology Assistants) 2001;2:44–5.
48. Microscopic screening coverage of Pap smears. Adv Med Lab Prof. 2001;13(9):12, 14–6.
49. The brown artifact – "cornflaking". ASC Bull. 2001;38(5):75, 78.
50. Papanicolaou stain – quality control and quality assurance. SCAN. 2001;12(2):33–6.
51. Microscopic screening coverage of Pap smears – a screening reticle to optimize and standardize slide step-size movement. Adv Med Lab Prof. 2001;13(9):12, 14–6.
52. Rapid review – is it effective? Is it efficient? SCAN. 2002;13(1):8–11.
53. 'PapPourri'. Adv Med Lab Prof. 2003;15(16):22–4.
54. Holladay EB, Gill GW, Clark JJ. Recuse yourself—you're biased! Diagn Cytopathol. 2003;29(5):303–4.
55. Chhieng DC, Roberson J, Gill GW. How to stay hip with HIPAA Privacy Rule?—A HIPAA checklist for the cytopathology laboratory. ASC Bull. 2004;41(4):57, 59–60, 72–3.
56. Q&A – Eating and drinking at the microscope is OK. Voice. 2004;1(1):4. Legislative Affairs.
57. Q&A – HIPAA does not require slide destruction before disposal. Voice. 2004;1(1):4. Legislative Affairs.
58. A vision of excellence. Adv Admin Lab. 2004;13(9):31, 34, 36, 38, 42–4, 46, 48. Cover story.
59. Roberson J, Gill GW, Chhieng DC. Creating a "HIPAA friendly" cytology laboratory: Part I. Focus. 2004;11(2):6–7. Newsletter of the Papanicolaou Society of Cytopathology.
60. The Pap test—Lab methods/instruments, then and now. LabMedicine. 2004;35(12):731.
61. Point: Why I am "pro" federally-mandated cytology proficiency testing. LabMedicine. 2004;36(4):215–8. Point-Counterpoint.
62. Roberson J, Gill GW, Chhieng DC. Creating a "HIPAA friendly" cytology laboratory: Part II. Focus. 2005;12(1):8–10. Newsletter of the Papanicolaou Society of Cytopathology.
63. The cytological thicket. Voice. 2005;2(3):6–7. ASCT Legislative Affairs.
64. PT is FT. Voice. 2005;2(5):4–5, 9. ASCT Legislative Affairs.
65. Ogden-Grable H, Gill GW. Phlebotomy puncture juncture: preventing phlebotomy articles—potential for harming your patients. LabMedicine. 2005;36(7):430–3.
66. Microscope cleaning tips, techniques, and timing. LabMedicine. 2005;36(8):460.
67. It's about MIME. ASC Bull. 2005;42(5):89, 92, 106.
68. Austin RM, Holladay EB, Gill GW. Blinded litigated Pap slides reviews gain acceptance as key legal test: a report of the Coalition for Objective

Blinded Rescreening Assessment (COBRA). ASC Bull. 2005;42(5):99, 102, 110.

69. Köhler illumination. LabMedicine. 2005;36(9):530.
70. CLIA '88: are we there yet? LabMedicine. 2005;36(11):694, 732. Quality Qorner.
71. Q&A: is there a standard of practice for screening Pap smears, and if so, what is its significance? LabMedicine. 2006;37(1):40.
72. Q&A: Köhler illumination. LabMedicine. 2006;37(1):42–3.
73. Enviro-Pap: an environmentally friendly, economical, and effective Pap stain. LabMedicine. 2006;37(2):105–8.
74. Monitoring cytotechnologist cytopathologist discrepancy in nongynecologic cytopathology. Diagn Cytopathol. 2006;34(4):270–1. Counterpoint.
75. The Pap test. DeMay RM, ASCP Press, Chicago, 2005. LabMedicine. 2006;37(5):315. Media review.
76. Opportunities and challenges in GYN cytology. Adv Admin Lab 2006;15(5):52–3.
77. Legislative and Regulatory Update. ASC Bull. 2006;43(6):128.
78. Screening and CPR. Micro Views. 2007;16(1):20–1. (SurgiPath publication).
79. Cytopathology Review Guide, 3rd edition, Holladay EB, ASCP Press, Chicago, 2009. LabMedicine. 2009;40(1):60. Media review.
80. Elgert PA, Gill GW. George N. Papanicolaou. LabMedicine. 2009;40(4):245–6. (Luminary in Cytopathology).

Letters to the Editor

1. Restoration of faded microscopical slides. Cytotechnol Bull. 1971;8(5):5, 11.
2. Unsuitability of liquid substitutes for cover glasses. Cytotechnol Bull. 1971;8(6):5–6.
3. Directions for finding an antioxidant. Cytotechnol Bull. 1972;9(1):7.
4. Controlling microscopical illumination. Cytotechnol Bull. 1972;9(3):2.
5. Carbowax in fixatives. Cytotechnol Bull. 1973;10(1):19–2.
6. Microscopy, filters, and mounting media. Cytotechnol Bull. 1973;10(2):6–7.
7. A new hematoxylin formula. Cytotechnol Bull. 1973;10(4):6.
8. Comparative filter techniques. Acta Cytol. 1975;19:207–9.
9. Xylene fumes and safety hoods. LabMedicine. 1982;13:60–1.
10. More on fume hoods. LabMedicine. 1982;13:447.
11. Bis-chloro-methyl ether (BCME): a response. Histo-Logic. 1982;12:176–7.
12. Gill GW, Gupta PK: Microscope objectives and cytopathology screening – another viewpoint. Hum Pathol. 1984;15:1100–1.

13. "Spaghetti and meatballs": more similes in cytopathology. Diagn Cytopathol. 1995;12(2):194.
14. Viewpoint: on CT workload overload and doing the job right the first time. ASCT News. 1997;18(4):57–61.
15. Complete imaging. CAP Today. 1997;11(7):25–6.
16. Screening cytologic slides. LabMedicine. 1997;28(10):621–2.
17. Pap piecework pay problematical still... Diagn Cytopathol. 1998;18(3):246.
18. Air-dried/rehydrated CV smears are different. Diagn Cytopathol. 1998;18(5):381–2.
19. Comments on which to use: No. 1 or No. 1-1/2 cover glasses? Micros Today. 1999;7:43.
20. More on cleaning a microscope's optical surfaces. Micros Today. 2001;7:36.
21. Opinions on GYN-cytology proficiency testing. LabMedicine. 2005;36(8):506.
22. Austin RM, Birdsong GG, Sidawy MK, Kaminsky DB, Benstein BB, Cibas ES, Mody DR, Guidos BJ, Moriarty AT, Rimm DL, Ehya H, Facik MS, Howell LP, Gill GW, Kurtycz DF, Powers CN, Young NA. Fine needle aspiration is a feasible and accurate technique in the diagnosis of lymphoma. J Clin Oncol. 2005;23(35):9029–30.

Vendor Literature

1. Clearium – a xylene (substitute) alternative, Micro Views, Surgipath Medical Industries Inc., 1986; 1(2):1, 7.
2. The cytotect triangle. MicroGram (EM Science). 1994;1(1):1.
3. What is the Papanicolaou stain anyway? MicroGram (EM Science). 1994;1(1):2.
4. Common cent$–"a penny saved is two pennies earned." MicroGram (EM Science). 1994;1(1):3.
5. Diamonds in the rough. MicroGram (EM Science). 1994;1(1):3.
6. EA stains – will the "real" EA please stand up? MicroGram (EM Science). 1994;1(2):1.
7. Diamonds in the rough – II. semordnilaps. MicroGram (EM Science). 1994;1(2):4.
8. EA solution solutions. MicroGram (EM Science). 1995;2(1):1–2.
9. Light green SF yellowish shortage. MicroGram (EM Science). 1995;2(1):3.
10. Common cent$ – "waste not, want not." MicroGram (EM Science). 1995;2(1):3.
11. Diamonds in the rough–III. compound fractures. MicroGram (EM Science). 1995;2(1):4.

12. Troubleshooting EA staining problems. MicroGram (EM Science). 1995;2(2):1, 3.
13. Bismarck brown Y and phosphotungstic acid in EA. MicroGram (EM Science). 1995;2(2):2.
14. Common cent$ – what do cleaning microscopes and front end alignments have in common? MicroGram (EM Science). 1995;2(2):3.
15. Diamonds in the rough – IV. spoonerisms. MicroGram (EM Science). 1997;2(2):4.
16. OG-6. MicroGram (EM Science). 1995;2(3):1, 3.
17. The post-OG/EA alcohol rinses. MicroGram (EM Science). 1995;2(3):2.
18. Diamonds in the rough – V. shades of black and white… MicroGram (EM Science). 1995;2(3):3.
19. Graded alcohols of hydration and dehydration. MicroGram (EM Science). 1995;2(4):1.
20. OG-6. MicroGram (EM Science). 1995;2(4):2.
21. Common cent$ – panes, of sorts… MicroGram (EM Science). 1995;2(4):3.
22. Diamonds in the rough – VI: oxymorons and a panel game. MicroGram (EM Science). 1995;2(4):4.
23. Differentiation and bluing – what's the difference? MicroGram (EM Science). 1995;2(5):1–3.
24. Common cent$ – the true impact of the lost raise, or "compound interest in reverse." MicroGram (EM Science). 1995;2(5):3.
25. Harris hematoxylin – an overview. MicroGram (EM Science). 1997;3(1):1–2.
26. Gill hematoxylin – background. MicroGram (EM Science). 1997;3(2):1–2.
27. Gill hematoxylin – composition and implications. MicroGram (EM Science). 1997;3(3):1–2.
28. Troubleshooting hematoxylin staining problems. MicroGram (EM Science). 1998;4(1):1–2.
29. H&E staining: oversight and insights. Dako Connection—Special Stains and H&E. 2010;14:104–14. http://www.dako.com/us/28829_connection_14.pdf
30. Ask the expert: H&E staining. Dako Connection—Special Stains and H&E. 2010;14:137–43. http://www.dako.com/us/28829_connection_14.pdf
31. Kumar GL, Gill GW. Introduction to special stains. Dako Connection—Special Stains and H&E. 2010;14:16–42. http://www.dako.com/us/28829_connection_14.pdf
32. Gill GW, Wood DM. Ask the experts: Biological Microtechnique. Dako Connection—Special Stains and H&E. 2010;14:144–50. http://www.dako.com/us/28829_connection_14.pdf

Index

G.W. Gill, *Cytopreparation: Principles & Practice*,
Essentials in Cytopathology 12, DOI 10.1007/978-1-4614-4933-1,
© Springer Science+Business Media New York 2013

GPSR Compliance
The European Union's (EU) General Product Safety Regulation (GPSR) is a set of rules that requires consumer products to be safe and our obligations to ensure this.

If you have any concerns about our products, you can contact us on

ProductSafety@springernature.com

In case Publisher is established outside the EU, the EU authorized representative is:

Springer Nature Customer Service Center GmbH
Europaplatz 3
69115 Heidelberg, Germany

www.ingramcontent.com/pod-product-compliance
Lightning Source LLC
Chambersburg PA
CBHW051149180926
51611CB00001B/5
* 9 7 8 1 4 6 1 4 4 9 3 2 4 *